T0253647

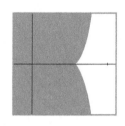

ANALYTIC NUMBER THEORY

An Introductory Course

ANALYTIC NUMBER THEORY

An Introductory Course

Paul T Bateman
Harold G Diamond

University of Illinois at Urbana-Champaign

 World Scientific

NEW JERSEY • LONDON • SINGAPORE • BEIJING • SHANGHAI • HONG KONG • TAIPEI • CHENNAI

Published by

World Scientific Publishing Co. Pte. Ltd.

5 Toh Tuck Link, Singapore 596224

USA office: 27 Warren Street, Suite 401-402, Hackensack, NJ 07601

UK office: 57 Shelton Street, Covent Garden, London WC2H 9HE

Library of Congress Cataloging-in-Publication Data
Bateman, P. T.
    Analytic number theory : an introductory course / Paul T. Bateman, Harold G. Diamond.
      p. cm. -- (Monographs in Number Theory ; v. 1)
    Includes bibliographical references and indexes.
    ISBN-13 978-981-238-938-1
    ISBN-10 981-238-938-5
    ISBN-13 978-981-256-080-3 (pbk)
    ISBN-10 981-256-080-7 (pbk)
    1. Number theory. 2. Nombres, Théorie des. I. Diamond, Harold G., 1940–
    II. Title. III. Series
    QA241 .B319 2004
    512.73--dc22

                                           2007297756

British Library Cataloguing-in-Publication Data
A catalogue record for this book is available from the British Library.

First published 2004
Reprinted 2009

Printed in Singapore

To our wives, Felice and Nancy

Preface

Number theory holds a distinguished position in mathematics for its many results which are at once profound and yet easy to state. It is a beautiful subject, and we hope this book will invite students to its study.

Our theme is the use of analysis to treat multiplicative problems in number theory. We study several of the principal methods and results in this area, particularly those involving reasonably stable arithmetical entities. Typical examples include counts of integers having regularly occurring properties or summatory functions of arithmetic functions.

It seems paradoxical that analysis should be useful in number theory. The integers, the central objects of study in number theory, are the prototype of discreteness, while mathematical analysis, on the other hand, is concerned with continuous phenomena. Analysis is applied in two ways in this book: through direct real variable estimations, which we call "elementary" methods; and by using transforms, which put the apparatus of complex function theory at our disposal. Analysis serves both to establish results and to yield better understanding of the structure of problems.

This book is based on lecture notes we have given to generations of students in introductory graduate level courses on analytic number theory at the University of Illinois. We enjoyed teaching the material, and we hope that some of this enthusiasm comes through in our text.

A feature of our presentation is use of Riemann-Stieltjes integrals to unify and motivate arguments involving sums and integrals. We had previously hesitated to publish our notes out of a concern that some of the methodology might be unfamiliar to the intended audience. We are cautiously optimistic that now our formulation will be generally accepted. In an appendix, we have presented the integration theory and a few further results that may be less well known; other background material is commonly taught in undergraduate courses in real analysis, complex analysis, and algebra or number theory.

Problems appear in the text near relevant techniques for their solution. They generally illustrate some point and give substance to theory; we encourage readers to consider them. The problems vary considerably in their difficulty.

Along with other writers, we suffer from a lack of symbols. For example, φ is used here for Euler's function as well as for various other functions. We generally identify each function in case of possible ambiguity. Also, usage of symbols is not always consistent among authors and topics. For instance, the number of distinct prime factors of an integer n is generally denoted by $\omega(n)$; in the chapters on sieves this symbol has another customary usage, so $\nu(n)$ serves to denote the number of distinct prime factors there. In the Symbol Index, we provide thumbnail sketches of symbols as a quick reminder to readers; these are not full definitions!

We are pleased to acknowledge the contributions of many people to this book. Most of our subject matter comes from the lectures and writings of distinguished number theorists (K. Chandrasekharan, H. Halberstam, A. E. Ingham, E. Landau, H. Rademacher, C. L. Siegel, and E. C. Titchmarsh, to name a few). Many students and colleagues over the years have provided stimulation, suggestions, and corrections to our original notes. We received help on parts of the manuscript from S. Ullom and from the referee for W.S.P. We are very appreciative of the assistance of F. Bateman, H. Halberstam, and J. Steinig for their many mathematical, grammatical, and typographic suggestions; and of A. J. Hildebrand for mathematical and LaTeX advice. We thank H. Britt for typing the manuscript.

Finally, we request readers to advise us of errors or obscurities that they find.

Urbana, Illinois
June, 2004

Contents

Chapter 1

Introduction

1.1 Three problems

The rational integers play an important role in many parts of analysis, e.g. as periods of functions such as $\sin 2\pi z$. In the other direction, one might try to apply analysis to establish properties of integers. Analytic number theory can be described as the study of problems concerning integers by use of methods from analysis. These problems are often easy to state; however, this is a poor guide for deciding how difficult they are to solve. Many innocent sounding arithmetical problems have not yet been solved or have been solved only by sophisticated methods.

We shall pose three problems here, each readily understood, and begin work upon the last one. Our approach is necessarily *ad hoc* at this stage, for we have available no general theory. The object here is to meet some ideas which will occur again. Also, it is interesting to see what we can do "from scratch." After some more machinery has been developed, the first two problems will be taken up and the third will be treated more efficiently and systematically.

1.2 Asymmetric distribution of quadratic residues

Let p be a prime number. In the sequel *the symbols p, p', ..., p_1, p_2, ... will be reserved for primes* and n, n', ..., n_1, n_2, ... for positive integers. We say that an integer n is a *quadratic residue modulo p* if $p \nmid n$ and n is congruent to some square modulo p. For the first few primes $p \equiv 3 \pmod 4$ we list the least positive residues modulo p and underline the quadratic residues:

1

$p = 3:$	1 2
$p = 7:$	1 2 3 4 5 6
$p = 11:$	1 2 3 4 5 6 7 8 9 10
$p = 19:$	1 2 3 4 5 6 7 8 9 10 11 12 13 14 15 16 17 18
$p = 23:$	1 2 3 4 5 6 7 8 9 10 11 12 13 14 15 16 17 18 19 20 21 22

Table 1.1   QUADRATIC RESIDUES

This table suggests that, generally, residues occur near the beginning of each sequence and nonresidues occur near the end. We are led to conjecture

Theorem 1.1   *Let p be a prime, $p \equiv 3$ (mod 4). There are more quadratic residues modulo p between 0 and $p/2$ than between $p/2$ and p.*

This is a true theorem, and one obviously involving only integers. No "elementary" proof is presently known. This is not surprising, since the ordering of the least positive residues $r \equiv k^2$ (mod p) is connected in a subtle way with the ordering of the integers $1 \leq k < p$. All known proofs involve such analytic tools as Fourier series or functions of a complex variable.

The above table suggests (and this is a familiar fact from elementary number theory) that if $p \equiv 3$ (mod 4) and n is a quadratic residue modulo p, then $p - n$ is not a quadratic residue and conversely. For primes $p \equiv 1$ (mod 4), Theorem 1.1 cannot hold, for in that case (again by elementary number theory) n is a quadratic residue precisely when $p - n$ is.

1.3   The prime number theorem

It has been known since the time of Euclid that there are infinitely many primes. (A proof of this fact is sketched in §1.5.) For $x \geq 1$, let $\pi(x)$ denote the number of primes in the interval $[1, x]$. Mathematicians have long sought exact formulas for $\pi(x)$ or for the nth prime number p_n. Around 1800 Gauss and Legendre independently conjectured

Theorem 1.2 (The prime number theorem).

$$\lim_{x \to \infty} \frac{\pi(x)}{x/\log x} = 1.$$

This theorem, which we shall call the P.N.T., is perhaps the most famous result in analytic number theory. Its proof withstood the best efforts of 19th century mathematicians until the end of the century, when proofs were discovered independently by J. Hadamard and C. J. de la Vallée Poussin.

Although this theorem deals ultimately with integers, it is perhaps less surprising that analysis plays a role here than in the first example. Indeed, the very statement of the theorem contains the notions of limit and logarithm, both of which belong to the domain of analysis.

1.4 Density of squarefree integers

A positive integer is said to be *squarefree* if it is not divisible by the square of any prime. We denote the squarefree integers by Q. The first few elements of Q are 1, 2, 3, 5, 6, 7, 10. We ask: What proportion of the positive integers are squarefree? This question is rather vague, and it can be made more precise as follows: We first define $Q(x)$ to be the number of squarefree integers not exceeding x. Next, we ask whether $Q(x)/x$ tends to a limit as x tends to ∞, and finally what the value of this limit is, if it exists. In case the limit exists, it is called the (asymptotic) *density* of Q.

One can make a numerical experiment on a list of the positive integers by first deleting all multiples of 4, then all multiples of 9, then of 25, etc. The first operation leaves about 3/4 of the integers. The second leaves about 8/9 of those surviving the first operation, the third 24/25, etc. We claim that divisibility by p^2 and divisibility by p'^2 ($p \neq p'$) are, in some sense, independent events (cf. the proof of Theorem 1.3, below, and §12.2).

The heuristic reasoning suggests that as $x \to \infty$,

$$\frac{Q(x)}{x} \to \lim_{n \to \infty} (1 - 2^{-2})(1 - 3^{-2}) \cdots (1 - p_n^{-2}) =: \prod_p (1 - p^{-2})$$

and numerical experiments reveal that

$$\frac{Q(10)}{10} = .7, \quad \frac{Q(100)}{100} = .61, \quad \frac{Q(1000)}{1000} = .608, \quad \frac{Q(10,000)}{10,000} = .6083.$$

We shall answer the question about the proportion of squarefree integers by the following three theorems.

Theorem 1.3 *The squarefree integers have the density*

$$\lim_{x \to \infty} Q(x)/x = \prod_{p}(1 - p^{-2}).$$

Theorem 1.4 (Euler product formula).

$$\prod_{p}(1 - p^{-2})^{-1} = \sum_{n=1}^{\infty} n^{-2}.$$

Theorem 1.5 $\left\{\sum_{n=1}^{\infty} n^{-2}\right\}^{-1} = 6/\pi^2 = 0.607927\ldots.$

Corollary 1.6 *The density of squarefree integers is $6/\pi^2$.*

Proof of Theorem 1.3. Let r be any nonnegative integer and for $x \geq 1$, let $Q^{(r)}(x)$ be the number of positive integers $n \leq x$ such that n is not divisible by the square of any of the first r primes. For example, $Q^{(0)}(x) = [x]$ and $Q^{(1)}(x) = [x] - [x/4]$. Here $[u]$ denotes the greatest integer not exceeding u. Clearly,

$$Q^{(0)}(x) \geq Q^{(1)}(x) \geq Q^{(2)}(x) \geq \ldots \geq Q(x).$$

We shall first prove that if y is a multiple of $2^2 3^2 \cdots p_r^2$, then

$$Q^{(r)}(y) = y(1 - 2^{-2})(1 - 3^{-2}) \cdots (1 - p_r^{-2}).$$

An integer n is not divisible by the square of any of the first r primes precisely when n satisfies the simultaneous congruences

$$n \equiv a_i \pmod{p_i^2}, \quad 1 \leq i \leq r,$$

for an r-tuple of integers $(a_1, \ldots, a_r)$ with $0 < a_i < p_i^2$. For any fixed r-tuple $(a_1, \ldots, a_r)$ these simultaneous congruences have a unique solution among any $p_1^2 p_2^2 \cdots p_r^2$ consecutive integers (Chinese remainder theorem). There are $(p_1^2 - 1)(p_2^2 - 1) \cdots (p_r^2 - 1)$ r-tuples satisfying $0 < a_i < p_i^2$ for $1 \leq i \leq r$. Thus if a is a positive integer and $y = a p_1^2 \cdots p_r^2$, then

$$Q^{(r)}(y) = a(p_1^2 - 1) \cdots (p_r^2 - 1) = y(1 - p_1^{-2}) \cdots (1 - p_r^{-2}).$$

Incidentally, this reasoning makes precise the sense in which we regard divisibility by p^2 and divisibility by p'^2 as independent events.

For arbitrary positive x let $y = [xp_1^{-2} \cdots p_r^{-2}] p_1^2 \cdots p_r^2$. We have

$$0 \le Q^{(r)}(x) - Q^{(r)}(y) \le x - y < p_1^2 p_2^2 \cdots p_r^2$$

and also

$$0 \le (x - y) \prod_{\nu=1}^{r} (1 - p_\nu^{-2}) < x - y < p_1^2 p_2^2 \cdots p_r^2.$$

Thus

$$Q^{(r)}(x) = x \prod_{\nu=1}^{r} (1 - p_\nu^{-2}) - (x - y) \prod_{\nu=1}^{r} (1 - p_\nu^{-2}) + Q^{(r)}(x) - Q^{(r)}(y)$$

$$= x \prod_{\nu=1}^{r} (1 - p_\nu^{-2}) + \theta p_1^2 \cdots p_r^2,$$

where θ is a number of modulus at most 1. Hence

$$\varlimsup_{x \to \infty} \frac{Q(x)}{x} \le \varlimsup_{x \to \infty} \frac{Q^{(r)}(x)}{x} = \lim_{x \to \infty} \frac{Q^{(r)}(x)}{x} = \prod_{\nu=1}^{r} (1 - p_\nu^{-2}).$$

This inequality is valid for each r and thus

$$\varlimsup_{x \to \infty} \frac{Q(x)}{x} \le \prod_{p} (1 - p^{-2}),$$

where the product is interpreted as the limit of the preceding one as $r \to \infty$.

We next estimate $Q(x)$ from below. Let r be a positive integer. Then $Q^{(r)}(x) - Q(x)$ counts the number of integers $n \in [1, x]$ which contain no factor p^2 with $p \le p_r$ and at least one factor p^2 for some $p > p_r$. Thus

$$Q^{(r)}(x) - Q(x) \le \#\{n \le x : \exists \nu > r, \ p_\nu^2 \mid n\} \le \sum_{\nu=r+1}^{\infty} \#\{n \le x : p_\nu^2 \mid n\}$$

$$= \sum_{\nu=r+1}^{\infty} \left[\frac{x}{p_\nu^2} \right] < \sum_{\nu=r+1}^{\infty} \frac{x}{\nu^2} < \int_{r}^{\infty} \frac{x}{t^2} \, dt = \frac{x}{r},$$

whence

$$\varliminf_{x \to \infty} \frac{Q(x)}{x} \ge \varliminf_{x \to \infty} \frac{Q^{(r)}(x)}{x} - \frac{1}{r} = \prod_{\nu=1}^{r} (1 - p_\nu^{-2}) - \frac{1}{r}.$$

Letting $r \to \infty$, we obtain

$$\lim_{x \to \infty} \frac{Q(x)}{x} \geq \prod_p (1 - p^{-2}).$$

Thus the limit of $Q(x)/x$ exists and equals $\prod_p (1 - p^{-2})$.   □

The following problem gives a positive lower bound for the last infinite product.

PROBLEM 1.1   Show that $\sum_p p^{-2} < 1/2$.   Using this fact show that $Q(n) > n/2$ for any $n \geq 1$. (It can be shown that

$$\operatorname*{g.l.b.}_{n \geq 1} \frac{Q(n)}{n} = \frac{Q(176)}{176} = \frac{106}{176} = .602272\ldots,$$

but this is more involved.)

PROBLEM 1.2   Let n be a positive integer. Show that

(a) $\displaystyle\prod_{j=2}^{n} (1 - j^{-2}) = (n+1)/(2n)$,

(b) $\displaystyle\prod_{j=n+1}^{\infty} (1 - j^{-2}) = 1 - 1/(n+1)$.

(c) Give an upper bound for ϵ such that

$$\prod_p (1 - p^{-2}) = (1 - \epsilon) \prod_{p<100} (1 - p^{-2}).$$

Proof of Theorem 1.4.   Multiplying the first two factors, we get

$$(1 - 2^{-2})^{-1}(1 - 3^{-2})^{-1}$$

$$= (1 + 2^{-2} + 4^{-2} + \cdots)(1 + 3^{-2} + 9^{-2} + \cdots) = {\sum}' n^{-2},$$

where Σ' denotes the sum over all positive integers n divisible by no prime greater than 3. The rearrangement of terms is valid by absolute convergence of each geometric series. By the unique factorization theorem we know that if an integer n appears in Σ', it appears precisely once.

Using similar reasoning we next write

$$\prod_{p \leq x} (1 - p^{-2})^{-1} = \prod_{p \leq x} (1 + p^{-2} + p^{-2 \cdot 2} + p^{-3 \cdot 2} + \cdots) = {\sum}'' n^{-2},$$

where Σ'' extends over all positive integers n divisible by no prime greater than x. Now

$$0 \le \sum_{n=1}^{\infty} n^{-2} - {\sum}'' n^{-2} = {\sum}''' n^{-2},$$

where Σ''' extends over all integers n divisible by at least one prime greater than x. It is clear that

$${\sum}''' n^{-2} < \sum_{n=[x]+1}^{\infty} n^{-2},$$

and the last sum tends to zero as $x \to \infty$, since $\sum n^{-2}$ converges. Thus,

$$\lim_{x \to \infty} \prod_{p \le x} (1 - p^{-2})^{-1} = \sum n^{-2}.$$

We note that the limit of the partial products is nonzero, and thus the infinite product converges. $\qquad\Box$

Proof of Theorem 1.5. The following proof uses the method of *deus ex machina*. It is motivated by the fact that the numbers n^{-2} are related to the Fourier coefficients of a tractable function.

Let $f(x) = x^2$, $|x| \le \pi$, and continue f as a 2π periodic function. Since f is a continuous function with a piecewise continuous derivative, f is represented by its Fourier series

$$f(x) = \frac{\pi^2}{3} + 4 \sum_{n=1}^{\infty} (-1)^n n^{-2} \cos n x.$$

If we evaluate f at $x = \pi$, we find that $\sum n^{-2} = \pi^2/6$. $\qquad\Box$

PROBLEM 1.3 (An alternative proof of Th. 1.5.) Integrate the function $z^{-2} \cot \pi z$ along the perimeter of a rectangle with vertices $\pm(N+1/2)\pm Ni$, where N is a positive integer which is allowed to tend toward ∞. Note that $\cot \pi z$ is bounded on this contour and apply the residue theorem.

One can ask also how swiftly $Q(x)/x$ converges to $6/\pi^2$ as x tends toward ∞. We shall consider this question again in Chapters 3, 8, and 10.

1.5   The Riemann zeta function

For s a real number, $s > 1$, define the zeta function by $\zeta(s) = \sum_{n=1}^{\infty} n^{-s}$. We note that the series for zeta diverges for any $s \le 1$ and converges uniformly on any interval $[a, \infty)$ for $a > 1$ (Weierstrass' M test). The proof of Theorem 1.4 adapts to show that the Euler product formula

$$\zeta(s) = \prod_p (1 - p^{-s})^{-1}$$

is valid for all $s > 1$. This relation is the analytic equivalent of the unique factorization theorem for integers, and it suggests that the zeta function should be useful in investigations of multiplicative properties of integers.

As $s \to 1+$, $\zeta(s) \to +\infty$, since for any finite number $X > 1$ we have

$$\lim_{s \to 1+} \zeta(s) \ge \lim_{s \to 1+} \sum_{n \le X} n^{-s} = \sum_{n \le X} n^{-1} > \log X,$$

and $\log X \to \infty$ as $X \to \infty$. Euler used this fact and the product formula to give a proof of the infinitude of primes. He noted (essentially) that if there were only finitely many primes, then $\prod_p (1 - p^{-s})^{-1}$ would have a finite limit as $s \to 1+$, whereas in fact the limit is infinite.

We remark that related reasoning yields the stronger assertion that *the sum of the reciprocals of the primes diverges.* Indeed, for $x \ge 2$,

$$\sum_{n \le x} \frac{1}{n} < \prod_{p \le x} \left(1 - \frac{1}{p}\right)^{-1} < \prod_{p \le x} e^{2/p} = \exp\left\{2 \sum_{p \le x} \frac{1}{p}\right\}.$$

Since the harmonic series diverges, we have

$$\sum_p 1/p = \infty. \tag{1.1}$$

We shall later extend the zeta function into the complex plane as a meromorphic function whose only singularity is a simple pole at $s = 1$. Here we shall determine the residue at that point.

Lemma 1.7 $\lim_{s \to 1+} (s - 1)\zeta(s) = 1.$

Proof.   We have for $s > 0$ and for $n \ge 1$ and $n \ge 2$ respectively

$$\int_n^{n+1} x^{-s}\, dx < n^{-s} < \int_{n-1}^n x^{-s}\, dx.$$

Thus for $s > 1$,

$$\sum_{n=1}^{\infty} n^{-s} > \sum_{n=1}^{\infty} \int_{n}^{n+1} x^{-s} dx = \int_{1}^{\infty} x^{-s} dx = \frac{1}{s-1}$$

and

$$\sum_{n=1}^{\infty} n^{-s} < 1 + \sum_{n=2}^{\infty} \int_{n-1}^{n} x^{-s} dx = 1 + \frac{1}{s-1},$$

and the result follows. $\square$

We now consider the function $\widehat{Q}$ defined for real values of $s > 1$ by $\widehat{Q}(s) := \sum' n^{-s}$, where $\sum'$ extends over all positive squarefree integers n. The series converges for $s > 1$. (Why?) The function $\widehat{Q}$ is called the *generating function* associated with the set of squarefree integers.

We want to express $\widehat{Q}(s)$ in terms of a product formula. By formally multiplying out the factors we see that $\prod_p (1+p^{-s})$ is expressible as $\sum' n^{-s}$. Thus we are led to conjecture the formula

$$\widehat{Q}(s) = \prod_p (1 + p^{-s}), \quad s > 1. \tag{1.2}$$

PROBLEM 1.4   Prove (1.2) using the reasoning which established the Euler product formula. ((1.2) will also follow from a general principle described in Corollary 6.4.)

If we write $1 + p^{-s} = (1 - p^{-2s})/(1 - p^{-s})$, form the product over all $p \leq X$, and let $X \to \infty$, we obtain the identity (for $s > 1$)

$$\prod_p (1 + p^{-s}) = \prod_p (1 - p^{-2s}) \prod_p (1 - p^{-s})^{-1}.$$

Thus $\widehat{Q}(s) = \zeta(s)/\zeta(2s)$ for $s > 1$.

PROBLEM 1.5   A positive integer n is called *cubefree* if it is not divisible by the cube of any prime. Express the generating function associated with cubefree integers in terms of the Riemann zeta function.

We now consider $\lim_{s \to 1+} (s-1)\widehat{Q}(s)$. The function $s \mapsto \zeta(2s)$ is continuous at $s = 1$, since it is given by a uniformly convergent series of functions continuous in a neighborhood of $s = 1$. Thus

$$\lim_{s \to 1+} \zeta(2s) = \zeta(2) = \pi^2/6 = 1.644934\ldots.$$

Also, we have seen that $(s - 1)\zeta(s) \to 1$ as $s \to 1+$, and hence

$$\lim_{s \to 1+} (s - 1)\widehat{Q}(s) = 6/\pi^2 = \lim_{x \to \infty} \frac{Q(x)}{x}.$$

A somewhat similar relation is the following:

$$\lim_{s \to 1+} (s - 1)\zeta(s) = 1 = \lim_{x \to \infty} [x]/x.$$

In these examples the density of a set of positive integers equals the residue of the associated generating function at $s = 1$. The next theorem shows that such relations are not accidental.

Theorem 1.8   (Dirichlet-Dedekind). *Let $\mathcal{A}$ be a set of (distinct) positive integers and for $x \geq 0$ let $A(x)$ be the number of elements of $\mathcal{A}$ not exceeding x. Suppose that $\mathcal{A}$ has density α, i.e. that $A(x)/x \to \alpha$ as $x \to \infty$. Then*

$$\lim_{s \to 1+} (s - 1) \sum_{n \in \mathcal{A}} n^{-s} = \alpha.$$

Proof.   For $s > 1$ and N any positive integer we have (since $A(0) = 0$)

$$\sum_{n \in \mathcal{A} \cap [1,N]} n^{-s} = \sum_{n=1}^{N} n^{-s}[A(n) - A(n - 1)]$$

$$= \sum_{n=1}^{N-1} A(n)(n^{-s} - (n + 1)^{-s}) + A(N)N^{-s}$$

$$= \sum_{n=1}^{N-1} A(n) \int_{n}^{n+1} s\, x^{-s-1} dx + A(N)N^{-s}$$

$$= s \int_{1}^{N} x^{-s-1} A(x) dx + A(N)N^{-s}.$$

Now let $N \to \infty$. Since $A(N) \leq N$, the last term tends to zero. Thus we have $\sum_{n \in \mathcal{A}} n^{-s} = s \int_{1}^{\infty} x^{-s-1} A(x) dx$ or

$$(s - 1) \sum_{n \in \mathcal{A}} n^{-s} - \alpha = s(s - 1) \int_{1}^{\infty} x^{-s} \left(\frac{A(x)}{x} - \alpha\right) dx + \alpha(s - 1).$$

Given $\epsilon > 0$, choose X large enough that

$$|x^{-1} A(x) - \alpha| < \epsilon \text{ for } X \leq x < \infty.$$

Then, since the integral over $[1, X]$ is a bounded function of s as $s \to 1+$, we have

$$\varlimsup_{s \to 1+} |(s - 1) \sum_{n \in \mathcal{A}} n^{-s} - \alpha| \leq \varlimsup_{s \to 1+} s(s - 1) \int_X^\infty x^{-s} \epsilon \, dx = \epsilon. \qquad \square$$

The converse of the above theorem is false. As an example, consider

$$\mathcal{A} := \bigcup_{j=0}^\infty \{n \in \mathbb{Z}^+ : 4^j \leq n < 2 \cdot 4^j\}.$$

It is not hard to show (Do it!) that

$$\lim_{s \to 1+} (s - 1) \sum_{n \in \mathcal{A}} n^{-s} = 1/2$$

but that

$$\varlimsup_{x \to \infty} \frac{A(x)}{x} = \frac{2}{3}, \quad \varliminf_{x \to \infty} \frac{A(x)}{x} = \frac{1}{3}.$$

We can restate these relations by saying that the set $\mathcal{A}$ has *upper density* 2/3 and *lower density* 1/3. A set of positive integers always has an upper and a lower density; the density exists precisely when the two are equal.

One of our main goals is to develop ways to extract number theoretic information from the associated generating function. We shall generally make use of the powerful apparatus of function theory for this purpose.

PROBLEM 1.6 Let $\mathcal{A}$ be a set of positive integers for which $\sum_{n \in \mathcal{A}} \frac{1}{n}$ converges. Show that $\mathcal{A}$ has a density and that its value is 0.

PROBLEM 1.7 Assuming that the odd squarefree integers have a density, determine this density.

1.6 Notes

1.2. Theorem 1.1 was first proved by G. Lejeune-Dirichlet, J. reine angew. Math., vol. 18 (1838), pp. 259–274; also in Werke, vol. 1, Berlin 1889 (reprinted by Chelsea Publishing Co, 1969), pp. 357–374. Dirichlet is generally regarded as the founding father of analytic number theory.

1.3. The first person to conjecture the prime number theorem explicitly in a published work was A. M. Legendre in Théorie des nombres, 2nd ed., Courcier, Paris, 1808, p. 394, and in Théorie des nombres, 3rd ed., Didot, Paris, 1830, vol. 2, p. 65. In the early 1790's, when C. F. Gauss was in his teens, he inserted the following note (here translated from German and expressed in modern notation) on a blank page of his personal copy of vol. 1 of J. C. Schulze's table of logarithms:

$$\text{Prime numbers less than } a \quad (a \to \infty) \qquad a/\log a. \qquad (1.3)$$

He recalled this assertion over a half a century later in a letter about the distribution of primes written to the astronomer J. F. Encke. This letter is reproduced in Werke, vol. 2, Göttingen, 1863 & 1876, pp. 444–447. The conjecture (1.3) is given in Werke, vol. 10, part 1, Göttingen, 1917, p. 11. The conjectures of both Legendre and Gauss seem to have been based entirely on empirical grounds.

The P.N.T. was proved independently by J. Hadamard, Bull. Soc. Math. France, vol. 24 (1896), pp. 199–220 (also in Hadamard's Selecta, Gauthier-Villars, Paris, 1935, pp. 111–132 and in Oeuvres, vol. 1, CNRS, Paris, 1968, pp. 189–210) and Ch. J. de la Vallée Poussin, Ann. Soc. sci. Bruxelles, Sér. I, vol. 20 (1896), pp. 183–256 and 281–397. For a brief history of the subject, see our article in Amer. Math. Monthly, vol. 103 (1996), pp. 729–741. For a more extended history, see [Nark].

1.4. Theorem 1.3 and Corollary 1.6 go back to L. Gegenbauer, Denkschriften Akad. Wiss. Wien, mat.-natur. Klasse, vol. 49 part 1 (1885), pp. 37–80. Theorem 1.3 (even in the more precise version given as Theorem 3.5) is much less profound than Theorems 1.1 and 1.2.

The inequality $Q(x)/x \geq 106/176$ of Problem 1.1 was established by K. Rogers in Proc. Amer. Math. Soc., vol. 15 (1964), pp. 515–516.

Theorems 1.4 and 1.5 are due to L. Euler.

1.5. For real values of s, the Riemann zeta function had already been considered by Euler in the eighteenth century. G. F. B. Riemann took the very important step of considering zeta for complex values of s as well, in Monatsberichte Akad. Wiss. Berlin (1859), pp. 671–680; also in Collected Papers, R. Narasimhan, ed., Springer-Verlag, Berlin, 1990, pp. 145–153. Accounts of Riemann's work are given in [Edw], [LanH], and [Mat].

Theorem 1.8 is the subject of Supplement II in Dirichlet, Vorlesungen über Zahlentheorie, R. Dedekind, ed., 4th ed., Braunschweig, Vieweg, 1894.

Chapter 2

Calculus of Arithmetic Functions

2.1 Arithmetic functions and convolution

Here we shall develop a calculus of arithmetic functions that is useful in treating problems involving multiplicative structure. A complex valued function defined on $\mathbb{Z}^+$ is called an *arithmetic function*. An arithmetic function f corresponds to a sequence $\{c_n\}_{n=1}^{\infty}$ with $c_n = f(n)$. We shall occasionally use the sequence notation. The set $\{n \in \mathbb{Z}^+ : f(n) \neq 0\}$ is called the *support* of f. If S is a set of positive integers, the function f satisfying $f(n) = 1$ if $n \in S$, and $f(n) = 0$ if $n \notin S$ is called the *indicator function* of S and is generally denoted by 1_S.

The integers we consider are usually positive, and if we write expressions such as $\{n \leq x : n$ satisfies property $P\}$, this is to be understood as $\{n \in \mathbb{Z} \cap [1, x] : n$ satisfies property $P\}$.

Let f and g be arithmetic functions. We write $f = g$ if $f(n) = g(n)$ for all $n \geq 1$ and we define $f \leq g$ similarly. We define addition of f and g and scalar multiplication by $\lambda \in \mathbb{C}$ by setting, for all $n \in \mathbb{Z}^+$,

$$(f + g)(n) = f(n) + g(n), \quad (\lambda f)(n) = \lambda f(n).$$

Equipped with the operations of addition and scalar multiplication, the arithmetic functions form a vector space over $\mathbb{C}$.

Example 2.1 We define the following arithmetic functions.

(1) $0(n) := 0$, all $n \in \mathbb{Z}^+$,

(2) $1(n) := 1_{\mathbb{Z}^+}(n) = 1$, all $n \in \mathbb{Z}^+$,

13

(3) $e_j(n) := 1_{\{j\}}(n) := \begin{cases} 1 & \text{if } n = j, \\ 0 & \text{if } n \neq j, \end{cases}$

(4) $e := e_1,$

(5) $\tau(n) := \sum_{d|n} 1 = $ the number of positive divisors of n,

(6) $\sigma(n) := \sum_{d|n} d = $ the sum of the positive divisors of n,

(7) $\mu_2 := e_1 - e_2,$

(8) $\mu(n) := \begin{cases} 1 & \text{if } n = 1, \\ (-1)^j & \text{if } n \text{ is the product of } j \text{ distinct primes,} \\ 0 & \text{if } n \text{ is divisible by the square of a prime,} \end{cases}$

(9) $|\mu| = 1_Q = $ the indicator function of squarefree integers,

(10) $\omega(n) := \sum_{p|n} 1 = $ the number of distinct prime divisors of n,

(11) $\Omega(n) := $ the total number of prime divisors of n (counting multiplicity).

PROBLEM 2.1   Find all integers $n > 1$ such that $\prod_{d|n} d = n^2$.

Let us recall the rule for multiplying two polynomials or power series. Given

$$f(x) = \sum_{j \geq 0} f_j x^j, \quad g(x) = \sum_{k \geq 0} g_k x^k,$$

we have

$$h(x) = f(x)g(x) = \sum_{i \geq 0} h_i x^i,$$

where $h_i = \sum_{j+k=i} f_j g_k$. The rule that creates the sequence $\{h_i\}_{i \geq 0}$ from $\{f_j\}_{j \geq 0}$ and $\{g_k\}_{k \geq 0}$ is called the Cauchy multiplication or additive convolution of the two sequences. This convolution is essential in the study of additive problems. Here we shall investigate only multiplicative problems, which usually require so-called Dirichlet series rather than power series, and the appropriate rule of composition is multiplicative convolution.

Definition 2.2   Given arithmetic functions f and g, define a new arithmetic function $f * g$, the Dirichlet product or the *multiplicative convolution* of f and g, by setting, for all $n \in \mathbb{Z}^+$,

$$(f * g)(n) := \sum_{ij=n} f(i)g(j).$$

The above sum extends over all ordered pairs of positive integers (i, j) whose product $ij = n$. The sum can also be expressed as

$$\sum_{i|n} f(i)g(n/i) \quad \text{or} \quad \sum_{j|n} f(n/j)g(j).$$

In particular we have

$$(f * g)(1) = f(1)g(1), \quad (f * g)(2) = f(1)g(2) + f(2)g(1),$$

$$(f * g)(4) = f(1)g(4) + f(2)g(2) + f(4)g(1).$$

Unless otherwise noted, convolutions will be multiplicative.

PROBLEM 2.2 Find expressions for $(f * g)(n)$ when $n = p^k$, or $n = pp'$ $(p \neq p')$, or $n = pp'p''$ (p, p', p'' distinct).

As we mentioned before, the arithmetic functions form a vector space over $\mathbb{C}$. When the convolution operation is admitted, the arithmetic functions form an algebra $\mathcal{A}$ over $\mathbb{C}$. This means that the convolution operation is associative, that it is distributive with respect to addition, and that for $\lambda \in \mathbb{C}$ and $f, g \in \mathcal{A}$,

$$\lambda(f * g) = (\lambda f) * g = f * (\lambda g).$$

The associativity of convolution is the only property which requires calculation to establish. Some otherwise messy computations amount to a verification of associativity. The proof of the Möbius inversion formula, which will be established below, is such an example. Associativity of convolution and a useful formula can be established by showing that

$$\{f * (g * h)\}(n) = \sum_{ijk=n} f(i)g(j)h(k) = \{(f * g) * h\}(n),$$

where the sum extends over all ordered triples of positive integers whose product is n.

We call a pair of elements f and g in an algebra *zero divisors* if $f \neq 0$, $g \neq 0$, but f times g is identically 0. It is useful to indicate the first nonzero value of an arithmetic function. For $f \in \mathcal{A}$, $f \neq 0$, let $\mathrm{fnz}(f) := \min\{j : f(j) \neq 0\}$. In case $f = 0$, we set $\mathrm{fnz}(0) := \infty$. For example $\mathrm{fnz}(e_3) = 3$.

PROBLEM 2.3   Show that $\mathrm{fnz}(f * g) = \mathrm{fnz}(f) \cdot \mathrm{fnz}(g)$. Deduce that $\mathcal{A}$ has no zero divisors, and hence cancellation in convolution equations is valid.

PROBLEM 2.4   Find all solutions $f \in \mathcal{A}$ of the convolution equation $f * f = e$. Same question for $f * f = f$.

PROBLEM 2.5   Verify each of the following convolution relations:

(1) $f * g = g * f$, i.e., convolution is commutative,

(2) $f * e = f$, i.e., e is a unity element for $\mathcal{A}$,

(3) $1 * 1 = \tau$,

(4) $(1 * |\mu|)(n) = 2^{\omega(n)}$,

(5) $(\mu_2 * 1)(n) = \begin{cases} 1 & \text{if } n \text{ is odd,} \\ 0 & \text{if } n \text{ is even,} \end{cases}$

(6) $|(f * g)(n)| \leq (|f| * |g|)(n)$ for all $n \in \mathbb{Z}^+$,

(7) $e_j * e_k = e_{jk}$.

PROBLEM 2.6   Express ω and Ω in the form $1 * f$ and $1 * F$ for suitable functions f and F.

Definition 2.3   For $f \in \mathcal{A}$, set $f^{*0} := e$ and $f^{*n} := f^{*(n-1)} * f$ for $n \geq 1$.

We shall have use for two mappings of $\mathcal{A}$ into itself. For k any nonnegative integer define L^k by setting

$$(L^k f)(n) := (\log n)^k f(n).$$

For any $\alpha \in \mathbb{C}$ define T^α by setting

$$(T^\alpha f)(n) := n^\alpha f(n).$$

We write L for L^1 and T for T^1. Define $L^0 = T^0$ (= identity operator). A linear map of an algebra into itself which obeys the product rule of differentiation is called a *derivation* of the algebra.

Lemma 2.4   *L is a derivation of $\mathcal{A}$ and, for each fixed α, T^α is an isomorphism of $\mathcal{A}$.*

Proof. L and T^α are clearly linear. Given $f, g \in \mathcal{A}$ we have

$$\{L(f * g)\}(n) = \sum_{ij=n} (\log i + \log j) f(i) g(j)$$

$$= \{(Lf) * g\}(n) + \{f * (Lg)\}(n),$$

and

$$\{T^\alpha(f * g)\}(n) = \sum_{ij=n} (ij)^\alpha f(i) g(j) = \{(T^\alpha f) * (T^\alpha g)\}(n).$$

It is easy to see that T^α is one-to-one and onto and that the composition $T^{-\alpha} \circ T^\alpha = T^0$. $\qquad\square$

Remark 2.5 As a derivation, L satisfies the familiar power formula and Leibniz formula

$$L(f^{*n}) = n f^{*(n-1)} * Lf, \tag{2.1}$$

$$L^n(f * g) = \sum_{j=0}^{n} \binom{n}{j} L^j f * L^{n-j} g. \tag{2.2}$$

PROBLEM 2.7 Verify each of the following relations:

(1) $(T1)(n) = n$, if $n \in \mathbb{Z}^+$,
(2) $T1 * T1 = T\tau$,
(3) $Lf = 0$ if and only if $f = ce$ for some constant c,
(4) $(T1) * 1 = \sigma$.

PROBLEM 2.8 Let $f, g \in \mathcal{A}$, $f \neq 0$, $g \neq 0$, and $(Lf) * g = f * (Lg)$. Show that $g = cf$ for some constant c. Hint. First show that $\text{fnz}(f) = \text{fnz}(g)$.

2.2 Inverses

Definition 2.6 Let $f \in \mathcal{A}$. We say that f is *invertible* if there exists a $g \in \mathcal{A}$ such that $f * g = e$. In this case we call g an *inverse* of f. Inverses are unique, when they exist: If g and g' are both inverses of f, then

$$g = g * (f * g') = (g * f) * g' = g'.$$

If $f * g = e$, we shall write $g = f^{*-1}$ and refer to g as *the* inverse of f.

PROBLEM 2.9   Let $S = \{2^n : 0 \leq n < \infty\}$, 1_S be the indicator function of S, and let $\mu_2 = e - e_2$. Show that $1_S * \mu_2 = e$.

We shall investigate when $f \in \mathcal{A}$ is invertible. The equation $f * g = e$ is equivalent to the infinite system of equations

$$1 = f(1)g(1),$$

$$0 = f(2)g(1) + f(1)g(2),$$

$$0 = f(3)g(1) + f(1)g(3),$$

$$0 = f(4)g(1) + f(2)g(2) + f(1)g(4),$$

$$\cdots .$$

For f to have an inverse it is clearly necessary that $f(1) \neq 0$. This condition is also sufficient, as we shall now see. The first equation in the above system gives the value of $g(1)$. Knowing $f(1)$, $f(2)$, and $g(1)$, we can find $g(2)$ from the next equation. If we have found $g(1), \ldots, g(n-1)$, then we can find $g(n)$ from the nth equation:

$$g(n) = -(f(1))^{-1} \sum_{\substack{jk=n, \\ j>1}} f(j)g(k).$$

Thus we have proved

Theorem 2.7   *$f \in \mathcal{A}$ is invertible if and only if $f(1) \neq 0$.*

We shall give a second proof of the theorem in terms of power series in the next section.

The function 1 is invertible and, as we shall show, its inverse is μ. This assertion is familiar from elementary number theory as the formula $\sum_{d|n} \mu(d) = e(n)$ and can be established by a combinatorial argument. However, we prefer to defer the proof until the next section. Accepting for the moment the formula $\mu * 1 = e$, we can easily prove

Lemma 2.8   (Möbius inversion formulas). *Let f and $F \in \mathcal{A}$. Then $F = 1 * f$ if and only if $f = F * \mu$. More generally, if g is any invertible element of $\mathcal{A}$, then $F = g * f$ if and only if $f = g^{*-1} * F$.*

Proof.   Let $F = g * f$. Then

$$g^{*-1} * F = g^{*-1} * g * f = f.$$

The converse is proved similarly. □

Example 2.9 The Euler φ function. We define $\varphi \in \mathcal{A}$ by setting

$$\varphi(n) := \#\{j \leq n : (n, j) = 1\}.$$

Thus, $\varphi(1) = \varphi(2) = 1$, $\varphi(3) = \varphi(4) = 2$, etc. Given $n, d \in \mathbb{Z}^+$ with $d \mid n$, define $S_d = \#\{j \leq n : (n, j) = d\}$. We have

$$S_d = \#\{k \leq n/d : (n/d, k) = 1\} = \varphi(n/d);$$

also, every integer in $[1, n]$ is counted exactly once by $\sum_{d \mid n} S_d$. Thus

$$n = \sum_{d \mid n} \varphi(n/d) \quad \text{or} \quad \varphi * 1 = T1.$$

By the Möbius inversion formula we obtain the representation

$$\varphi = T1 * \mu. \tag{2.3}$$

2.3 Convergence

Definition 2.10 Given a sequence $\{f_\nu\}_{\nu=1}^\infty$ with each $f_\nu \in \mathcal{A}$, we say that $\{f_\nu\}$ converges (in $\mathcal{A}$) if $\lim_{\nu \to \infty} f_\nu(n)$ exists and is a finite number for each $n \in \mathbb{Z}^+$. If the limit function is denoted by f, we shall write $f_\nu \to f$. This is a pointwise notion of convergence. If we are given a sequence of complex numbers $\{a_n\}_{n=0}^\infty$ and $f \in \mathcal{A}$, define $\sum_{n=0}^\infty a_n f^{*n}$ as the limit of the sequence of partial sums of the arithmetic functions $a_n f^{*n}$. We say that the infinite product $f_1 * f_2 * \cdots$ exists provided that (i) $\prod_{i=1}^\infty \{f_i(1)\}$ converges as an infinite product of complex numbers, i.e. there exist at most a finite number of indices i for which $f_i(1) = 0$ and with these factors omitted, the remaining product converges to a finite nonzero limit, and (ii) the sequence of partial products $f_1 * \cdots * f_N$ converges as $N \to \infty$.

PROBLEM 2.10 Show that, as $\nu \to \infty$,

(1) $f_\nu(n) := e(n\nu) \to 0$,
(2) $g_\nu(n) := \exp(\exp[-(\nu - n)^2]) \to 1$.

PROBLEM 2.11 Write out $(f_1 * \cdots * f_5)(2)$. Does this example suggest the special role that $\prod_{i=1}^\infty \{f_i(1)\}$ plays?

PROBLEM 2.12   Let $\{f_i\}_{i=1}^{\infty}$ be a sequence of arithmetic functions, none of which is identically zero. Also assume that $f_1 * f_2 * \cdots$ converges. Prove that $f_1 * f_2 * \cdots \neq 0$. Hint. Consider the f_i for which $f_i(1) = 0$.

Lemma 2.11   *Let $\{h_j\}_{j=1}^{\infty}$ be a sequence of arithmetic functions with $h_j(1) = 0$ for all j. Then, for all indices $\nu > \log n / \log 2$,*

$$(h_1 * \cdots * h_\nu)(n) = 0.$$

Proof.   Suppose $\nu > \log n / \log 2$. In

$$(h_1 * \cdots * h_\nu)(n) = \sum_{n_1 n_2 \cdots n_\nu = n} h_1(n_1) h_2(n_2) \cdots h_\nu(n_\nu)$$

at least one n_i satisfies $n_i \leq n^{1/\nu} < 2$. Thus $n_i = 1$, $h_i(n_i) = 0$, and so $(h_1 * \cdots * h_\nu)(n) = 0$. $\square$

This lemma implies that *any* power series of any arithmetic function that vanishes at $n = 1$ is convergent:

Lemma 2.12   *Let $h \in \mathcal{A}$ satisfy $h(1) = 0$, and let $\{a_j\}_{j \geq 0}$ be an arbitrary sequence of complex numbers. Then $\sum_{j \geq 0} a_j h^{*j}$ converges.*

The proof consists in observing that for each positive n the expression

$$\left(\sum_{0 \leq j \leq N} a_j h^{*j} \right)(n)$$

is constant for all $N > \log n / \log 2$.

Remark 2.13   A little more effort shows that if $\sum_{\nu=0}^{\infty} |a_\nu| r^\nu < \infty$ and if $h \in \mathcal{A}$ with $|h(1)| < r$, then $\sum a_\nu h^{*\nu}$ converges in $\mathcal{A}$. Most of the later results on power series that are stated for $\mathcal{A}_0$ are in fact valid in this more general setting. However, Lemma 2.12 suffices for our intended applications.

Lemma 2.14   *Let $\{f_j\}_{j=1}^{\infty}$ be a sequence of arithmetic functions with $f_j(1) = 0$ for all j. Suppose that*

$$\sum_{j=1}^{\infty} |f_j(n)| < \infty, \quad all \ \ n \in \mathbb{Z}^+.$$

Then the convolution $(e + f_1) * (e + f_2) * \cdots$ converges.

Proof. Clearly, $\prod_{j=1}^{\infty} e(1)$ converges. Let n be any positive integer and let $\nu = [\log n / \log 2]$. By formally multiplying out the product and applying Lemma 2.11 we have for any $N \geq \nu$,

$$\{(e + f_1) * \cdots * (e + f_N)\}(n)$$

$$= e(n) + \sum_{j=1}^{N} f_j(n) + \sum_{1 \leq i < j \leq N} (f_i * f_j)(n) + \cdots$$

$$+ \sum_{1 \leq i_1 < i_2 < \cdots < i_\nu \leq N} (f_{i_1} * f_{i_2} * \cdots * f_{i_\nu})(n).$$

There are at most ν convolution factors in each term. As $N \to \infty$, each series converges absolutely. The second one, for example, is dominated by

$$\sum_{1 \leq i < j < \infty} \sum_{\ell m = n} |f_i(\ell)| \, |f_j(m)| \leq \sum_{\ell m = n} \sum_{i=1}^{\infty} |f_i(\ell)| \sum_{j=1}^{\infty} |f_j(m)|$$

$$\leq \tau(n) \max_{1 \leq \ell \leq n} \{\sum_{i=1}^{\infty} |f_i(\ell)|\}^2 < \infty.$$

The other series are estimated similarly. Thus the product converges for each fixed n. $\qquad \square$

Second proof of Theorem 2.7. Again, $f(1) \neq 0$ if f has an inverse. Now suppose that $f(1) = 1$. Let $h = e - f \in \mathcal{A}$. Then h vanishes at 1, and $\sum_{i=0}^{\infty} h^{*i}$ converges by Lemma 2.12. Now let n be any positive integer and let r be an integer greater than $\log n / \log 2$. We have

$$\left(f * \sum_{i=0}^{\infty} h^{*i}\right)(n) = \sum_{d|n}(e - h)\left(\frac{n}{d}\right) \sum_{i=0}^{\infty} h^{*i}(d) = \sum_{d|n}(e - h)\left(\frac{n}{d}\right) \sum_{i=0}^{r-1} h^{*i}(d)$$

$$= \left\{(e - h) * \sum_{i=0}^{r-1} h^{*i}\right\}(n) = (e - h^{*r})(n) = e(n).$$

Thus, if $f(1) = 1$, then

$$f^{*-1} = \sum_{i=0}^{\infty} h^{*i} = \sum_{i=0}^{\infty} (e - f)^{*i}. \tag{2.4}$$

If $f(1) = \alpha \neq 0$, define $f_1 = \alpha^{-1}f$. Then $f_1(1) = 1$ and f_1 has an inverse g_1. Let $g = \alpha^{-1}g_1$ and note that

$$f * g = \alpha^{-1}f * \alpha g = f_1 * g_1 = e. \qquad \square$$

Lemma 2.15 *Let $\{f_j\}$, $\{g_j\}$, $\{F_j\}$ and $\{G_j\}$ be sequences of arithmetic functions with the properties $f_j \to f$, $g_j \to g$, $F_j \to 0$, and $\sup_j |G_j(n)| < \infty$ for each $n \in \mathbb{Z}^+$. Also, let α, β be complex constants. Then*

(1) $\alpha f_j + \beta g_j \to \alpha f + \beta g$,

(2) $T^\alpha f_j \to T^\alpha f$,

(3) $L f_j \to L f$,

(4) $F_j * G_j \to 0$,

(5) $f_j * g_j \to f * g$.

Proof. (1), (2), and (3) are obvious. (2) and (3) assert that T^α and L are continuous. Let $M(n) = \sup_j |G_j(n)|$. Then for each $n \in \mathbb{Z}^+$ we have

$$|(F_j * G_j)(n)| \leq \sum_{k|n} M(n/k)|F_j(k)| \leq \max_{1 \leq \ell \leq n} M(\ell) \sum_{k|n} |F_j(k)| \to 0$$

as $j \to \infty$, proving (4). We establish (5) by writing

$$f * g - f_j * g_j = f * (g - g_j) + g_j * (f - f_j).$$

For each n, $g_j(n)$ is a bounded sequence of numbers. Noting that $g - g_j \to 0$, $f - f_j \to 0$, we apply (4) to obtain (5). $\qquad \square$

We apply the above results on convergence to show that 1 and μ are inverses. Our proof is based on the arithmetic analogue of the Euler product for the Riemann zeta function (§1.5).

Theorem 2.16 $\mu * 1 = e$.

Proof. For p a prime let $\mu_p := e - e_p$ (cf. Example 2.1). Also let

$$1_p = \sum_{j=0}^{\infty} e_{p^j} = \text{indicator function of } \{p^j : j = 0, 1, 2, \dots\}.$$

In Problem 2.9, we stated (in different notation) that $1_2 * \mu_2 = e$. For any prime p we have $1_p * \mu_p = e$. One way of seeing this is to apply formula (2.4):

$$\mu_p^{*-1} = \sum_{j=0}^{\infty} e_p^{*j} = \sum_{j=0}^{\infty} e_{p^j} = 1_p .$$

We now show that

$$1_2 * 1_3 * 1_5 * \cdots = 1, \quad \mu_2 * \mu_3 * \mu_5 * \cdots = \mu,$$

i.e. that the partial products over primes $p \leq X$ converge to the limits 1 and μ respectively as $X \to \infty$. Since $1_p(1) = \mu_p(1) = 1$, the condition of convergence of the numerical infinite product is trivial, and we have

$$1_2 * 1_3 * 1_5 * \cdots (1) = 1, \quad \mu_2 * \mu_3 * \mu_5 * \cdots (1) = 1.$$

For $n \geq 2$, let $n = p_1^{\alpha_1} p_2^{\alpha_2} \cdots p_r^{\alpha_r}$, where $p_1, p_2, \ldots$ are the primes in their usual order and the α_i are nonnegative integers. For any $N > r$ we have

$$(1_{p_1} * \cdots * 1_{p_N})(n) = \sum_{k_1 k_2 \cdots k_N = n} 1_{p_1}(k_1) 1_{p_2}(k_2) \cdots 1_{p_N}(k_N)$$

$$= \{1_{p_1}(p_1^{\alpha_1}) \cdots 1_{p_r}(p_r^{\alpha_r})\}\{1_{p_{r+1}}(1) \cdots 1_{p_N}(1)\}$$

$$= 1.$$

All other terms of $\sum_{k_1 \cdots k_N = n}$, corresponding to other decompositions of n, necessarily vanish. Thus we have proved that the partial products converge.

Also, we have

$$(\mu_{p_1} * \cdots * \mu_{p_N})(n) = \sum_{k_1 k_2 \cdots k_N = n} \mu_{p_1}(k_1') \mu_{p_2}(k_2) \cdots \mu_{p_N}(k_N)$$

$$= \{\mu_{p_1}(p_1^{\alpha_1}) \cdots \mu_{p_r}(p_r^{\alpha_r})\}\{\mu_{p_{r+1}}(1) \cdots \mu_{p_N}(1)\}$$

$$= \begin{cases} 0, & \text{if some } \alpha_i > 1 \\ (-1)^s, & \text{if } n \text{ is the product of } s \text{ distinct primes} \end{cases}$$

$$= \mu(n).$$

By Lemma 2.15 we now have

$$\mu * 1 = \left(\lim_{N \to \infty} \mu_2 * \cdots * \mu_{p_N} \right) * \left(\lim_{N \to \infty} 1_2 * \cdots * 1_{p_N} \right)$$

$$= \lim_{N \to \infty} \left\{ \mu_2 * \cdots * \mu_{p_N} * 1_2 * \cdots * 1_{p_N} \right\}$$

$$= \lim_{N \to \infty} e = e. \qquad \square$$

We conclude this section by establishing two results, arithmetic analogues of familiar theorems of analysis, which will be of use in studying the exponential of an arithmetic function: a convergent power series in an arithmetic function is continuous and termwise differentiation of such a series is valid.

Lemma 2.17 *Let $g_j \in \mathcal{A}$, $g_j(1) = 0$ for $j = 1, 2, \ldots$, and $g_j \to g \in \mathcal{A}$ as $j \to \infty$. Let $a_0, a_1, \ldots$ be any sequence of complex numbers. Then*

$$\lim_{j \to \infty} \sum_{i=0}^{\infty} a_i g_j^{*i} = \sum_{i=0}^{\infty} a_i g^{*i}.$$

Proof. We fix $n \geq 1$ and show the two sides of the formula agree at n. For any $r \geq \lceil \log n / \log 2 \rceil$ we have

$$\left\{ \sum_{i=0}^{\infty} a_i g_j^{*i} \right\}(n) = \sum_{i=0}^{r} a_i g_j^{*i}(n) \to \sum_{i=0}^{r} a_i g^{*i}(n) = \left\{ \sum_{i=0}^{\infty} a_i g^{*i} \right\}(n).$$

The limiting process is valid since the summation involves only a finite number of nonzero terms. (For $n = 1$ both sides equal a_0.) $\qquad \square$

Lemma 2.18 *Let $g \in \mathcal{A}$, $g(1) = 0$ and let $\sum_{j=0}^{\infty} a_j g^{*j}$ be a power series. Then*

$$L\left(\sum_{j=0}^{\infty} a_j g^{*j} \right) = \left(\sum_{j=1}^{\infty} j a_j g^{*(j-1)} \right) * Lg.$$

Proof. Fix $n \geq 1$. We evaluate each side of the formula at n. The summation on each side extends over at most $1 + \lceil \log n / \log 2 \rceil$ indices j. Now L is linear and for each fixed j we apply the power formula (2.1). $\qquad \square$

An example of the last lemma is

$$L\left(\sum_{j=0}^{\infty} g^{*j}/j!\right) = \left(\sum_{j=0}^{\infty} g^{*j}/j!\right) * Lg. \qquad (2.5)$$

The series occurring in (2.5) is an analogue of the familiar exponential series. In the next section we shall investigate some properties of this series and show its use in multiplicative number theory.

2.4 Exponential mapping

Let

$$\mathcal{U} := \{f \in \mathcal{A} : f(1) \neq 0\},$$
$$\mathcal{A}_i := \{f \in \mathcal{A} : f(1) = i\}, \quad i = 0, 1.$$

Recall that the invertible elements of $\mathcal{A}$ are precisely the elements of $\mathcal{U}$. It is easy to see that $\mathcal{U}$ and $\mathcal{A}_1$ are groups under $*$ and that $\mathcal{A}_0$ is a group under addition. We remark that $\mathcal{A}_0$ is the unique maximal ideal of $\mathcal{A}$.

By analogy with real numbers, we seek a continuous map which transforms multiplication into addition. In this section we define an exponential map on $\mathcal{A}_0$ and show that it does what we expect of an exponential. We define exp by setting

$$\exp \nu = e + \nu + \frac{\nu * \nu}{2!} + \frac{\nu * \nu * \nu}{3!} + \cdots.$$

It is clear that $\exp \nu \in \mathcal{A}_1$. We shall generally use Greek letters to denote generic elements of the domain of exp and Latin letters for generic elements of the range.

Example 2.19

$$(\exp e_2)(n) = \begin{cases} 1/\alpha!, & \text{if } n = 2^{\alpha} \text{ for some } \alpha \geq 0, \\ 0, & \text{if } n \neq 2^{\alpha} \text{ for all } \alpha \geq 0. \end{cases}$$

By the chain rule of calculus, if $f = \exp g$, then $f' = f \cdot g'$. Recalling that L is a derivation, in the following theorem we establish a similar formula for arithmetic functions. Also, we show that an analogue of the power series

$$\log x = \sum_{j=1}^{\infty} (-1)^{j-1} (x-1)^j / j, \quad |x-1| < 1,$$

holds here as well.

Theorem 2.20 *The exponential map is a one-to-one mapping from $\mathcal{A}_0$ onto $\mathcal{A}_1$ which is continuous in both directions and is a group isomorphism. If $\lambda \in \mathcal{A}_0$ and $f \in \mathcal{A}_1$, then the following three equations are equivalent:*

$$f = \exp \lambda, \tag{2.6}$$

$$L\lambda = Lf * f^{*-1}, \tag{2.7}$$

$$\lambda = \sum_{j=1}^{\infty} (-1)^{j-1} (f-e)^{*j} / j. \tag{2.8}$$

Proof. (1) **Homomorphism property.** Let $\lambda, \nu \in \mathcal{A}_0$ and let n be a given positive integer. Consider the rectangular array

$$\left\{ \left(\frac{\lambda^{*j}}{j!} * \frac{\nu^{*k}}{k!} \right)(n) \right\}_{j,k=0}^{\infty}.$$

By Lemma 2.11, all terms with $j + k > \log n / \log 2$ are zero. If the array is summed by rows and then columns we obtain $(\exp \lambda * \exp \nu)(n)$. On the other hand, if the matrix is summed along a diagonal $\{(j,k) : j + k = \ell\}$ we obtain

$$\frac{1}{\ell!} \sum_{j+k=\ell} \frac{\ell!}{j!k!} (\lambda^{*j} * \nu^{*k})(n) = \frac{1}{\ell!} (\lambda + \nu)^{*\ell}(n),$$

and if the last expression is then summed over all nonnegative values of ℓ, we obtain $\{\exp(\lambda + \nu)\}(n)$. Thus

$$\exp(\lambda + \nu) = (\exp \lambda) * (\exp \nu).$$

(2) **Differentiation identity.** If $f = \exp \lambda$, then by (2.5), $Lf = f * L\lambda$.

(3) **One-to-one property.** First we prove that $\exp(-\lambda) = (\exp \lambda)^{*-1}$. Indeed, $\exp \lambda \in \mathcal{A}_1$ and thus is invertible. We have

$$(\exp \lambda) * (\exp \lambda)^{*-1} = e = \exp(\lambda + (-\lambda)) = (\exp \lambda) * (\exp(-\lambda)).$$

The assertion follows from the uniqueness of inverses.

Now assume $\lambda_1, \lambda_2 \in \mathcal{A}_0$ with $\exp \lambda_1 = \exp \lambda_2$. Since $e = \exp(\lambda_1 - \lambda_2)$, we have

$$0 = Le = L \exp(\lambda_1 - \lambda_2) = \exp(\lambda_1 - \lambda_2) * L(\lambda_1 - \lambda_2) = L(\lambda_1 - \lambda_2).$$

Thus $\lambda_1(n) = \lambda_2(n)$ for all $n \geq 2$ and $\lambda_1(1) = \lambda_2(1) = 0$.

(4) **Equivalence of (2.7) and (2.8).** Lemma 2.18 and the formula (2.4) for inverses yield the following identity for $f \in \mathcal{A}_1$:

$$L\left\{ \sum_{j=1}^{\infty} \frac{(-1)^{j-1}}{j}(f - e)^{*j} \right\} = \left\{ \sum_{j=1}^{\infty} (-1)^{j-1}(f - e)^{*(j-1)} \right\} * Lf = f^{*-1} * Lf.$$

Now if $\lambda(1) = 0$, we have

$$L\lambda = f^{*-1} * Lf \iff L\lambda = L\left\{ \sum_{j=1}^{\infty} \frac{(-1)^{j-1}}{j}(f - e)^{*j} \right\}$$

$$\iff \lambda = \sum_{j=1}^{\infty} \frac{(-1)^{j-1}}{j}(f - e)^{*j}.$$

(5) **Equivalence of (2.6) and (2.7).** If $f = \exp \lambda$, then (2) implies that $Lf = f * L\lambda$, and we immediately obtain the desired formula for $L\lambda$. For the converse, we need the following identity:

$$0 = L(f * f^{*-1}) = Lf * f^{*-1} + f * Lf^{*-1}.$$

Now we assume that $L\lambda = Lf * f^{*-1}$ and write

$$L(f^{*-1} * \exp \lambda) = (\exp \lambda) * (Lf^{*-1} + f^{*-1} * L\lambda)$$

$$= (\exp \lambda) * f^{*-1} * (L\lambda + f * Lf^{*-1})$$

$$= (\exp \lambda) * f^{*-1} * (L\lambda - f^{*-1} * Lf) = 0.$$

Thus $f^{*-1} * \exp \lambda = ke$ for some constant k and so $\exp \lambda = kf$. Evaluation at 1 shows that $k = 1$.

(6) **exp maps onto $\mathcal{A}_1$.** Let $f \in \mathcal{A}_1$ be given. There exists a. $\lambda \in \mathcal{A}_0$ satisfying (2.8). By the equivalence of (2.6) and (2.8), $\exp \lambda = f$, and hence f is in the range of exp. Since exp is a one-to-one map of $\mathcal{A}_0$ onto

$\mathcal{A}_1$, there is a well defined inverse map, which we call log, defined on $\mathcal{A}_1$. The relations established in (4) and (5) give the following formula for log:

$$\log f := \sum_{j=1}^{\infty} (-1)^{j-1}(f-e)^{*j}/j.$$

(7) **Continuity.** Both exp and log are given as power series — in terms of $\lambda \in \mathcal{A}_0$ in the first case and of $f - e \in \mathcal{A}_0$ in the second. By Lemma 2.17 each mapping is continuous. $\square$

2.4.1   The 1 function as an exponential

Since $1 \in \mathcal{A}_1$, it is representable as an exponential. We now find the logarithm of 1, using the decomposition occurring in Theorem 2.20. We work with μ, the inverse function of 1, for simplicity in calculations.

For p a prime, let $\mu_p = e - e_p$. By the power series for log, we have

$$-\log \mu_p = \sum_{j=1}^{\infty} \frac{1}{j} e_p^{*j} = \sum_{j=1}^{\infty} \frac{1}{j} e_{p^j}.$$

Thus, $1_p = \exp f_p$, where f_p is the sum of the last series. By the homomorphism property of exp, for any prime P, we have

$$1_2 * \cdots * 1_P = \exp\left\{ \sum_{p \leq P} f_p \right\}.$$

The left side of the last equation converges to 1 as $P \to \infty$ (cf. proof of Theorem 2.20). If we define κ as $f_2 + f_3 + f_5 + \cdots$, then

$$\kappa(n) := \begin{cases} 1/j, & \text{if } n = p^j, \text{ some } p \text{ and some } j \geq 1, \\ 0, & \text{if } n = 1 \text{ or } n \text{ is divisible by more than one prime.} \end{cases} \qquad (2.9)$$

Now exp is continuous and hence we deduce that

$$1 = \exp \kappa. \qquad (2.10)$$

Of course, the Möbius function μ satisfies $\mu = \exp(-\kappa)$.

Formula (2.10) is remarkable for the fact that the left hand side is connected with the set of positive integers, while the right hand side involves in a simple way the primes and their powers. It is not too surprising then that this equation should be useful in the study of prime numbers. To this end we introduce von Mangoldt's Λ function, a weighted prime (and prime power) counting function, defined by setting $\Lambda = L\kappa$, i.e.,

$$\Lambda(n) := \begin{cases} \log p, & \text{if } n = p^j \text{ for some } p \text{ and some } j \geq 1, \\ 0, & \text{if } n = 1 \text{ or if } n \text{ is divisible by more than one prime.} \end{cases}$$

By Theorem 2.20, equation (2.10) is equivalent to the following equation:

Lemma 2.21 $L1 = 1 * \Lambda$.

This formula was first exploited by Chebyshev; it served as the starting point of his investigation of the distribution of primes (see §4.2). It is desirable to have a direct proof of Lemma 2.21. This in turn will give another proof of the validity of equation (2.10).

Proof. We have $(L1)(1) = 0 = (\Lambda * 1)(1)$. For $n \geq 2$, set

$$n = \prod_{p|n} p^{\alpha_p(n)}.$$

Then

$$(\Lambda * 1)(n) = \sum_{p^\beta | n} \log p = \sum_{p|n} \alpha_p(n) \log p = \log n = (L1)(n). \qquad \square$$

2.4.2 Powers and roots

It is easy to see that if $f = \exp \lambda$, then $f^{*n} = \exp n\lambda$. In particular, we have $\exp 2\kappa = 1 * 1 =: \tau$. An arithmetic function f with the property that $f * f = g$ is called a (convolution) *square root* of g. If f is a square root of g, then of course $-f$ is one also. If $g \in \mathcal{A}_1$ there is a unique square root $f \in \mathcal{A}_1$, determined inductively by the infinite set of equations

$$f(1) = 1,$$

$$g(2) = (f * f)(2) = f(2)f(1) + f(1)f(2),$$

$$g(3) = (f * f)(3) = f(3)f(1) + f(1)f(3),$$

$$g(4) = (f * f)(4) = f(4)f(1) + f(2)f(2) + f(1)f(4),$$

PROBLEM 2.13   Let $f \in \mathcal{A}_1$ satisfy $f * f = 1$. Show that f is given by $f = \exp(\kappa/2)$. Compute $f(n)$ for $n = 1, 2, \ldots, 10$.

PROBLEM 2.14   Let f be as in the preceding problem. Show that

$$f = (e + (1 - e))^{*1/2} = \sum_{j=0}^{\infty} \binom{1/2}{j} (1 - e)^{*j}.$$

Hint. The binomial expansion and uniqueness of the Maclaurin series give a useful combinatorial identity.

PROBLEM 2.15   Show that an arithmetic function has at most two convolution square roots. Hint. $\mathcal{A}$ has no zero divisors.

We have seen in Lemmas 2.4 and 2.15 that each mapping T^α is linear, continuous, and satisfies $T^\alpha(f^{*n}) = (T^\alpha f)^{*n}$ for any $f \in \mathcal{A}$. It follows that *each mapping T^α commutes with* exp. That is, for any $\lambda \in \mathcal{A}_0$ we have

$$T^\alpha(\exp \lambda) = \exp(T^\alpha \lambda).$$

As an example, the sum-of-divisors function σ satisfies

$$\sigma := T1 * 1 = \exp(T\kappa + \kappa).$$

What is a similar representation for Euler's function φ?

The support of the arithmetic function $\kappa = \log 1$ is the set of prime powers. Since we are most frequently interested in the primes themselves, we might ask what would happen if the powers of primes were "deleted from" κ. Precisely, we take ν as the indicator function of the primes and compute $\exp \nu$. Since $0 \leq \nu \leq \kappa$ and the series for exp has only positive terms, we have $e \leq \exp \nu \leq \exp \kappa = 1$. (Recall that for f, $g \in \mathcal{A}$, "$f \leq g$" means that $f(n) \leq g(n)$ for all $n \in \mathbb{Z}^+$.) The exact formula for $\exp \nu$ is

$$(\exp \nu)(p_1^{\alpha_1} \cdots p_r^{\alpha_r}) = \frac{1}{\alpha_1! \cdots \alpha_r!}. \tag{2.11}$$

We shall sketch three different proofs of this formula and ask the reader to give the details.

For one proof we set $n = p_1^{\alpha_1} \cdots p_r^{\alpha_r}$ and $\beta = \alpha_1 + \cdots + \alpha_r$ and note that in the exponential series all terms except the βth vanish at n. We have

$$(\exp \nu)(n) = \frac{1}{\beta!} \nu^{*\beta}(p_1^{\alpha_1} \cdots p_r^{\alpha_r}),$$

and the last expression can be evaluated combinatorially to yield (2.11).

Another proof of equation (2.11) can be given by using Theorem 2.20. If $f = \exp \nu$, then by (2.5),

$$f(n) \log n = \sum_{j \mid n} f(n/j) \nu(j) \log j = \sum_{p \mid n} f(n/p) \log p.$$

The last formula enables us to verify (2.11) by induction on n.

For a third proof of (2.11), recall how we evaluated $\exp e_2$ near the beginning of §2.4. For each p we can similarly evaluate $\exp e_p$, then express $\exp \nu$ as

$$(\exp e_2) * (\exp e_3) * (\exp e_5) * \cdots$$

to obtain (2.11). An argument of this type occurred in the proof of Theorem 2.16. The technique of resolving a function into a convolution product is effective for certain problems, and we shall study it systematically in the next section.

2.5 Multiplicative functions

Let

$$\mathcal{M} := \{ f \in \mathcal{A} : f \neq 0, f(mn) = f(m)f(n) \text{ if } (m, n) = 1 \}.$$

The elements of $\mathcal{M}$ are called *multiplicative* arithmetic functions. It is easy to see that a multiplicative function is determined by its values on prime powers. If $p_1, p_2, \ldots, p_r$ are distinct primes and $\alpha_1, \alpha_2, \ldots, \alpha_r$ are any nonnegative integers, then we have

$$f(p_1^{\alpha_1} p_2^{\alpha_2} \cdots p_r^{\alpha_r}) = f(p_1^{\alpha_1}) f(p_2^{\alpha_2}) \cdots f(p_r^{\alpha_r}).$$

Conversely, a function $f \neq 0$ satisfying this equality for all finite sets of distinct primes and all nonnegative integer exponents is multiplicative.

If $g(mn) = g(m)g(n)$ for *all* m, $n \in \mathbb{Z}^+$ and if $g \neq 0$, then g is called *completely multiplicative*. Such functions are determined by their values on primes. We have

$$g(p_1^{\alpha_1} p_2^{\alpha_2} \cdots p_r^{\alpha_r}) = g(p_1)^{\alpha_1} g(p_2)^{\alpha_2} \cdots g(p_r)^{\alpha_r}.$$

Conversely, a function $g \neq 0$ satisfying the last relation for all primes and nonnegative integer exponents is completely multiplicative.

As the examples below suggest, many significant arithmetic functions are multiplicative. Elements of $\mathcal{M}$ necessarily possess a certain degree of regularity, and this quality is useful in studying an arithmetic function. In this section we shall characterize elements of $\mathcal{M}$ in terms of their convolution logarithms (in the sense of §2.4), and from this deduce some familiar theorems about multiplicative functions.

Example 2.22 The following functions are multiplicative: τ; μ; σ; $f(n) = c^{\omega(n)}$ (c a nonzero constant); $\exp(e_2)$; $|\mu|$; $e + f$, where f is a function whose support is contained among the powers of a fixed prime.

Example 2.23 The following functions are completely multiplicative: $T^{\alpha}1$; $1_p = \sum_{j=0}^{\infty} e_{p^j}$; $1_p * 1_{p'}$ ($p \neq p'$); $f(n) = c^{\Omega(n)}$ (c a nonzero constant).

Example 2.24 Let f, g, and h be multiplicative and assume $h(n) \neq 0$ for all $n \in \mathbb{Z}^+$. Then the following functions are multiplicative: $|f|$; $1/h$; $f \cdot g$. Here $(f \cdot g)(n) = f(n)g(n)$, the pointwise product. If f, g, and h are completely multiplicative, then so too are $|f|$, $1/h$, and $f \cdot g$.

PROBLEM 2.16 Let $f \in \mathcal{M}$. Show that

$$f(m)f(n) = f((m,n))f([m,n]), \tag{2.12}$$

where (m,n) is the greatest common divisor of m and n and $[m,n]$ is their least common multiple. Conversely, if $f \in \mathcal{A}_1$ and (2.12) holds, show that $f \in \mathcal{M}$.

Lemma 2.25 $\mathcal{M} \subset \mathcal{A}_1$.

Proof. There is an $n_0 \geq 1$ such that $f(n_0) \neq 0$. Since $(n_0, 1) = 1$, we have $f(n_0) = f(n_0)f(1)$ and hence $f(1) = 1$. $\square$

Every multiplicative function can be represented in a canonical way as a convolution product. We have applied this idea earlier in the proof of Theorem 2.16, where we factored 1 and μ.

Lemma 2.26 Let $f \in \mathcal{A}_1$ and let $p_1, p_2, \ldots,$ denote the primes with their usual order. For $j = 1, 2, \ldots$ define functions $f_j \in \mathcal{M}$ by setting

$$f_j(m) = \begin{cases} f(m), & \text{if } m = p_j^{\alpha} \text{ for some } \alpha \geq 0, \\ 0, & \text{otherwise.} \end{cases}$$

Then

$$f = f_1 * f_2 * \cdots \iff f \in \mathcal{M}.$$

Proof. It is easy to verify that each $f_j \in \mathcal{M}$. Given $n = p_1^{\alpha_1} p_2^{\alpha_2} \cdots p_r^{\alpha_r}$ (each $\alpha_j \geq 0$), we take $\nu \geq r$ and observe that

$$(f_1 * \cdots * f_\nu)(n) = \prod_{j=1}^{r} f_j(p_j^{\alpha_j}) \prod_{j=r+1}^{\nu} f_j(1) = \prod_{j=1}^{r} f_j(p_j^{\alpha_j}).$$

The infinite convolution product $f_1 * f_2 * \cdots$ converges because (1) $f_j(1) = 1$ for all j and (2) if $n = p_1^{\alpha_1} \cdots p_r^{\alpha_r}$, then $(f_1 * \cdots * f_\nu)(n)$ remains constant for all $\nu \geq r$. If $f = f_1 * f_2 * \cdots$, then with the preceding identity,

$$f(n) = (f_1 * \cdots * f_\nu)(n) = \prod_{j=1}^{r} f_j(p_j^{\alpha_j}) = \prod_{j=1}^{r} f(p_j^{\alpha_j})$$

and f is multiplicative. Conversely, if f is multiplicative, then using the identity again,

$$f(n) = \prod_{j=1}^{r} f(p_j^{\alpha_j}) = \prod_{j=1}^{r} f_j(p_j^{\alpha_j}) = (f_1 * \cdots * f_\nu)(n)$$

for all $\nu \geq r$, and thus $f = f_1 * f_2 * \cdots$. $\qquad\square$

Since $\mathcal{M} \subset \mathcal{A}_1$, we can express each multiplicative function f as the exponential of a function $\lambda \in \mathcal{A}_0$. We now show that $f \in \mathcal{A}_1$ is multiplicative if and only if its logarithm λ has support contained in the set of prime powers.

Theorem 2.27 *Let $f = \exp \lambda$ for some $\lambda \in \mathcal{A}_0$. Then $f \in \mathcal{M} \Longleftrightarrow \lambda(n) = 0$ for $n = 1$ and for all n divisible by (at least) two different primes.*

Proof. Suppose first that λ is zero except on primes and prime powers. Let $p_1, p_2, \ldots$ be the primes in their usual order. For each $j \in \mathbb{Z}^+$ define $\lambda_j \in \mathcal{A}_0$ by

$$\lambda_j(m) = \begin{cases} \lambda(m), & \text{if } m = p_j^\beta \text{ for some } \beta \geq 1, \\ 0, & \text{otherwise,} \end{cases}$$

and for r a positive integer define $\rho_r \in \mathcal{A}_0$ by setting

$$\lambda = \lambda_1 + \lambda_2 + \cdots + \lambda_r + \rho_r.$$

The support of ρ_r is contained in the set $\{p^\alpha : p > p_r, \alpha \in \mathbb{Z}^+\}$.

Note that $(\exp \lambda_j)(m) = 0$ if $m \neq p_j^\beta$ for some $\beta \geq 1$, since in this case $\lambda_j(m) = (\lambda_j * \lambda_j)(m) = \cdots = 0$. Similarly, $(\exp \rho_r)(m) = 0$ if $m \neq 1$ and all prime divisors of m are at most equal to p_r, because $\rho_r(m) = (\rho_r * \rho_r)(m) = \cdots = 0$. Of course $(\exp \rho_r)(1) = 1$.

In view of these remarks, for $n = p_1^{\alpha_1} \cdots p_r^{\alpha_r}$ we have

$$f(n) = (\exp \lambda_1 * \exp \lambda_2 * \cdots * \exp \lambda_r * \exp \rho_r)(n)$$

$$= \left\{ \prod_{i=1}^{r} (\exp \lambda_i)(p_i^{\alpha_i}) \right\} (\exp \rho_r)(1).$$

If we specialize n to $p_i^{\alpha_i}$, we see that

$$f(p_i^{\alpha_i}) = (\exp \lambda_i)(p_i^{\alpha_i}).$$

Returning to the general case we have

$$f(n) = f(p_1^{\alpha_1}) f(p_2^{\alpha_2}) \cdots f(p_r^{\alpha_r}),$$

and hence f is multiplicative.

Now suppose f is multiplicative. We form the canonical representation of f as $f_1 * f_2 * \cdots$ as described in the preceding lemma. Each $f_j \in \mathcal{A}_1$ and thus we can write $f_j = \exp \gamma_j$ for some $\gamma_j \in \mathcal{A}_0$. We claim that if $\gamma_j(m) \neq 0$, then $m = p_j^\alpha$ for some $\alpha \geq 1$. Indeed, by Theorem 2.20

$$\gamma_j = \sum_{k=1}^{\infty} (-1)^{k-1} (f_j - e)^{*k} / k,$$

and $f_j - e$ and all its powers have support contained in the set of integers $\{ p_j^\alpha : \alpha \geq 1 \}$. We set $\lambda = \sum_{j=1}^{\infty} \gamma_j$ (a convergent series since $\mathrm{fnz}(\gamma_j) \geq p_j$), and verify that

$$f = \lim_{\nu \to \infty} f_1 * \cdots * f_\nu = \lim_{\nu \to \infty} \exp (\gamma_1 + \cdots + \gamma_\nu) = \exp \lambda.$$

It is clear from the construction that the support of λ is contained in the set of prime powers. By Theorem 2.20, λ is unique. $\qquad\square$

Theorem 2.27 exhibits a one-to-one correspondence between elements of $\mathcal{M}$ and elements of $\mathcal{A}_0$ whose support lies in the set of prime powers. The last set is clearly a group under addition. By the isomorphic property of exp, $\mathcal{M}$ is a group under convolution.

Theorem 2.28 *$\mathcal{M}$ is a subgroup of $\mathcal{A}_1$ under convolution.*

Proof. We spell out the preceding remarks explicitly. If $f = \exp \lambda$ and $g = \exp \nu$ are elements of $\mathcal{M}$, then λ and ν have support in the set of prime powers. The arithmetic functions $-\lambda$, $\lambda + \nu$, and 0 have their supports contained in the same set, and thus $\exp(-\lambda) = f^{*-1}$, $\exp(\lambda + \nu) = f * g$, and $\exp(0) = e$ are multiplicative. $\square$

PROBLEM 2.17 Let $f = \exp \lambda$, f multiplicative. Compute $\lambda(p)$ and $\lambda(p^2)$ in terms of values of f.

PROBLEM 2.18 Show directly from the definition of $*$ that the convolution of two multiplicative functions is multiplicative.

PROBLEM 2.19 Show that τ, σ, and $\varphi \in \mathcal{M}$. Verify the following formulas for $n = \prod_{i=1}^{r} p_i^{\alpha_i}$, with $p_1, \ldots, p_r$ distinct primes and each $\alpha_i \geq 1$:

$$\tau(n) = \prod_{i=1}^{r} (\alpha_i + 1),$$

$$\sigma(n) = \prod_{i=1}^{r} (1 + p_i + \cdots + p_i^{\alpha_i}),$$

$$\varphi(n) = n \prod_{i=1}^{r} (1 - p_i^{-1}),$$

$$1 * 1 * 1(n) = \prod_{i=1}^{r} (\alpha_i + 1)(\alpha_i + 2)/2.$$

PROBLEM 2.20 Let $f \in \mathcal{A}_1$, $f * f = 1$. Show that f is multiplicative.

PROBLEM 2.21 Let α be an arbitrary real number. Show that T^α maps $\mathcal{M}$ into itself.

PROBLEM 2.22 Show that a multiplicative function $f = \exp \lambda$ is completely multiplicative if and only if $\lambda(p^\alpha) = \lambda(p)^\alpha / \alpha$ for each prime p and $\alpha \geq 1$.

PROBLEM 2.23 Let f be any completely multiplicative function except e. Show that $f * f$ and f^{*-1} are *not* completely multiplicative.

PROBLEM 2.24 Let $f \in \mathcal{A}_1$. Show that f is completely multiplicative if and only if $f * f = f \cdot \tau$.

PROBLEM 2.25 Let $f \in \mathcal{A}_1$. Show that f is completely multiplicative if and only if for each g and $h \in \mathcal{A}$ we have $f \cdot (g * h) = (f \cdot g) * (f \cdot h)$.

PROBLEM 2.26 Show that $|\mu|$ is multiplicative but not completely multiplicative. Express $|\mu|$ as $f_1 * f_2 * \cdots$ as in Lemma 2.26. Describe the functions f_i. Find $\log |\mu|$.

PROBLEM 2.27 Liouville's λ function is defined by $\lambda(n) = (-1)^{\Omega(n)}$. Show that λ is completely multiplicative. Find $\log \lambda$. What is the relation between λ and $|\mu|$?

PROBLEM 2.28 For $b \in \mathbb{Z}^+$ define $f_b(n) = (n, b)$, the g.c.d. of n and b. Prove that f_b is multiplicative.

PROBLEM 2.29 Let $f = \exp \lambda$, where $\lambda \in \mathcal{A}_0$, and let g be completely multiplicative. Show that $f \cdot g = \exp(\lambda \cdot g)$.

PROBLEM 2.30 For $b \in \mathbb{Z}^+$ define $g_b(n) = 1$ if $(n, b) = 1$, and $g_b(n) = 0$ if $(n, b) > 1$. Show that g_b is completely multiplicative.

PROBLEM 2.31 Let S be the set of squares. Prove that

$$\tau^2 = 1^{*4} * 1_S^{*-1}.$$

In the beginning of this section we mentioned that multiplicative functions possess a certain amount of regularity. As one example of this regularity we shall characterize real valued monotone multiplicative functions.

Theorem 2.29 (P. Erdős). *Let f be a real valued monotone multiplicative function. If f is nondecreasing, then $f = T^\alpha 1$ for some real nonnegative α. If f is nonincreasing then either $f = T^\alpha 1$ for some real nonpositive α or $f = e + ce_2$ for some $c \in [0, 1]$.*

Proof. Suppose first that f is nondecreasing, so that $f(n) \geq f(1) = 1$ for all n. Let a be a fixed positive integer greater than 1. We claim that for any positive integer v

$$f(a^v - 1) \geq f(a)^{v-1}, \quad f(a^v + 1) \leq f(a)^{v-1} f(a + 1). \tag{2.13}$$

For $v = 1$ these inequalities are trivial. If (2.13) holds for a given value of v, then by monotonicity and multiplicativity

$$f(a^{v+1} - 1) \geq f(a^{v+1} - a) = f(a)f(a^v - 1) \geq f(a)^v$$

and

$$f(a^{v+1} + 1) \leq f(a^{v+1} + a) = f(a)f(a^v + 1) \leq f(a)^v f(a + 1).$$

Thus (2.13) holds for all $v \in \mathbb{Z}^+$. It follows that

$$f(a)^{v-1} \leq f(a^v) \leq f(a)^{v-1}f(a + 1)$$

for all integers $v \geq 1$.

Now let n be any integer exceeding a (still fixed), and define the integer r by $a^r \leq n < a^{r+1}$. Then $r \geq 1$ and $f(a^r) \leq f(n) \leq f(a^{r+1})$, whence

$$f(a)^{r-1} \leq f(n) \leq f(a)^r f(a + 1),$$

and therefore

$$f(a)^{(1/\log a)-(2/\log n)} \leq f(n)^{1/\log n} \leq f(a)^{1/\log a} f(a + 1)^{1/\log n}.$$

Letting $n \to \infty$, we get

$$\lim_{n\to\infty} f(n)^{1/\log n} = f(a)^{1/\log a}$$

for each $a \geq 2$. It follows that $f(a)^{1/\log a} = e^\alpha$, where α is a nonnegative constant, i.e., $f(a) = a^\alpha$ for each $a \geq 2$.

Now we consider the case in which f is nonincreasing. Suppose there were an $n_0 \in \mathbb{Z}^+$ for which $f(n_0) < 0$. We have $(n_0, n_0 + 1) = 1$ and $f(n_0 + 1) \leq f(n_0) < 0$. Thus

$$0 > f(n_0) \geq f(n_0(n_0 + 1)) = f(n_0)f(n_0 + 1) > 0.$$

This is impossible and hence $f \geq 0$ on $\mathbb{Z}^+$.

Suppose there exists an $n \in \mathbb{Z}^+$ such that $f(n) = 0$. By the monotonicity and nonnegativity of f there exists an $n_1 \in \mathbb{Z}^+$ such that $f(n) > 0$ for $1 \leq n \leq n_1$ and $f(n) = 0$ for $n > n_1$. Suppose that $n_1 \geq 3$. Then

$$f(n_1(n_1 - 1)) = f(n_1)f(n_1 - 1) > 0.$$

But this is impossible, since $n_1(n_1 - 1) > n_1$ for $n_1 \geq 3$. It follows that either

    (a) $1 = f(1) \geq f(2) \geq 0 = f(3) = f(4) = \cdots$

or

    (b) $f > 0$ on $\mathbb{Z}^+$.

In case (a) we have $f = e_1 + ce_2$ with some $c \in [0, 1]$; this function is multiplicative and is nonincreasing. In case (b) $1/f$ is multiplicative and monotone nondecreasing. By the first part of the theorem, $1/f(n) = n^\alpha$ for some $\alpha \geq 0$. Thus, if f is nonincreasing and $f(3) > 0$, then

$$f(n) = n^{-\alpha} = (T^{-\alpha}1)(n). \qquad \square$$

2.6 Notes

2.1. The function μ is named for A. F. Möbius, who explored its properties in J. reine angew. Math., vol. 9 (1832), pp. 105–123.

2.5. Theorem 2.29 was first proved by P. Erdős in Ann. of Math. (2), vol. 47 (1946), pp. 1–20. A shorter proof was given by J. Lambek and L. Moser in Proc. Amer. Math. Soc., vol. 4 (1953), pp. 544–545. We have given here a variant of that argument based on suggestions of M. Nair and J. Steinig.

Chapter 3

Summatory Functions

3.1 Generalities

Given $f \in \mathcal{A}$, define a function $F : \mathbb{R} \to \mathbb{C}$ by setting $F(x) = \sum_{n \leq x} f(n)$ for $x \geq 1$ and $F(x) = 0$ for $x < 1$. F is called the *summatory function* of f. Examples of summatory functions include the following, which we define explicitly only on $[1, \infty)$; the functions are always taken as 0 on $(-\infty, 1)$. For g a real valued function on some set, let $g^+(x) = \max\{g(x), 0\}$.

1. $N(x) := [x]^+ = \displaystyle\sum_{n \leq x} 1$,

2. $Q(x) := \displaystyle\sum_{n \leq x} |\mu(n)|$,

3. $\pi(x) := \displaystyle\sum_{n \leq x} 1_{\mathcal{P}}(n)$, $\mathcal{P} :=$ set of primes,

4. $M(x) := \displaystyle\sum_{n \leq x} \mu(n)$,

5. $N_2(x) := \displaystyle\sum_{n \leq x} \tau(n) = \sum_{n \leq x} (1 * 1)(n)$.

Except for the function $f = 1$, all of the preceding arithmetic functions are rather irregular. The τ function, for example, satisfies $\tau(p^\alpha) = \alpha + 1$, from which we see that τ is unbounded but assumes the value 2 infinitely often. On the other hand, a summatory function involves a large number of values of the associated arithmetic function. Thus we might hope that fluctuations of the arithmetic function are somehow smoothed out, enabling us to make statements about its average behavior. We shall see that in all of the above examples this hope is—to a greater or lesser degree—realized.

In this chapter we develop techniques for estimating various summatory functions and apply these methods to some problems of number theoretic interest. Also, we are going to introduce Riemann-Stieltjes integrators as a generalization of arithmetic functions. This will enable us to treat a broader class of problems and make efficient use of techniques from analysis.

It is often convenient to express estimates in terms of the O- and o- symbols. Let F be a real or complex valued function defined on a set I, usually a real half line $[a, \infty)$ for a suitable value of a, and G a positive valued function also defined on I. We say that $F = O(G)$ (on I) if the function F/G is bounded on I. Equivalent notations are $F \ll G$ and $G \gg F$. For $I = [a, \infty)$, we say that $F = o(G)$ if $\lim_{x \to \infty} F(x)/G(x) = 0$. In case F depends also upon one or more parameters, one must check whether estimates hold uniformly in the parameters; in nonuniform cases, we usually mention the parameter in a subscript.

For F a function on $\mathbb{R}$ we define supp F, the *support* of F, to be the closure of the set of points $x \in \mathbb{R}$ at which $F(x) \neq 0$. We say that F is *supported in a set* $E \subset \mathbb{R}$ if supp $F \subset E$. For example, the support of the summatory function of the arithmetic function $e - e_2$ is the interval $[1, 2]$.

The summatory function of the convolution of two arithmetic functions f and g can be represented as a sum involving the summatory function of f or that of g.

Lemma 3.1   *Let f and $g \in \mathcal{A}$ and have summatory functions F and G respectively. Then*

$$\sum_{n \leq x} (f * g)(n) = \sum_{m \leq x} F\left(\frac{x}{m}\right) g(m) = \sum_{\ell \leq x} G\left(\frac{x}{\ell}\right) f(\ell).$$

Proof.   The left hand side equals

$$\sum_{n \leq x} \sum_{\ell m = n} f(\ell) g(m) = \sum_{\ell m \leq x} f(\ell) g(m).$$

The last sum extends over all *lattice points* (ℓ, m), i.e. points in the cartesian plane having integer coordinates, which satisfy $\ell \geq 1$, $m \geq 1$, and $\ell m \leq x$. For a fixed integer $m \in [1, x]$, the index ℓ can range between 1 and x/m. Thus the last sum can be expressed as $\sum_{m \leq x} F(x/m) g(m)$. If we sum first on m and then on ℓ, we obtain $\sum_{\ell \leq x} G(x/\ell) f(\ell)$. $\square$

PROBLEM 3.1   Show that

$$\sum_{n \leq x} \tau(n) = \sum_{m \leq x} [x/m],$$

$$\sum_{n \leq x} \sigma(n) = \frac{1}{2} \sum_{\ell \leq x} \left[\frac{x}{\ell}\right] \left[\frac{x}{\ell} + 1\right] = \sum_{m \leq x} m \left[\frac{x}{m}\right].$$

PROBLEM 3.2   Suppose that f, $g \in \mathcal{A}$ and f, $g \geq 0$. Let $h = f * g$ and F, G, H be the respective summatory functions. Show that $H(x) \leq F(x)G(x)$ holds for all $x \geq 1$. Find conditions on f and/or g for which equality holds (a) for a particular $x \geq 2$, (b) for all $x \geq 2$.

PROBLEM 3.3   Show that $1 = |\mu| * 1_S$, where S is the set of squares. Show that

$$\sum_{n \leq x} \frac{1}{n} \leq \sum_{\ell=1}^{\infty} \frac{1}{\ell^2} \sum_{m \leq x} \frac{|\mu(m)|}{m}$$

holds for all $x \geq 1$. Deduce that $\sum_{m=1}^{\infty} |\mu(m)|/m = \infty$.

Let f, $g \in \mathcal{A}$ and F be the summatory function of f. If we have a "good" approximation of F by a "smooth" function, we often can estimate the summatory function of $f * g$ even though the values of g are quite erratically distributed. As an example, we show that $\mu(n)/n$ has a bounded summatory function. This estimate is not completely obvious, because μ assumes the values 0, -1, 1 in a seemingly irregular manner and $\Sigma|\mu(n)|/n = \infty$.

Lemma 3.2 $\sum_{n \leq x} \mu(n)/n = O(1)$ *on* $[1, \infty)$.

Proof.   We shall show that $\sum_{n \leq x} x\mu(n)/n = O(x)$ on $[1, \infty)$. Recall that $1 * \mu = e$ and by Lemma 3.1,

$$\sum_{n \leq x} \left[\frac{x}{n}\right] \mu(n) = \sum_{n \leq x} e(n) = 1 \quad \text{for } x \geq 1.$$

Now

$$\sum_{n \leq x} \frac{x}{n} \mu(n) = \sum_{n \leq x} \left[\frac{x}{n}\right] \mu(n) + \sum_{n \leq x} \left(\frac{x}{n} - \left[\frac{x}{n}\right]\right) \mu(n),$$

and the last sum is less than x in absolute value. Thus

$$\left|\sum_{n \leq x} \frac{\mu(n)}{n}\right| < 1 + \frac{1}{x}. \qquad \square$$

Remarks 3.3   The inequalities $0 \leq x/n - [x/n] < 1$ of the preceding proof will be of frequent use in the sequel. Also, it will turn out that $\Sigma \mu(n)/n$ converges and has the value 0; however, this is a much deeper fact than boundedness. In §5.4 we shall show that the convergence of $\Sigma \mu(n)/n$ is closely related to the P.N.T.

3.2   Estimate of $Q(x) - 6x/\pi^2$

In §1.4 we saw that $Q(x) := \sum_{n \leq x} |\mu(n)| \sim 6x/\pi^2$ as $x \to \infty$. The $\sim$ notation here is read as "is asymptotic to" and means that

$$Q(x)/\{6x/\pi^2\} \to 1 \quad (x \to \infty).$$

Now we investigate the rate of convergence in the last expression.

Lemma 3.4 $|\mu| = 1 * f$, where $f(n) := \mu(\sqrt{n})$ if n is a square and $f(n) := 0$ otherwise.

Proof.   Let $g := \mu * |\mu|$. Since μ and $|\mu|$ are multiplicative, so is g. An easy calculation shows that $g(1) = 1$ and that for any p, $g(p^2) = -1$ and $g(p^\alpha) = 0$ if $\alpha \neq 0, 2$. Also, f is multiplicative. Thus $g = f$, and we have (by Möbius inversion) $|\mu| = 1 * \mu * |\mu| = 1 * f$.   □

PROBLEM 3.4   Give another proof of Lemma 3.4 using the factorization of $|\mu|$ from Problem 2.26 and an analogue of the polynomial identity

$$1 + x = (1 - x^2)/(1 - x).$$

Theorem 3.5   *If* $x \geq 1$, *then*

$$|Q(x) - 6x/\pi^2| < 3x^{1/2}.$$

Proof.   Let f be as in the preceding lemma. We have

$$Q(x) = \sum_{n \leq x} (f * 1)(n) = \sum_{n \leq x} \left[\frac{x}{n}\right] f(n) = \sum_{m \leq \sqrt{x}} \left[\frac{x}{m^2}\right] \mu(m)$$

$$= \sum_{m \leq \sqrt{x}} \frac{x}{m^2} \mu(m) + \sum_{m \leq \sqrt{x}} \left\{\left[\frac{x}{m^2}\right] - \frac{x}{m^2}\right\} \mu(m) =: I + J.$$

Clearly, $|J| \le \sqrt{x}$, since each summand is bounded by 1. Now set

$$I = x \sum_{m=1}^{\infty} m^{-2}\mu(m) - x \sum_{m>\sqrt{x}} m^{-2}\mu(m) =: I_1 + I_2.$$

This representation is valid since each series converges. Now

$$m^{-2} < \int_{m-\frac{1}{2}}^{m+\frac{1}{2}} t^{-2} dt$$

by the midpoint approximation of a convex function, and hence

$$|I_2| \le x \sum_{m>\sqrt{x}} m^{-2} < x \int_{\sqrt{x}-\frac{1}{2}}^{\infty} t^{-2} dt = \frac{x}{\sqrt{x}-\frac{1}{2}} \le \sqrt{x}+1$$

for $x \ge 1$. Let us denote $\Sigma\mu(m)m^{-2}$ by c. Then $I_1 = cx$ and

$$|Q(x) - cx| < 2\sqrt{x}+1 \le 3\sqrt{x}.$$

Hence $Q(x)/x \to c$ as $x \to \infty$, and recalling Corollary 1.6, we have $c = 1/\zeta(2) = 6/\pi^2$. $\qquad\square$

Remarks 3.6 The preceding proof showed incidentally that

$$\{\Sigma\mu(m)m^{-2}\} \cdot \{\Sigma n^{-2}\} = 1.$$

Also, we have the formula $1 * \mu = e$. It is not hard to see that these identities are related. We shall return to this relationship in §6.1.

PROBLEM 3.5 Using a similar analysis, show that the number of cube-free integers in $[1,x]$ is $x/\zeta(3) + O(x^{1/3})$. You may assume the fact that $\Sigma\mu(n)n^{-3} = 1/\zeta(3)$ (cf. Problem 1.5).

PROBLEM 3.6 Show that for $x \ge 1$,

$$\sum_{n\le x} \frac{\varphi(n)}{n} = \sum_{n\le x} \left[\frac{x}{n}\right] \frac{\mu(n)}{n} = \frac{6}{\pi^2}x + O(\log ex).$$

PROBLEM 3.7 Show that if (i) $g \in \mathcal{A}$, (ii) $\sum_{n=1}^{\infty} |g(n)|/n$ converges, and (iii) $f = 1 * g$, then

$$\lim_{N\to\infty} \frac{1}{N} \sum_{n=1}^{N} f(n) = \sum_{n=1}^{\infty} \frac{g(n)}{n}.$$

x	$Q(x)$	$6x/\pi^2$	$Q(x) - 6x/\pi^2$	$x^{1/2}$	$x^{2/5}$	$x^{1/4}$
20	13	12.158	0.841	4.472	3.314	2.114
50	31	30.396	0.603	7.071	4.781	2.659
100	61	60.792	0.207	10.000	6.309	3.162
176	106	106.995	-0.995	13.266	7.910	3.642
200	122	121.585	0.414	14.142	8.325	3.760
500	306	303.963	2.036	22.360	12.011	4.728
1000	608	607.927	0.072	31.622	15.848	5.623
2000	1217	1215.854	1.145	44.721	20.912	6.687

Table 3.1 $Q(x)$ DATA

PROBLEM 3.8   Define Liouville's function by $\lambda(n) = (-1)^{\Omega(n)}$. Show that $\sum_{n \le x} \lambda(n)/n = O(1)$

(a) by using $\lambda * 1 = 1_S$, S the set of squares,
(b) by using $\lambda * |\mu| = e$.

The data of Table 3.1 suggest that the error estimate we have given for $Q(x) - 6x/\pi^2$ is quite conservative. In the proof of Theorem 3.5 we used the relations

$$[x/m^2] = x/m^2 + O(1) \quad \text{and} \quad \sum_{m \le \sqrt{x}} |\mu(m)| \le \sqrt{x}.$$

It is possible to reorganize our argument to employ a better estimate of $M(y) := \sum_{n \le y} \mu(n)$ than the trivial absolute value bound. Such estimates can be deduced from the P.N.T. (see Theorem 8.17). Further, if the famous Riemann hypothesis (see §8.3) is true, then $M(y) = O(y^{\frac{1}{2}+\epsilon})$ (cf. Problem 8.15) and $Q(x) - 6x/\pi^2 = O(x^{9/28+\epsilon})$ for any fixed $\epsilon > 0$.

3.3   Riemann-Stieltjes integrals

We shall have use for a more general integral than that of Riemann. Either a Riemann-Stieltjes (R.S.) or Lebesgue-Stieltjes integral would suffice for our work. We have chosen the R.S. integral, since it is adequate for our purposes and probably familiar to most readers. All statements we make about R.S. integrals can be interpreted in terms of Lebesgue theory. We shall assume knowledge of the basic facts about the R.S. integral as presented e.g. in [Apos]. Here and in the Appendix we present some results about integrals that we shall need which are not discussed by Apostol.

Let F be a complex valued function defined on an interval (a, ∞), where $a \in \mathbb{R}$ or $a = -\infty$. We say that F is *locally of bounded variation* (loc. B.V.) on (a, ∞) if the variation of F on each compact subinterval $[b, c] \subset (a, \infty)$ is finite. For example, the functions $x \mapsto \sin 1/x$ and $x \mapsto x^2 \sin x$ are loc. B.V. on $(0, \infty)$. If F is defined on $\mathbb{R}$ and $F(x) = 0$ for all $x < 1$, then we define F_v, the *total variation function* of F, as an extended real valued function on $\mathbb{R}$ by setting $F_v(x)$ equal to the total variation of F on $[0, x]$. Familiar facts about F_v include the following:

(1) F_v is monotone nondecreasing (write briefly: "$\uparrow$" or "increasing"),

(2) $|F(b) - F(a)| \le |F_v(b) - F_v(a)|$ for all $a, b \in \mathbb{R}$,

(3) if $F \uparrow$, then $F_v = F$.

A function F supported in $[1, \infty)$ is loc. B.V. iff $F_v(x) < \infty$ for each real x. For example, if $F(x) = [x] - x + 1$ for $x \ge 1$ and $F(x) = 0$ for $x < 1$, then $F_v(x) = [x] + x - 1$ for $x \ge 1$.

We define the R.S. integral following S. Pollard (cf. Notes) by using refinements of partitions. There is another definition, the "uniform R.S. integral," in which the R.S. sums are required to tend to a limit as the distance between consecutive partition points tends to zero. We are using the Pollard integral because it allows us to integrate a larger class of functions than the uniform R.S. integral does. In particular, the following useful theorem is valid for the Pollard integral but not for the uniform integral.

Theorem 3.7 *Let F and g be functions of bounded variation on $[a, b]$ and assume that F is continuous from the right at each point of $[a, b)$ and g is continuous from the left at each point of $(a, b]$. Then $\int_a^b g \, dF$ exists.*

The proof is given in the Appendix. If a R.S. integral is regarded as a function of a variable upper or lower limit, it satisfies a one sided continuity condition. Precisely, we have

Lemma 3.8 *Let F and g satisfy the hypotheses of the preceding theorem. For $a < x < b$ let $\varphi(x) = \int_a^x g \, dF$. Then φ is continuous from the right and $\varphi(x) - \varphi(x-) = g(x)\{F(x) - F(x-)\}$.*

This lemma also is proved in the Appendix. The proof of the last lemma shows incidentally that a discontinuity of F at the right limit b is reflected in the value of the integral, but a discontinuity of F at the left limit a does not affect the value of the integral.

If F satisfies the hypotheses of Theorem 3.7 on $[1, x]$, then

$$\int_1^x 1 \, dF = F(x) - F(1).$$

Usually we want to include the contribution arising from a possible discontinuity of F at 1. For example, if $f \in \mathcal{A}$ and F is its summatory function, then F has a jump of $f(1)$ at the point 1. If F and g satisfy the hypotheses of Theorem 3.7 on $[1 - \epsilon, x]$ for some $\epsilon > 0$, then we define

$$\int_{1-}^x g \, dF := \lim_{\delta \to 0+} \int_{1-\delta}^x g \, dF.$$

If $F(u) = 0$ for $1 - \epsilon \le u < 1$, as is the case when F is the summatory function of an arithmetic function, the limit operation is unnecessary, provided of course, that $\delta \le \epsilon$.

The following two examples will be the most important instances of R.S. integrals for our purposes. Proofs of the assertions are not difficult to supply. Also, they can be found in [Apos], Th. 7.11 and Th. 7.35.

Example 3.9 (Conversion of a sum into an R.S. integral). Let $f \in \mathcal{A}$ with summatory function F and let $0 < a < b$ be given. Let g be continuous from the left at all integers in $(a, b]$. Then

$$\int_a^b g \, dF = \sum_{a < n \le b} g(n) f(n). \tag{3.1}$$

Example 3.10 (Conversion to a Riemann integral). Let f and g be functions which are (properly) Riemann integrable on any bounded interval $[1, X]$. Let $F(x) = \int_1^x f(t) \, dt$ for all $x \ge 1$. Let $1 \le a < b < \infty$. Then

$$\int_a^b g \, dF = \int_a^b g(x) f(x) \, dx. \tag{3.2}$$

Riemann-Stieltjes integrals can be integrated by parts: if $\int_a^b g \, dF$ exists, then $\int_a^b F \, dg$ exists and we have the familiar formula

$$\int_a^b g \, dF + \int_a^b F \, dg = g(x) F(x) \Big|_a^b.$$

PROBLEM 3.9 Estimate $\sum_{n \le x} |\mu(n)| \log n$ using Theorem 3.5 and an integration by parts of an appropriate Stieltjes integral.

Integration by parts yields the following integral inequality, which will be very useful in the sequel.

Lemma 3.11 (Comparison of integrators). *Let g be a nonnegative, left continuous, decreasing function on $(1 - \epsilon, \infty)$. Let F and φ be functions on $\mathbb{R}$ which are supported in $[1, \infty)$, are right continuous, and are locally of bounded variation. Assume $F(x) \le \varphi(x)$ for all x. Then*

$$\int_{1-}^{x} g\,dF \le \int_{1-}^{x} g\,d\varphi.$$

Proof. We have

$$\int_{1-}^{x} g\,(d\varphi - dF) = g(x)\{\varphi(x) - F(x)\} - \int_{1-}^{x} (\varphi - F)\,dg.$$

The right hand side is nonnegative since $g \downarrow$ and g is nonnegative. □

PROBLEM 3.10 Using the estimate $Q(x) > x/2$ for all $x \ge 1$, show that

$$\sum_{n \le x} |\mu(n)|/n > \frac{1}{2} + \frac{1}{2} \log x, \quad x \ge 1.$$

PROBLEM 3.11 By iterating the method of the preceding lemma (or otherwise) show that the series

$$1 - 2 \cdot 2^{-s} + 3^{-s} + 4^{-s} - 2 \cdot 5^{-s} + 6^{-s} + 7^{-s} - 2 \cdot 8^{-s} + 9^{-s} + \cdots$$

converges and is nonnegative for any real positive s.

Let f be a function on $\mathbb{R}$ which does not oscillate too much on an interval I. Then we expect sums and integrals of f over I to be fairly close. Precisely, we have

Lemma 3.12 (Euler's summation formula). *Let f be a continuous function on $[a, b]$ with a piecewise continuous derivative and let c be a constant. Then*

$$\sum_{a < n \le b} f(n) = \int_{a}^{b} f(t)\,dt - (t - [t] - c)\,f(t)\Big|_{a}^{b} + \int_{a}^{b} (t - [t] - c)\,f'(t)\,dt.$$

Proof.   Write the given sum as a Stieltjes integral as in (3.1), insert a comparison term and integrate by parts:

$$\sum_{a<n\leq b} f(n) = \int_a^b f(t)\,d[t] = \int_a^b f(t)\,dt + \int_a^b f(t)\,(d[t] - dt)$$

$$= \int_a^b f(t)\,dt + f(t)\,([t] - t + c)\Big|_a^b - \int_a^b ([t] - t + c)\,f'(t)\,dt.$$

We have used equation (3.2) to replace df by $f'(t)dt$.   □

The above formula is a special case of the Euler-Maclaurin summation formula; cf. Notes. If the function f has two (or more) continuous derivatives on $[a, b]$ then it may be advantageous to integrate by parts again (or several times). As the first additional step, we define φ on $[0, \infty)$ by setting

$$\varphi(x) := \int_0^x \left([t] - t + \frac{1}{2}\right) dt;$$

the choice $c = 1/2$ makes φ bounded. We now have

$$\sum_{a<n\leq b} f(n) = \int_a^b f(t)\,dt + \left([t] - t + \frac{1}{2}\right)f(t)\Big|_a^b$$

$$- \varphi(t)f'(t)\Big|_a^b + \int_a^b \varphi(t)f''(t)\,dt. \tag{3.3}$$

As an example of the application of Euler's formula we estimate the summatory function of $T^{-\alpha}1$ for real positive values of α.

Lemma 3.13   *If $x \geq 1$, then*

$$\sum_{n\leq x} n^{-1} = \log x + \gamma + \theta/x, \quad \text{where } |\theta| \leq 1, \tag{3.4}$$

$$\sum_{n\leq x} n^{-\alpha} = \frac{x^{1-\alpha}}{1-\alpha} + \zeta(\alpha) + O(x^{-\alpha}), \tag{3.5}$$

if $0 < \alpha < 1$ or $\alpha > 1$. Here γ is Euler's constant, $.577215\ldots$, and ζ is defined as in §1.5 if $\alpha > 1$; if $0 < \alpha < 1$, then ζ is defined by

$$\zeta(\alpha) = -\frac{\alpha}{1-\alpha} - \alpha \int_1^\infty t^{-\alpha-1}(t - [t])\,dt. \tag{3.6}$$

The constant implied by the O-symbol in (3.5) is absolute.

Proof. To show (3.4), take $f(t) = t^{-1}$ in Lemma 3.12 and let $a = 1$, $b = x$, and $c = 0$. We obtain

$$\sum_{1 < n \leq x} \frac{1}{n} = \log x - \frac{x - [x]}{x} - \int_1^x (t - [t]) \, t^{-2} \, dt.$$

Since the integral converges as $x \to \infty$, we can write

$$\sum_{1 \leq n \leq x} \frac{1}{n} = \log x + \gamma - \frac{x - [x]}{x} + \int_x^\infty (t - [t]) \, t^{-2} \, dt,$$

where we have set $\gamma = 1 - \int_1^\infty (t - [t]) \, t^{-2} \, dt$. Also, we have

$$\left| \frac{[x] - x}{x} + \int_x^\infty (t - [t]) \, t^{-2} \, dt \right| \leq \frac{1}{x}$$

and thus (3.4) holds. We see from (3.4) that γ satisfies

$$\gamma = \lim_{x \to \infty} \left(\sum_{n \leq x} \frac{1}{n} - \log x \right).$$

If $0 < \alpha < 1$, we take $f(t) = t^{-\alpha}$ and apply Euler's formula as in the case $\alpha = 1$. Here we take $\zeta(\alpha)$ as in equation (3.6). If $\alpha > 1$, we write

$$\zeta(\alpha) = \sum_{n \leq x} n^{-\alpha} + \lim_{y \to \infty} \sum_{x < n \leq y} n^{-\alpha}$$

and apply the Euler formula to the second sum with $f(t) = t^{-\alpha}$, and then let $y \to \infty$. The error estimates in the cases $0 < \alpha < 1$ and $\alpha > 1$ are uniform in α (as well as in x, of course). $\quad\square$

Remarks 3.14 The error term in (3.4) cannot be changed to $o(1/x)$ and in (3.5) cannot be changed to $o(x^{-\alpha})$. Euler's constant turns up in various areas of mathematics; it is a famous unsolved problem whether γ is rational or irrational.

Note carefully how we estimated $\int_1^x (t - [t]) t^{-2} dt$ by writing the integral as $\int_1^\infty - \int_x^\infty$, obtaining a constant plus an error term. Similar reasoning was applied earlier in the proof of Theorem 3.5, where we wrote

$$I := \sum_{m \le \sqrt{x}} \frac{x}{m^2} \mu(m) = \left\{ \sum_{m=1}^{\infty} - \sum_{m > \sqrt{x}} \right\} \frac{x}{m^2} \mu(m)$$

$$= \frac{x}{\zeta(2)} + O\left(x \sum_{m > \sqrt{x}} m^{-2} \right).$$

PROBLEM 3.12   Let $a_n = -1$ if n is not a square and $a_n = 2\sqrt{n} - 1$ if n is a square. Prove that $\sum_{n=1}^{\infty} a_n/n = \gamma$.

PROBLEM 3.13   For each $k \in \mathbb{Z}^+$ show that

$$\sum_{n \le x} (\log n)^k = \int_1^x \log^k t\, dt + O(\log^k x).$$

Note that the sum is $\log([x]!)$ when $k = 1$. The constant implied by the O-symbol is uniform with respect to k.

PROBLEM 3.14   For each $k \in \mathbb{Z}^+$ show that

$$\gamma_k := \lim_{N \to \infty} \left\{ \sum_{n=1}^{N} \frac{(\log n)^k}{n} - \frac{(\log N)^{k+1}}{k+1} \right\}$$

exists and that $\gamma_k = O(k!)$ as $k \to \infty$. Assuming the fact (cf. Th. 8.1) that $\zeta(s) - 1/(s-1)$ is entire, show that $\gamma_k = O_R(R^{-k} k!)$ holds for any $R > 0$.

3.4   Riemann-Stieltjes integrators

Up to this point we have considered arithmetic functions, sums, and occasionally Riemann or R.S. integrals. We shall now introduce an extension of the notion of arithmetic function based on R.S. integrators. The idea underlying the generalization is to view an arithmetic function as a collection of "mass points" located at positive integer points on the real line. Then it becomes reasonable to admit more general types of "mass distributions" of a continuous or discrete character. This change will lead to a theory involving more analysis than we have previously used. On the other hand, it will enable us to treat a larger class of problems, unify many arguments that are valid for both sums and integrals, and allow us to pass easily between sums and integrals. The material of this section, like that of §3.3, also could have been developed in terms of Lebesgue-Stieltjes theory.

Let $\mathcal{V}$ denote the class of complex valued functions on $\mathbb{R}$ which are

(1) zero in $(-\infty, 1)$,
(2) continuous from the right,
(3) loc. B.V.

Given $f \in \mathcal{A}$, its summatory function F is a member of $\mathcal{V}$ and the total variation function satisfies $F_v(x) = \sum_{n \leq x} |f(n)|$. An element of $\mathcal{V}$ is representable as $F_1 - F_2 + iF_3 - iF_4$ where each F_j is monotone ([Apos], Th. 6.13), and hence ([Apos], Th. 4.51) has limits from the left also.

Definition 3.15 Associated with every $F \in \mathcal{V}$ is a complex valued function dF defined on sets that are finite unions of half open intervals $(a, b] \subset \mathbb{R}$. This function, called the *Riemann-Stieltjes integrator of F*, is defined by the following three conditions:

(1) $dF(\phi) := 0$, where ϕ is the empty set,
(2) $dF((a, b]) := F(b) - F(a)$,
(3) if $E_1, \ldots, E_n$ are each of the form $(a, b]$ and if $E_i \cap E_j = \phi$ for $i \neq j$ then

$$dF(\bigcup_{i=1}^{n} E_i) := \sum_{i=1}^{n} dF(E_i).$$

The last property is called *finite additivity*. It is easy to verify that the preceding three properties are consistent. Also, the symbol dF denoting an integrator is consistent with the dF occurring in R.S. integrals in that $dF(E) = \int_E 1 \, dF$ for any set E which is a finite union of intervals $(a, b] \subset \mathbb{R}$.

Let us see how arithmetic functions generate R.S. integrators. For any $f \in \mathcal{A}$, its summatory function F lies in $\mathcal{V}$. For any $a < b < \infty$ we take

$$dF((a, b]) := F(b) - F(a) = \sum_{a < n \leq b} f(n).$$

For example, if $f = 1$, $F(x) = [x]^+ = N(x)$, and E is the union of half open intervals $(a, b]$, then

$$dN(E) = \sum_{n \in E} 1(n) = \sum_{n=1}^{\infty} 1_E(n).$$

Dirac point measure at c is a similar kind of integrator which is not (usually) generated by an arithmetic function. Given $c \geq 1$, let $F(x) = 0$

for $-\infty < x < c$ and $F(x) = 1$ for $c \le x < \infty$. It is clear that $F \in \mathcal{V}$ and that if S is any finite union of half open intervals, then $dF(S) = 1$ or 0 according as c belongs to S or not. This integrator is usually denoted by δ_c. In case $c \in \mathbb{Z}^+$, then δ_c arises from the arithmetic function e_c.

Continuous analogues of the summatory function of an arithmetic function provide another important class of examples. Let f be a function supported in $[1, \infty)$ which is Riemann integrable on each bounded interval $[1, X]$ and let

$$F(x) := \int_1^x f(t)dt.$$

It is well known that F is continuous. Also, we have

$$F_v(x) = \int_1^x |f(t)|\, dt.$$

By the integral analogue of the triangle inequality, $F_v(x) \le \int_1^x |f(t)|\, dt$, and hence $F \in \mathcal{V}$; for the opposite inequality we approximate f from above and below by step functions. The integrator dF is defined by

$$dF((a, b]) := F(b) - F(a) = \int_a^b f(t)dt$$

and for E a finite union of half open intervals, $dF(E) = \int_E f(t)dt$.

An important special case arises upon taking $f(x) := 1$ for $x \ge 1$ and $f(x) := 0$ for $x < 1$. Then $F(x) = x - 1$ for $x \ge 1$ and $F \in \mathcal{V}$. If S denotes any finite union of half open intervals on $\mathbb{R}$, then $dF(S) = $ length of $(S \cap [1, \infty))$. This integrator, which we usually denote by dx, dt, or du, will be called the *Riemann integrator*. The approximation of expressions involving dN by ones involving dt will be a recurring theme.

3.4.1   Convolution of integrators

In §2.1 we defined an arithmetic function h as the convolution product of given arithmetic functions f and g by the formula

$$h(n) := (f * g)(n) := \sum_{ij=n} f(i)g(j) \tag{3.7}$$

and we established summation relations such as

$$\sum_{n \le x} h(n) = \sum_{i \le x} \left\{ \sum_{j \le x/i} g(j) \right\} f(i). \tag{3.8}$$

Here we extend these ideas to define a convolution of integrators. Speaking informally for the moment, we shall develop an analog of (3.7) by taking

$$dH(E) = \iint_{st \in E} dF(s)\, dG(t) \tag{3.9}$$

for F and G functions in $\mathcal{V}$ with associated integrators dF and dG and E a finite union of half open intervals. Also, as an extension of (3.8), we define the cumulative function $H(x) := 0$ for $x < 1$ and

$$H(x) := \int_{1-}^{x} G(x/t)\, dF(t) \qquad (x \ge 1). \tag{3.10}$$

The remainder of this subsection is devoted to developing and explaining these formulas. Among other things, we shall

(1) show that the integral (3.10) in fact exists,
(2) show that H as defined by (3.10) is in $\mathcal{V}$,
(3) define integrators on a product space,
(4) define $dF * dG$, and
(5) show that dH as given in (3.9) can be identified with $dF * dG$.

The integral in (3.10) is well defined. For any fixed $x \ge 1$, set $g(t) := G(x/t)$ for $t > 0$, and note that g is continuous from the *left* and loc. B.V. By Theorem 3.7, (3.10) exists as a Riemann-Stieltjes integral for each $x \ge 1$. We take (3.10) as the basis of the convolution construction.

Integration by parts and a change of variable $(x/t \mapsto u)$ yield

$$H(x) = G(1)F(x) + \int_{1}^{x+} F(x/u)\, dG(u).$$

We can change the $x+$ to an x in the upper limit of the integral by using Lemma 3.8. The last formula can be rewritten as

$$H(x) = \int_{1-}^{x} F(x/u)\, dG(u), \tag{3.11}$$

where we have included $F(x)\, G(1)$ as the contribution to the integral at $u = 1$. (Recall that $G(1-) = 0$.)

Note that if F and G are the respective summatory functions of f, $g \in$ $\mathcal{A}$, then H is the summatory function of $f * g$ and dH is the R.S. integrator corresponding to the arithmetic function $f * g$. (Verify this!)

Lemma 3.16 *Let* F, $G \in \mathcal{V}$ *and let* H *be defined by* (3.10). *Then*

$$H_v(x) \leq \int_{1-}^{x} G_v(x/t)\, dF_v(t) \leq G_v(x) F_v(x)$$

and hence H *is loc. B.V.*

A proof of this lemma is given in the Appendix.

We define a *rectangle* to be a set of the form

$$\{(x,y) \in \mathbb{R} : a < x \leq b,\ c < y \leq d\} = (a,b] \times (c,d].$$

Note that the sides are parallel to the coordinate axes, the top and right sides belong to the rectangle, the other two sides do not belong, and the upper right corner is the only vertex that is included in the rectangle. Suppose we are given a finite union of rectangles. By the extension of horizontal and vertical lines we can express the given set as a finite union of mutually disjoint rectangles. Also, we see that if A and B are each finite unions of rectangles, then so too are the sets $A \cap B$ and $A \smallsetminus B$.

Definition 3.17 Given integrators dF and dG we introduce a *product integrator* $dF \times dG$ as a complex valued function on sets which are finite unions of rectangles. It is defined by the following conditions:

(1) $(dF \times dG)(\phi) := 0$ for ϕ the empty set,

(2) $(dF \times dG)\left((a,b] \times (c,d]\right) := \left(F(b) - F(a)\right)\left(G(d) - G(c)\right)$,

(3) $dF \times dG$ is finitely additive,

i.e. if $R_1, \ldots, R_n$ are mutually disjoint rectangles, then

$$(dF \times dG)\left(\bigcup_{i=1}^{n} R_i\right) := \sum_{i=1}^{n} (dF \times dG)(R_i).$$

If S is any finite union of rectangles, $(dF \times dG)(S)$ depends only on the set S and not on any particular representation as a union of rectangles. This can be seen by superimposing two different representations of S and extending all horizontal and vertical line segments that occur. Then $dF \times dG$ is evaluated on the "refined" representation of S and shown to be the same as $dF \times dG$ on each of the original representations.

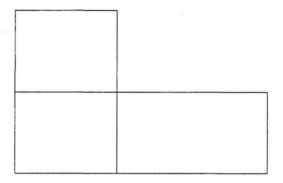

Fig. 3.1   A REFINED REPRESENTATION

We now extend product integrators to a larger class of sets than finite unions of rectangles. We say that a bounded set $S \subset \mathbb{R} \times \mathbb{R}$ *has content* if for each pair of monotone nondecreasing functions F and G in $\mathcal{V}$ there exist sequences $\{A_n\}_{n=1}^{\infty}$ and $\{B_n\}_{n=1}^{\infty}$ with each A_n and each B_n a finite union of rectangles such that $A_n \subset S \subset B_n$ for all n and

$$\lim_{n \to \infty} (dF \times dG)(A_n) = \lim_{n \to \infty} (dF \times dG)(B_n).$$

We define the common value of the limit to be $(dF \times dG)(S)$. A rectangle clearly has content.

If $S \subset \mathbb{R} \times \mathbb{R}$ is a bounded set and $S \cap ([a, \infty) \times [a, \infty)) = \phi$, the empty set, for some $a < 1$, then S has content and $(dF \times dG)(S) = 0$ for any F, $G \in \mathcal{V}$. Also, if S and S' have content, then so do the sets $S \cup S'$, $S \cap S'$ and $S \smallsetminus S'$. In particular, a finite union of rectangles has content.

Given an arbitrary function $F \in \mathcal{V}$ we can express F as $\sum_{j=1}^{4} c_j F_j$, $c_1 = 1$, $c_2 = -1$, $c_3 = i$, $c_4 = -i$, with each F_j a monotone function in $\mathcal{V}$. Suppose that $S \subset \mathbb{R} \times \mathbb{R}$ is a set having content and that $F, G \in \mathcal{V}$. We express F and G each in terms of monotone functions and set

$$(dF \times dG)(S) = \sum_{j,k=1}^{4} c_j c_k (dF_j \times dG_k)(S)$$

(which is independent of a particular choice of F_j and G_k). With these definitions we see that any integrator $dF \times dG$ is finitely additive on the class of sets having content.

Various nonrectangular sets in $\mathbb{R}^2$ have content, but there is a particular collection, which we call hyperbolic regions, that we shall need in the sequel. We show that these sets have content and that the product integrator can be evaluated on them by iterated integration. For any $x < 1$, let $S_x = \phi$, the empty set, and for any $x \geq 1$ let

$$S_x := \{(u, v) \in (\tfrac{1}{2}, \infty) \times (\tfrac{1}{2}, \infty) : uv \leq x\}.$$

By setting the left side and bottom of the sets to be $1/2$, we include possible contributions of the product integrator along the lines $x = 1$ and $y = 1$.

Lemma 3.18   *Each set S_x has content. Let F and $G \in \mathcal{V}$ and let H be defined on $[1, \infty)$ by equation (3.10) and $H(x) = 0$ for $x < 1$. Then*

$$(dF \times dG)(S_x) = H(x).$$

A proof of this lemma is given in the Appendix.

Remarks 3.19   If we combine (3.10), (3.11) and Lemma 3.18, we see that

$$(dF \times dG)(S_x) = (dG \times dF)(S_x).$$

Alternatively, this equation can be established from first principles by noting that the set S_x is symmetric with respect to the line $s = t$ in the s–t plane. This gives an independent verification of equation (3.11). Obviously, the factors of a product integrator cannot be interchanged in general.

We can use the representation of $H(x)$ in Lemma 3.18 to establish

Lemma 3.20   *Let F, $G \in \mathcal{V}$, let $H(x) = 0$ for $x < 1$, and let $H(x)$ be defined by equation (3.10) for $x \geq 1$. Then H is continuous from the right and $H \in \mathcal{V}$.*

A proof of this lemma also is given in the Appendix.

Definition 3.21   Finally, we define the multiplicative convolution of two integrators and identify it with a third one. Let $F, G \in \mathcal{V}$ and let H be the function given by (3.10). We define $dF * dG$, the *multiplicative convolution* of dF and dG, on finite unions of intervals of the form $(a, b]$ by taking

(1) $(dF * dG)(\phi) := 0$,
(2) $(dF * dG)((a, b]) := (dF \times dG)(S_b \setminus S_a)$,
(3) $dF * dG$ is finitely additive.

We define convolution powers of an integrator dF by setting $dF^{*0} := \delta_1$ and $dF^{*n} := dF^{*n-1} * dF$ for $n = 1, 2, \ldots$.

With the aid of Lemma 3.18 we see that

$$(dF * dG)((a, b]) = H(b) - H(a).$$

Remark 3.19 implies that convolution is commutative. Thus we have

$$dF * dG = dG * dF = dH, \tag{3.12}$$

$$\int_{1-}^{x} dF * dG = \int_{1-}^{x} G(x/t)\, dF(t) = \int_{1-}^{x} F(x/t)\, dG(t). \tag{3.13}$$

It is often easiest to find the integrator dH by first computing $H(x)$ using the preceding representation.

Example 3.22 We compute $dN * dt$ by setting

$$H(x) = \int_{1-}^{x} dN * dt = \int_{1-}^{x} \left(\frac{x}{t} - 1\right) dN(t), \quad x \geq 1.$$

H is continuous and for $x \notin \mathbb{Z}^+$, H has a continuous derivative given by

$$H'(x) = \int_{1-}^{x} \frac{dN(u)}{u} = \sum_{n \leq x} \frac{1}{n}.$$

Thus

$$dN * dt = dH = \left\{ \int_{1-}^{t} u^{-1} dN(u) \right\} dt.$$

PROBLEM 3.15   Let S denote a finite union of intervals of the form $(a, b]$ and dA any integrator. Establish the following useful convolution identities:

1. $\delta_1 * dA = dA$,

2. $dA * t^{-1} dt = A(t)\, t^{-1}\, dt$,

3. $(dN * dN)(S) = \sum_{n \in S} \tau(n) =: dN_2(S)$,

4. $(dN * dQ)(S) = \sum_{n \in S} 2^{\omega(n)}$,

5. $dt * dt = \log t\, dt$,

6. $dN * dM = \delta_1$,

7. $(\delta_1 + dt) * (\delta_1 - t^{-1} dt) = \delta_1$,

8. $\int_{1-}^{x} (t^{-1} dN) * (t^{-1} dt) = \frac{1}{2} \log^2 x + \gamma \log x - \gamma_1 + O(x^{-1})$. (For the definition of γ_1, see Problem 3.14.)

Let φ be a loc. B.V. function which is continuous from the left on $(0, \infty)$, let F and $G \in \mathcal{V}$ and let $0 < a < b < \infty$. We define

$$\int_a^b \varphi \, (dF * dG) := \int_a^b \varphi \, dH,$$

where $dH = dF * dG$ is the integrator associated with the function $H \in \mathcal{V}$ given in (3.10). We establish a formula for this integral in

Lemma 3.23   *Let F, G, φ, a, and b be as above. Then*

$$\int_a^b \varphi \, (dF * dG) = \int_{s=1-}^b \left\{ \int_{t=a/s}^{b/s} \varphi(st) \, dG(t) \right\} dF(s). \qquad (3.14)$$

A proof of this lemma is given in the Appendix.

Finally, we have

Lemma 3.24   *Multiplicative convolution of integrators is associative.*

A proof of this lemma also is given in the Appendix.

We give now a convolution inequality that will be useful in the sequel.

Lemma 3.25   *Let A, B, C, $D \in \mathcal{V}$ and suppose that $A(x) \le B(x)$ and $C(x) \le D(x)$ for all $x \ge 1$. Further, suppose that B and C (or A and D) are monotone increasing. Then for all $x \ge 1$,*

$$\int_{1-}^x dA * dC \le \int_{1-}^x dB * dD.$$

Proof.   Suppose that B and C are monotone increasing. We have

$$\int_{1-}^x dD * dB - \int_{1-}^x dA * dC$$

$$= \int_{1-}^x (dD - dC) * dB + \int_{1-}^x dC * (dB - dA)$$

$$= \int_{1-}^x (D - C)\left(\frac{x}{t}\right) dB(t) + \int_{1-}^x (B - A)\left(\frac{x}{t}\right) dC(t) \ge 0.$$

The other case is established by an analogous argument. $\square$

Corollary 3.26   *Let A, $B \in \mathcal{V}$ and suppose that both A and B are monotone increasing and $A(x) \le B(x)$ for all $x \ge 1$. Then for $j = 1, 2, \ldots,$*

$$\int_{1-}^x dA^{*j} \le \int_{1-}^x dB^{*j} \qquad (x \ge 1.)$$

Proof. For $F = A, B$, define $F_j(x) := \int_{1-}^{x} dF^{*j}$. Note that $F \uparrow$ and $dF_j = dF^{*j}$. We show that $B_j(x) \geq A_j(x)$ by induction on j. The case $j = 1$ is given to be true. Assuming $B_j(x) \geq A_j(x)$ for all $x \geq 1$ and some $j \geq 1$, we apply the preceding lemma with $A = A$, $B = B$ and $C = A_j$ and $D = B_j$. $\qquad\square$

PROBLEM 3.16 Show by example that the preceding corollary can fail to hold if the monotonicity condition for A is dropped.

3.4.2 Generalization of results on arithmetic functions

It is now possible to generalize most of the results of §§2.2–2.4. We shall present here a brief indication of this extension (cf. Notes). If dA is an integrator and $n \in \mathbb{Z}^+$, we define dA^{*n} to be the n fold multiplicative convolution of dA with itself and we set $dA^{*0} = \delta_1$, the unity element of integrators under convolution.

We can define operators L^n $(n \in \mathbb{Z}^+)$ and T^α $(\alpha \in \mathbb{C})$ from integrators to integrators, which are analogs of the corresponding operators on $\mathcal{A}$:

$$(L^n dF)(E) := \int_{t \in E} (\log t)^n dF(t),$$

$$(T^\alpha dF)(E) := \int_{t \in E} t^\alpha dF(t)$$

for E a finite union of intervals $(a, b]$. We set $L^1 := L$ and $T^1 := T$. By analogy with Lemma 2.4, we can assert that L is a derivation on integrators and that for each $\alpha \in \mathbb{C}$, T^α is an isomorphism of the algebra of integrators. We have the following formulas, valid for all integrators dF and dG, $n \in \mathbb{Z}^+$, and $\alpha \in \mathbb{C}$:

$$T^\alpha(dF * dG) = (T^\alpha dF) * (T^\alpha dG), \tag{3.15}$$

$$L(dF * dG) = (L\, dF) * dG + dF * (L\, dG), \tag{3.16}$$

$$L^n(dF * dG) = \sum_{j=0}^{n} \binom{n}{j} (L^j dF) * (L^{n-j} dG), \tag{3.17}$$

$$L(dF^{*n}) = n\, dF^{*(n-1)} * L\, dF. \tag{3.18}$$

Equations (3.15) and (3.16) are proved using Lemma 3.23 with appropriate choices of φ. Equations (3.17) and (3.18) follow from (3.16) by induction arguments. (In fact the same arguments work for any derivation of an algebra.)

Example 3.27   With Λ as in §2.4.1, let

$$\psi(x) := \sum_{n \le x} \Lambda(n) = \sum_{p^\alpha \le x} \log p.$$

Then Lemma 2.21 can be rephrased as $LdN = dN * d\psi$.

Example 3.28   Let

$$S(x) := \sum_{n \le x} \sigma(n) = \sum_{n \le x} \sum_{d \mid n} d;$$

then $dS = (TdN) * dN$.

PROBLEM 3.17   Verify that

$$(dt)^{*n} = L^{n-1} dt / (n-1)!$$

for any $n \in \mathbb{Z}^+$. Hint. Show that

$$T^{-1}(dt)^{*n} = (T^{-1}dt)^{*n} = T^{-1}L^{n-1}dt/(n-1)!\,.$$

We say that an integrator dA is *invertible* if there exists an integrator dB such that $dA * dB = \delta_1$. The argument given in §2.2, showing that an arithmetic functions has at most one inverse, applies here as well. Thus, in this case, we say that dB is *the* inverse of dA. The examples preceding Lemma 3.23 include two pairs of inverses. There is a criterion for invertibility which generalizes Theorem 2.7: an integrator dF is invertible iff the associated function $F \in \mathcal{V}$ satisfies $F(1) \ne 0$.

Invertibility formulas can be established using power series techniques as in the proof of Theorem 2.7. Also, the results of §2.4 on exponentials can be extended to integrators. These ideas require a notion of convergence of integrators. We shall not develop these themes any further since we shall not use them in the sequel.

PROBLEM 3.18   Show that the convolution inverse of $\delta_1 + dt$ can be obtained formally from the series $\delta_1 - dt + (dt * dt) - (dt * dt * dt) + \cdots$.

PROBLEM 3.19   Find the convolution inverse of $\delta_1 + Ldt$. Verify that the integrator you have found is indeed the inverse.

PROBLEM 3.20 (Exponential of an integrator). If $F \in \mathcal{V}$ and $F(1) = 1$, then there exists a function $\varphi \in \mathcal{V}$ with $\varphi(1) = 0$ such that

$$dF = \exp d\varphi := \delta_1 + d\varphi + (d\varphi * d\varphi)/2! + \cdots$$

and $LdF = dF * Ld\varphi$ (cf. Theorem 2.20). Show that

$$\delta_1 + dt = \exp\left\{\frac{1 - t^{-1}}{\log t}dt\right\}.$$

PROBLEM 3.21 Show that if $F \in \mathcal{V}$ and $F(x) = \int_1^x dF * t^{-1}dt$ for all $x \geq 1$, then $F = 0$. Hint. Use the identity $(\delta_1 + dt) * (\delta_1 - t^{-1}dt) = \delta_1$.

3.5 Stability

In §3.2 we estimated $Q(x)$ by noting that $|\mu| = 1 * f$, where f as defined in Lemma 3.4 is a "small" function in the sense that $\sum_{n \leq x} |f(n)| = O(\sqrt{x})$, while $N(x)$, the summatory function of 1, is well estimated by the "larger" smooth function x. We found that

$$Q(x) \sim cN(x), \quad \text{where} \quad c = \sum f(n)/n = 1/\zeta(2).$$

Summations of the convolution of a "large" and a "small" function occur quite often; another example of this phenomenon is given in Problem 3.7.

Here we study convolution integrals $\int_{1_-}^x dF * dG$ in which $F(x)$ is "suitably close" to a function $x^b P(\log x)$ for some $b \in \mathbb{C}$ and some polynomial P and where $G_v(x)$ is "much smaller" than $|F(x)|$ as $x \to \infty$. We shall see that the value of the convolution integral is asymptotic to $cF(x)$ for a suitable constant c.

The extreme case of stability is given by $dG = \delta_1$. A less trivial case is given in the following

PROBLEM 3.22 Let $F(x) \sim ax^b$, where $a, b \in \mathbb{C}$, $a \neq 0$, and $\Re b > 0$. Show that

$$\int_1^x dF * \frac{dt}{t} \sim \frac{a}{b}x^b.$$

Theorem 3.29   (Stability). *Let F and $G \in \mathcal{V}$ and satisfy for all $x \geq 1$*

$$F(x) = x^s P(\log x) + O(x^\theta \log^K ex),$$

$$G_v(x) = O(x^\tau \log^H ex),$$

where $s \in \mathbb{C}$; P is a polynomial of degree m; K, $H \geq 0$; $0 \leq \tau < \Re s$; and $0 \leq \theta \leq \Re s$. Then

$$\int_{1-}^{x} dF * dG = x^s P^*(\log x) + R(x),$$

where P^* is a polynomial of degree $\leq m$ and

$$R(x) = \begin{cases} O\{x^\theta \log^K ex\}, & \text{if } \theta > \tau \\ O\{x^\theta (\log ex)^{H+\max(m,K+1)}\}, & \text{if } \theta = \tau \\ O\{x^\tau (\log ex)^{H+m}\}, & \text{if } \theta < \tau. \end{cases}$$

Remarks 3.30   The constants implied in the estimates of $R(x)$ may depend on any of the parameters except, of course, x. The logarithms that occur are an unpleasant reality. They occur in examples as simple as

$$\int_{1}^{x} dt * dt = x \log x - x + 1.$$

We have assumed for simplicity that the log factors occurring in the error terms have nonnegative exponents. If the hypothesis for F holds with a negative value of K, then *a fortiori* it holds with $K = 0$, which we would use in the theorem. This remark applies to $H < 0$ as well. We have written $\log ex$ in place of $\log x$ in the estimates to insure their validity for x near 1.

Proof of Theorem 3.29.   We express the convolution as an iterated integral, approximate $F(x/t)$ by $P(\log x - \log t)$, which we then expand by Taylor's formula:

$$\int_{1-}^{x} dF * dG = \int_{1-}^{x} F(x/t)\, dG(t)$$

$$= \int_{1-}^{x} \left(\frac{x}{t}\right)^s P\left(\log \frac{x}{t}\right) dG(t) + \int_{1-}^{x} O\left\{\left(\frac{x}{t}\right)^\theta \left(\log \frac{ex}{t}\right)^K\right\} dG(t)$$

$$=: I + J, \quad \text{say}.$$

Now

$$I = x^s \sum_{\ell=0}^{m} \frac{1}{\ell!} P^{(\ell)}(\log x) \int_{1-}^{x} (-\log t)^\ell t^{-s} \, dG(t).$$

The integrals in I converge as $x \to \infty$, as we shall see in a moment. Consequently, we can write

$$I = x^s \sum_{\ell=0}^{m} \frac{1}{\ell!} P^{(\ell)}(\log x) \left\{ \int_{1-}^{\infty} - \int_{x}^{\infty} \right\} (-\log t)^\ell t^{-s} \, dG(t).$$

If we set $\sigma = \Re s$, then the triangle inequality and integration by parts yield

$$\left| \int_{x}^{\infty} \frac{\log^\ell t}{t^s} \, dG(t) \right| \le \int_{x}^{\infty} \frac{\log^\ell t}{t^\sigma} \, dG_v(t)$$

$$= \left. \frac{\log^\ell t}{t^\sigma} G_v(t) \right|_{x}^{\infty} - \int_{x}^{\infty} G_v(t) \frac{(\log t)^{\ell-1}}{t^{\sigma+1}} \{\ell - \sigma \log t\} \, dt$$

$$= O\{x^{\tau-\sigma}(\log ex)^{\ell+H}\}, \tag{3.19}$$

and thus

$$I = x^s \sum_{\ell=0}^{m} \frac{1}{\ell!} P^{(\ell)}(\log x) c_\ell + O\{x^\tau (\log ex)^{m+H}\}.$$

Here we have taken

$$c_\ell = \int_{1-}^{\infty} (-\log t)^\ell t^{-s} dG(t),$$

which is finite by (3.19). Now

$$|J| \le c \int_{1-}^{x} \left(\frac{x}{t}\right)^\theta \left(\log \frac{ex}{t}\right)^K dG_v(t)$$

$$\le c' \int_{1-}^{x} \left(\frac{x}{t}\right)^\theta \left(\log \frac{ex}{t}\right)^K \{\delta_1 + t^{\tau-1}(\log et)^H dt\}$$

by Lemma 3.11.

We consider separately the cases $\theta > \tau$, $\theta = \tau$, and $\theta < \tau$:

$$\theta > \tau: \ |J| \le c' x^\theta (\log ex)^K \left\{ \int_{1}^{\infty} t^{\tau-\theta-1}(\log et)^H dt + 1 \right\}$$

$$= O\{x^\theta (\log ex)^K\},$$

$$\theta = \tau: \quad |J| \le c'x^\theta (\log ex)^K \left\{ \int_1^x t^{-1} (\log et)^H dt + 1 \right\}$$

$$= O\{x^\theta (\log ex)^{K+H+1}\},$$

$$\theta < \tau: \quad |J| \le c' \left\{ \int_{1-}^{\sqrt{x}} + \int_{\sqrt{x}}^x \right\} \left(\frac{x}{t}\right)^\theta \left(\log \frac{ex}{t}\right)^K \{\delta_1 + t^{\tau-1}(\log et)^H dt\}$$

$$\le c'x^\theta (\log ex)^K (\log e\sqrt{x})^H \int_1^{\sqrt{x}} t^{\tau-\theta-1} dt$$

$$+ c'x^\tau (\log ex)^H \int_1^{\sqrt{x}} (\log eu)^K u^{\theta-\tau-1} du + c'x^\theta (\log ex)^K.$$

(We replaced t by x/u in the integral $\int_{\sqrt{x}}^x$.) Now

$$\int_1^{\sqrt{x}} t^{\tau-\theta-1} dt = O\{x^{(\tau-\theta)/2}\}, \quad \int_1^{\sqrt{x}} (\log eu)^K u^{\theta-\tau-1} du = O(1),$$

and thus, in case $\theta < \tau$, we have $|J| = O(x^\tau (\log ex)^H)$. Combining the estimates in each of the three cases we obtain the stated error terms. The polynomial P^* is defined by

$$P^*(u) = \sum_{\ell=0}^m \frac{1}{\ell!} P^{(\ell)}(u) c_\ell$$

and is of degree m, unless $c_0 = \int_{1-}^\infty t^{-s} dG(t) = 0$, in which case P^* is of lower degree. $\qquad \square$

PROBLEM 3.23   Apply the stability theorem to establish the following relations:

(1) $\sum_{n \le x} \varphi(n) = \sum_{n \le x} (T1 * \mu)(n) = x^2/(2\zeta(2)) + O(x \log ex)$.

(2) Let $S = \{n^2 : n \in \mathbb{Z}^+\}$. Then, for some constant c,

$$\sum_{n \le x} (L1 * 1_S)(n) = \zeta(2) x \log x + cx + O(x^{1/2} \log ex).$$

(3) Let $k \in \mathbb{Z}^+$. Then $\sum_{n \le x} (\log x/n)^k = k! \, x + O\{(\log ex)^k\}$.

(4) Prove that $\sum_{n \le x} \tau^2(n) \sim \pi^{-2} x \log^3 x$. Hint. $\tau^2 = 1^{*4} * 1_S^{*-1}$ from Problem 2.31.

PROBLEM 3.24   (E. Cohen). Let M be a given nonempty set of integers, each exceeding 1. Let S be the set of positive integers n such that, if $p_1^{m_1} \cdots p_r^{m_r}$ is the canonical factorization of n, then none of the exponents $m_1, \ldots, m_r$ is in M. Let $S(x) := \#\{n \leq x : n \in S\}$. Prove that

$$S(x) = \alpha x + O(x^{1/2}),$$

where

$$\alpha = \prod_p \left\{ 1 - \sum_{m \in M} (p^{-m} - p^{-m-1}) \right\}.$$

Hint. Use ideas from §3.2 and Theorem 3.29. The weaker assertion that $\lim_{x \to \infty} S(x)/x = \alpha$ follows from Problem 3.7.

PROBLEM 3.25   Suppose that a is a fixed positive integer.

(1) Show that, for $\Re s > 1$,

$$\sum_{(n,a)=1} \frac{n\mu^2(n)}{\varphi(n)n^s} = \sum_{\ell=1}^{\infty} \frac{1}{\ell^s} \sum_{m=1}^{\infty} \frac{c_m}{m^s},$$

where

$$\sum_{m=1}^{\infty} \frac{c_m}{m^s} = \prod_{p \mid a} (1 - p^{-s}) \prod_{p \nmid a} \left(1 + \frac{1}{(p-1)p^s} - \frac{p}{(p-1)p^{2s}} \right)$$

and the series $\sum_m c_m m^{-s}$ converges absolutely for $\Re s > 1/2$.

(2) Show that

$$\sum_{\substack{n \leq x \\ (n,a)=1}} \frac{n\mu^2(n)}{\phi(n)} = \frac{\varphi(a)}{a} x + O(x^{1/2+\epsilon}).$$

(3) Show that

$$\sum_{\substack{n \leq x \\ (n,a)=1}} \frac{\mu^2(n)}{\varphi(n)} = \frac{\varphi(a)}{a} \log x + K(a) + O(x^{-1/2+\epsilon}),$$

where $K(a)$ is a number depending on a.

3.6   Dirichlet's hyperbola method

Suppose we apply Lemma 3.1 or Problem 3.1 to estimate

$$N_2(x) := \sum_{n \leq x}(1 * 1)(n) = \sum_{n \leq x} \tau(n).$$

We have

$$N_2(x) = \sum_{n \leq x}\left[\frac{x}{n}\right] = \sum_{n \leq x}\frac{x}{n} + \sum_{n \leq x}\left(\left[\frac{x}{n}\right] - \frac{x}{n}\right) =: I + J.$$

It is clear that $|J| < x$. Applying Lemma 3.13, we have

$$I = x \log x + \gamma x + O(1).$$

Thus we obtain the simple asymptotic estimate

$$N_2(x) = x \log x + O(x). \tag{3.20}$$

PROBLEM 3.26   Let $\tau_2 = 1 * 1$ and $\tau_k = \tau_{k-1} * 1$ for any integer $k > 2$. Prove that

$$\sum_{n \leq x} \tau_k(n) = \frac{1}{(k-1)!} x \log^{k-1} x + O(x \log^{k-2} x).$$

The sum $\sum_{n \leq x} 1 * 1(n)$ extends over the collection of ordered pairs of positive integers (ℓ, m) satisfying the inequality $\ell m \leq x$. This is the collection of lattice points lying in the set we called S_x in §3.4.1. Dirichlet showed that one could improve the error estimate in (3.20) by using parts of each of the formulas of Lemma 3.1. His idea is given in

Theorem 3.31   (Dirichlet's hyperbola method).   *Let F, $G \in \mathcal{V}$. For given $x > 1$, let $y \geq 1$ and $z \geq 1$ be any numbers whose product is x. Then*

$$\int_{1-}^{x} dF * dG = \int_{1-}^{z} F\left(\frac{x}{t}\right) dG(t) + \int_{1-}^{y} G\left(\frac{x}{s}\right) dF(s) - F(y)G(z).$$

Proof.   The formula holds trivially if either $y = 1$ or $z = 1$. We assume that $y > 1$ and $z > 1$ and set

$$G_1(u) = G(u) \text{ for } 1 \leq u \leq z, \quad G_1(u) = G(z) \text{ for } z \leq u < \infty,$$

$$G_2(u) = 0 \text{ for } 1 \leq u \leq z, \quad G_2(u) = G(u) - G(z) \text{ for } z \leq u < \infty,$$

so that $G = G_1 + G_2$, and we apply the iterated integration formula to each of G_1, G_2. We find that

$$
\int_{1-}^{x} dF * dG = \int_{1-}^{x} dF * dG_1 + \int_{1-}^{x} dF * dG_2
$$

$$
= \int_{1-}^{x} F\left(\frac{x}{t}\right) dG_1(t) + \int_{1-}^{x} G_2\left(\frac{x}{s}\right) dF(s)
$$

$$
= \int_{1-}^{z} F\left(\frac{x}{t}\right) dG(t) + \int_{1-}^{x/z} \left\{G\left(\frac{x}{s}\right) - G(z)\right\} dF(s)
$$

$$
= \int_{1-}^{z} F\left(\frac{x}{t}\right) dG(t) + \int_{1-}^{y} G\left(\frac{x}{s}\right) dF(s) - F(y)G(z). \quad \square
$$

In applications we treat each of $\int_{1-}^{z}$, $\int_{1-}^{y}$ by the method used in estimating the main term in the stability theorem. We write, e.g.,

$$
F(x/t) = A(x/t) + \epsilon(x/t)
$$

where A is a smooth approximating function and ϵ the error function. We have then

$$
\int_{1-}^{z} F\left(\frac{x}{t}\right) dG(t) = \int_{1-}^{z} A\left(\frac{x}{t}\right) dG(t) + \theta \int_{1-}^{z} \left|\epsilon\left(\frac{x}{t}\right)\right| dG_v(t),
$$

where $|\theta| \leq 1$. Note the use of G_v in the last integral.

There are two possible advantages which this formula can offer: (1) The percentage errors in estimating $G(x/s)$ and $F(x/t)$ are generally smaller for large values of the arguments, and the arguments x/s and x/t are both rather large in Dirichlet's method (provided neither y nor z is near to 1). (2) The integration intervals $1 \leq s \leq y$ and $1 \leq t \leq z$ are much shorter than $[1, x]$. On the other hand, we might have such large estimates for F_v or G_v that the use of Dirichlet's formula proves wasteful.

The most famous application of the hyperbola method is to the Dirichlet divisor problem:

Corollary 3.32 *For $x \in [1, \infty)$,*

$$
N_2(x) := \sum_{n \leq x} (1 * 1)(n) = x \log x + (2\gamma - 1)x + O(\sqrt{x}).
$$

Proof. We apply Dirichlet's formula to $1 * 1$ for some y and z presently to be chosen. We write

$$N_2(x) = I + II - III,$$

where

$$I = \sum_{m \leq y} \left[\frac{x}{m}\right] = \sum_{m \leq y} \frac{x}{m} - \sum_{m \leq y} \left(\frac{x}{m} - \left[\frac{x}{m}\right]\right)$$

$$= x\left(\log y + \gamma + O\left(y^{-1}\right)\right) + O(y);$$

$$II = \sum_{n \leq z} \left[\frac{x}{n}\right] = x\left(\log z + \gamma + O\left(z^{-1}\right)\right) + O(z)$$

by the same calculation; and

$$III = [y][z] = (y + O(1))(z + O(1)) = yz + O(y) + O(z) + O(1).$$

If we put the three parts together and recall that $yz = x$, we obtain

$$N_2(x) = x(\log x + 2\gamma - 1) + O(z) + O(y) + O(1).$$

Now we select $y = z = \sqrt{x}$ to obtain the claimed error term. $\qquad\square$

The exponent $1/2$ in the error term of the corollary is far from optimal, as we shall indicate in §11.8.

PROBLEM 3.27 Show that

$$\lim_{x \to \infty} \frac{1}{x} \sum_{n \leq x} \left(\frac{x}{n} - \left[\frac{x}{n}\right]\right) = 1 - \gamma.$$

PROBLEM 3.28 Show that there exist constants $\alpha, \beta \in \mathbb{R}$ such that

$$\sum_{n \leq x} (1 * 1 * 1)(n) = \frac{x \log^2 x}{2} + \alpha x \log x + \beta x + O(x^{2/3} \log ex).$$

Hint. Express $1 * 1 * 1$ as $\tau * 1$ and choose y and z judiciously.

PROBLEM 3.29 Let $b > a > 0$. Show that the number of lattice points lying in the region $\{(s,t) : s, t > 0, s^a t^b \leq x\}$ is

$$\zeta(a/b)x^{1/b} + \zeta(b/a)x^{1/a} + O(x^{1/(a+b)}).$$

($\zeta(a/b)$ is defined by equation (3.6).)

PROBLEM 3.30 Show that if $M(x) := \sum_{n \le x} \mu(n) = O(x^{1/2+\epsilon})$ holds for every $\epsilon > 0$, then $Q(x) - 6x/\pi^2 = O(x^{2/5+\epsilon})$ for every $\epsilon > 0$. Hint. Recall that

$$Q(x) = \sum_{m^2 n \le x} \mu(m),$$

where m and n run over positive integers.

3.7 Notes

3.2. Theorem 3.5 was first proved by L. Gegenbauer in the paper cited in the Notes for §1.4.

The estimate $M(x) = O(x^{1/2+\epsilon})$ on the assumption of the Riemann hypothesis was first made by J. E. Littlewood and is briefly discussed in §8.8. See also §11.7.

For further discussion of $Q(x)$, see Problem 3.30 and the note to §3.6.

3.3. The Pollard version of the Riemann-Stieltjes integral, which we use, was introduced in Quarterly J. Pure Appl. Math., vol. 49 (1923), pp. 73–138.

For the Euler-Maclaurin sum formula see K. Knopp, Theory and Application of Infinite Series, trans. by R. C. Young, Blackie, London, 1928, §64; Chapter 13 of [HarD]; Chapter 8 of [Olv]; or §7.21 of [WW].

3.4. A measure theoretic generalization of arithmetic functions, including a notion of convergence and exponentials, is given in H. G. Diamond, Illinois J. Math., vol. 14 (1970), pp. 12–28.

3.5. The stability theorem (Th. 3.29) is a part of mathematical folklore. A special case occurs as Lemma 1 in P. T. Bateman and E. Grosswald, Ill. J. Math., vol. 2 (1958), pp. 88–98. A more general result occurs as Theorem 2 of J. P. Tull, Duke Math. J., vol. 26 (1959), pp. 73–80.

3.6. The hyperbola method and Corollary 3.32 go back to Dirichlet's paper in Abhandlungen Preuss. Akad. Wiss. 1849, pp. 69–83. Also in Werke, vol. 2, Berlin, 1897 (reprinted by Chelsea Publishing Co., 1969), pp. 49–66. The name comes from the interpretation of $N_2(x)$ as the number of points with positive integer coordinates lying on or under the hyperbola $st = x$.

The result of Problem 3.30 is due to A. Axer, Prace Mat. Fiz., vol. 21 (1910), pp. 65–95 and Sitzungsber. Akad. Wiss. Wien (2a), vol. 120 (1911), pp. 1253–1298. It was proved with the use of Theorem 3.31. The stronger assertion

$$Q(x) - 6x/\pi^2 = O(x^{\alpha+\epsilon}) \tag{3.21}$$

with $\alpha = 9/28$ was established by H. L. Montgomery and R. C. Vaughan under the assumption of the Riemann hypothesis. This result appears in Recent Progress in Analytic Number Theory, vol. 1, London, Academic Press, 1981, pp. 247–256. In the other direction, we shall show in §11.4 that $Q(x) - 6x/\pi^2$ is not $o(x^{1/4})$.

Chapter 4

The Distribution of Prime Numbers

4.1 General remarks

Problems concerning primes have interested people since antiquity. One particularly difficult problem was to find a simple formula for $\pi(x)$, the number of primes not exceeding x. This function grows quite erratically, and these attempts proved fruitless. Around the beginning of the 19th century, A. M. Legendre and C. F. Gauss formed hypotheses on the statistical behavior of the primes by examining tables. Independently, they conjectured that $\pi(x)$ is well approximated, in some sense, by $x/\log x$.

P. L. Chebyshev was the first person to make significant progress on estimating $\pi(x)$. Around the middle of the 19th century he established that $x/\log x$ is the "exact order" of $\pi(x)$. We shall give some estimates of Chebyshev's type in the next section.

One of the outstanding achievements of 19th century mathematics was the establishment by J. Hadamard and Ch. J. de la Vallée Poussin of

Theorem 4.1 (The prime number theorem). *As $x \to \infty$,*

$$\pi(x) \sim x/\log x.$$

A heuristic argument for the P.N.T. can be based on the formulas

$$1 = \exp \kappa \quad \text{and} \quad \delta_1 + dt = \exp\{(1 - t^{-1})dt/\log t\}$$

of §2.4 and §3.4. The summatory functions

$$\sum_{n \leq x} 1 = [x] \quad \text{and} \quad \int_{1-}^{x} (\delta_1 + dt) = x$$

are nearly equal. Also, the summatory functions associated with κ and $(1 - t^{-1})dt/\log t$ satisfy

$$\sum_{n \leq x} \kappa(n) = \pi(x) + \frac{1}{2}\pi(x^{1/2}) + \frac{1}{3}\pi(x^{1/3}) + \cdots \sim \pi(x)$$

and

$$\int_1^x (1 - t^{-1})dt/\log t \sim x/\log x.$$

Thus it is plausible that $x/\log x$ should be a good approximation of $\pi(x)$.

The proofs of Hadamard and de la Vallée Poussin, as well as all other proofs of the P.N.T. of the following fifty years, utilized the Riemann zeta function and contour integration or Fourier analysis. In fact, it was believed by many that there was no other path to the P.N.T. It came as a stunning surprise when in 1948 an "elementary" proof of the P.N.T. (i.e. one not utilizing functions of a complex variable or Fourier analysis) was established by A. Selberg and by P. Erdős, using a formula of Selberg. In Chapter 5 we shall present an elementary proof of the P.N.T.

In §1.5 we gave a proof of Euler that there are an infinitude of primes. Here is another proof of this result based on the idea that a finite number of primes would create a number system with density zero.

Lemma 4.2 $\pi(x) \to \infty$ *as* $x \to \infty$.

Proof.   We establish a one–to–one correspondence between $\mathbb{Z}^+$ and the set of sequences of nonnegative integers of which all but a finite number are zero. The rule of correspondence is $n = 2^{\alpha_1} 3^{\alpha_2} 5^{\alpha_3} \cdots \longleftrightarrow (\alpha_1, \alpha_2, \alpha_3, \ldots)$, where the α_i's occur as exponents in the unique prime factorization of n. If $n = 2^{\alpha_1} 3^{\alpha_2} \cdots p_j^{\alpha_j} \leq x$, then

$$\alpha_1 \leq [\log x / \log 2], \ldots, \alpha_j \leq [\log x / \log p_j].$$

Thus

$$\#\{(\alpha_1, \alpha_2, \ldots) : \alpha_i \geq 0, \prod_i p_i^{\alpha_i} \leq x\} \leq \left[1 + \frac{\log x}{\log 2}\right] \cdots \left[1 + \frac{\log x}{\log p_r}\right], \quad (4.1)$$

where p_r is the largest prime not exceeding x. The left side of the last inequality is $[x]$. If there existed only r primes, the right side would be bounded by $(1 + \log x / \log 2)^r = O(\log^r x)$ for all $x \geq 1$. Of course, the inequality $x \leq c \log^r x$ is false for sufficiently large x, and hence there must be an infinite number of primes. $\qquad\qquad\qquad\qquad\qquad\qquad\square$

PROBLEM 4.1   Given $r \in \mathbb{Z}^+$, find some number $x_0 = x_0(r)$ such that $x > (\log x)^r$ for all $x \geq x_0$.

PROBLEM 4.2   Derive a lower bound for $\pi(x)$ from inequality (4.1). (This estimate is imprecise because (4.1) is very wasteful.)

Now we are going to show that the set of primes has density zero, in the sense that $\lim_{x \to \infty} \pi(x)/x$ exists and has the value 0.

Lemma 4.3 $\pi(x) = o(x)$.

Proof.   We use an elementary sieve argument. For r a positive integer let $S = S_r$ denote the set of positive integers that are relatively prime to each of the first r primes and let $S(x) = \#\{n \in S \cap [1, x]\}$. We have $\pi(x) \leq r + S(x)$ for all $x \geq 1$, since the set of primes is contained in $\{p_1, p_2, \ldots, p_r\} \cup S$. We shall now show that if r is sufficiently large then S has small density.

Let $N = 2 \cdot 3 \cdots p_r$. Then $S(N) = \varphi(N)$, where φ is Euler's function, since $(n, N) = 1$ if and only if $p_i \nmid n$ for $1 \leq i \leq r$. Also, for k any positive integer $S(kN) = k\varphi(N)$, since any pair of numbers congruent mod N are both in S or both out of S.

We can now show that

$$\text{density of } S = \lim_{x \to \infty} \frac{S(x)}{x} = \frac{\varphi(N)}{N} = \prod_{i=1}^{r}(1 - p_i^{-1})$$

by writing

$$S\left(\left[\frac{x}{N}\right]N\right) \leq S(x) < S\left(\left[\frac{x+N}{N}\right]N\right),$$

substituting $S(kN) = k\varphi(N)$, dividing by x, and taking limits as $x \to \infty$.

Let $\epsilon > 0$ be given. Choose r so large that $\log p_r > 1/\epsilon$. (There are infinitely many primes!) Now

$$\prod_{i=1}^{r}(1 - p_i^{-1})^{-1} = {\sum}' \frac{1}{n},$$

where $\sum'$ denotes the sum over all positive integers n none of whose prime factors exceeds p_r. The density of S is less than ϵ, since

$$\sideset{}{'}\sum \frac{1}{n} > \sum_{n \leq p_r} \frac{1}{n} > \log p_r > 1/\epsilon.$$

As we noted before, $\pi(x) \leq r + S(x)$ and hence

$$\limsup_{x \to \infty} \frac{\pi(x)}{x} \leq \limsup_{x \to \infty} \frac{S(x)}{x} = \prod_{i=1}^{r}(1 - p_i^{-1}) < \epsilon. \qquad \square$$

PROBLEM 4.3   Show that for any positive integer r there is a number c_r such that

$$\pi(x) \leq (x + c_r) \prod_{i=1}^{r}(1 - p_i^{-1})$$

for all positive x. Show that we may take $c_1 = 1$, $c_2 = 5$, $c_3 = 11$.

4.2   The Chebyshev ψ function

Von Mangoldt's arithmetic function Λ and its summatory function ψ are far more convenient than the prime counting function π to use for our technical work. This is already suggested by the identity $1 = \exp \kappa$ established in §2.4, where Λ is defined. Also, $\psi(x)$ can be interpreted as the logarithm of a product involving all primes not exceeding x (see Problem 4.4 below), and multiplication is the most natural operation on primes.

Recall that

$$\psi(x) = \sum_{n \leq x} \Lambda(n) = \sum_{p^\alpha \leq x} \log p,$$

where the last sum extends over all primes p and all positive integral exponents α satisfying the indicated inequality. The functions Λ and ψ satisfy the Chebyshev identity (Lemma 2.21), which we can write as

$$L1 = \Lambda * 1 \quad \text{or} \quad LdN = d\psi * dN.$$

This relation played a central role in Chebyshev's investigation of the distribution of primes and will be the basis of several arguments in this chapter.

PROBLEM 4.4   Show that $\psi(N)$ equals the logarithm of the least common multiple of the positive integers $1, 2, \ldots, N$.

We show the close connection of π and ψ in

Lemma 4.4 $\dfrac{\psi(x)}{x} = \dfrac{\pi(x)}{x/\log x} + o(1).$

Proof. We have

$$\frac{\psi(x)}{\log x} = \sum_{p \leq x} \sum_{\alpha \leq \log x/\log p} \frac{\log p}{\log x} = \sum_{p \leq x} \left[\frac{\log x}{\log p}\right] \frac{\log p}{\log x} \leq \sum_{p \leq x} 1 = \pi(x)$$

or

$$\frac{\psi(x)}{x} \leq \frac{\pi(x)}{x/\log x}. \tag{4.2}$$

On the other hand, by the definition of κ, we have $1_p \leq \kappa$ or

$$\pi(x) = \sum_{p \leq x} 1 \leq \sum_{p^\alpha \leq x} \frac{\log p}{\log p^\alpha} = \sum_{2 \leq n \leq x} \frac{\Lambda(n)}{\log n} = \int_{2-}^{x} \frac{d\psi(t)}{\log t}.$$

Integration by parts shows that

$$\pi(x) \leq \frac{\psi(x)}{\log x} + \int_{2}^{x} \frac{\psi(t)dt}{t \log^2 t}.$$

If we combine Lemma 4.3 with (4.2) we see that $\psi(x) = o(x \log x)$. Inserting this estimate in the last integral we obtain

$$\pi(x) \leq \frac{\psi(x)}{\log x} + \int_{2}^{x} o(\log^{-1} t)\, dt = \frac{\psi(x)}{\log x} + o(\frac{x}{\log x})$$

or

$$\frac{\pi(x)}{x/\log x} \leq \frac{\psi(x)}{x} + o(1).$$

This inequality and (4.2) imply the truth of the lemma. □

The problem of estimating the prime counting function π can now be rephrased in terms of estimating ψ. In particular, the P.N.T. is equivalent to the assertion that $\psi(x) \sim x$. The following three estimates of ψ were first proved by Chebyshev. We shall use his identity in establishing them.

Lemma 4.5 $\psi(x) = O(x).$

Proof.   We convolve each side of Chebyshev's identity $LdN = d\psi * dN$ with $\delta_1 - t^{-1}dt$. This integrator is an "approximate inverse" of dN in the sense that

$$R(x) := \int_{1-}^{x} dN * (\delta_1 - t^{-1}dt)$$

is "close" to 1.

Since $R(x) = N(x) - \int_1^x N(u)u^{-1}du$, this function is decreasing in each interval $n < x < n+1$. By adding and subtracting a term we obtain

$$R(x) = N(x) - (x - 1) + \int_1^x \frac{u - N(u)}{u}\, du.$$

Now $N(x) - (x-1)$ has period 1 and the integral is increasing; hence we have $R(x+1) > R(x)$ for all $x \geq 1$. It follows that

$$R(x) \geq R(2-) = 1 - \log 2 > 3/10$$

for all $x \geq 1$. Thus we have

$$\int_{1-}^{x} d\psi * dN * (\delta_1 - t^{-1}dt) = \int_1^x R(x/t)\, d\psi(t) > \frac{3}{10}\psi(x).$$

On the other hand,

$$\int_{1-}^{x} LdN * (\delta_1 - \frac{dt}{t}) = \int_1^x LdN - \int_1^x u^{-1}\left(\int_1^u LdN\right)du$$

$$= x \log x - x + O(\log ex)$$

$$- \int_1^x \{\log u - 1 + O(u^{-1}\log eu)\}\, du$$

$$= x + O(\log^2 ex).$$

The two estimates and Chebyshev's identity imply that

$$\psi(x) \leq \frac{10}{3}x + O(\log^2 ex) = O(x). \qquad \square$$

Lemma 4.6   *There is a positive number c such that*

$$\psi(x) \geq cx + o(x).$$

Proof. Again we start with the formula $d\psi * dN = LdN$ and take a suitable approximate inverse, this time $\delta_1 - 2\delta_2$. We have

$$R(x) := \int_{1-}^{x} dN * (\delta_1 - 2\delta_2) = N(x) - 2N(x/2) = 0 \text{ or } 1.$$

On the one hand, we have by Problem 3.13,

$$\int_{1-}^{x} LdN * (\delta_1 - 2\delta_2) = x \log x - x - 2\left\{\frac{x}{2} \log \frac{x}{2} - \frac{x}{2}\right\} + O(\log ex)$$

$$= x \log 2 + O(\log ex).$$

On the other hand,

$$\int_{1-}^{x} d\psi * dN * (\delta_1 - 2\delta_2) = \int_{1}^{x} R(x/t) \, d\psi(t) \le \psi(x).$$

It follows that $\psi(x) \ge x \log 2 + O(\log ex)$. $\square$

The bounds that Chebyshev found were $\limsup \psi(x)/x \le 1.10555\ldots$ and $\liminf \psi(x)/x \ge .92129\ldots$. He used $\delta_1 - \delta_2 - \delta_3 - \delta_5 + \delta_{30}$ as the approximate inverse of dN along with some auxiliary arguments.

PROBLEM 4.5 Let $\chi_{(2,\infty)}$ denote the indicator function of the interval $(2,\infty)$. Use the approximate inverse $\delta_1 - 2\chi_{(2,\infty)}(t)t^{-1}dt$ to show that

$$\limsup_{x \to \infty} \psi(x)/x \le 1 + \log 2.$$

PROBLEM 4.6 Obtain an upper estimate for $\psi(x) - \psi(x/2)$ by using the approximate inverse $\delta_1 - 2\delta_2$. Use the inequality you have found to obtain an upper estimate for $\psi(x)$.

Lemma 4.7 $\liminf_{x \to \infty} \psi(x)/x \le 1$; $\limsup_{x \to \infty} \psi(x)/x \ge 1$.

Proof. Let $c = \liminf_{x \to \infty} \psi(x)/x$. We apply Chebyshev's formula once more, recalling that $\int_{1}^{x} LdN \sim x \log x$. On the other hand, we have

$$\int_{1-}^{x} \psi(\frac{x}{t}) \, dN(t) \ge \int_{1-}^{x} (c + o(1))\frac{x}{t} \, dN(t) = (c + o(1))x \log x.$$

It follows that $c \le 1$.

The estimate of the limit superior is obtained in the same way. $\square$

The lemma implies that if $\psi(x)/x$ tends to a limit as $x \to \infty$, then the value of the limit is 1.

PROBLEM 4.7   (P. Erdős, L. Kalmár). For $n \in \mathbb{Z}^+$ put $L(n) := \exp \psi(n)$ and recall (cf. Problem 4.4) that this quantity equals the least common multiple of $\{1, 2, \ldots, n\}$.

(1) For $k \in \mathbb{Z}^+$ show that

$$L(2k) \leq 2L(2k-1),$$

$$L(2k+1) \leq \binom{2k+1}{k+1} L(k+1) < 2^{2k} L(k+1).$$

(2) Use (1) to show that $L(n) < 4^n$, i.e., $\psi(n) < n \log 4$, for $n \in \mathbb{Z}^+$.

PROBLEM 4.8   (M. Nair, H. Lee). Let $L(\cdot)$ be as in the preceding problem.

(1) For $k \in \mathbb{Z}^+$ show that $L(2k)$ is divisible by $k\binom{2k}{k}$.
(2) If $k \in \mathbb{Z}^+$ and $k \geq 4$, prove by induction that $k\binom{2k}{k} > 2^{2k}$.
(3) Use (i) and (ii) to prove that $L(n) > 2^n$ for $n \in \mathbb{Z}^+$, $n \geq 7$.

4.3   Mertens' estimates

We present here elementary estimates of some sums and products involving primes which appear first to have been discovered by F. Mertens. Relation (4.3) will be used in proving the P.N.T. in the next chapter.

Lemma 4.8

$$\int_1^x t^{-1} d\psi(t) = \sum_{n \leq x} \frac{\Lambda(n)}{n} = \log x + O(1), \tag{4.3}$$

$$\int_1^x \psi(t) t^{-2} dt = \log x + O(1), \tag{4.4}$$

$$\sum_{p \leq x} \frac{\log p}{p} = \log x + O(1). \tag{4.5}$$

Proof. In order to prove (4.3) it suffices to show that

$$\int_1^x \frac{x}{t} d\psi(t) = x \log x + O(x).$$

We establish the last formula by using Chebyshev's identity. We write

$$\int_1^x \frac{x}{t} d\psi(t) = \int_1^x \left[\frac{x}{t}\right] d\psi(t) + \int_1^x \left(\frac{x}{t} - \left[\frac{x}{t}\right]\right) d\psi(t)$$

$$= \int_1^x dN * d\psi + \int_1^x O(1) d\psi$$

$$= \int_1^x LdN + O(\psi(x)) = x \log x + O(x).$$

Relation (4.4) follows from (4.3) by an integration by parts and another application of the estimate $\psi(x) = O(x)$.

We can deduce (4.5) from (4.3) by noting that

$$0 \le \int_1^x \frac{d\psi(t)}{t} - \sum_{p \le x} \frac{\log p}{p} = \sum_{p^\alpha \le x; \alpha \ge 2} \frac{\log p}{p^\alpha}$$

$$< \sum_p \sum_{\alpha \ge 2} \frac{\log p}{p^\alpha} = \sum_p \frac{\log p}{p(p-1)} < \sum_{n=2}^\infty \frac{\log n}{n(n-1)} < \infty. \qquad \square$$

Remarks 4.9 It is interesting to note that a straightforward application of the strong relation $\psi(x) \sim x$ (by Lemma 4.4 this is equivalent to the P.N.T.) yields the weaker result

$$\int_1^x t^{-1} d\psi(t) = \frac{\psi(x)}{x} + \int_1^x \frac{\psi(t)}{t^2} dt \sim \log x.$$

On the other hand, a more delicate argument, to be given in §5.4, shows that the P.N.T. is in fact "equivalent" to the relation

$$\int_1^x t^{-1} d\psi(t) = \log x + c + o(1),$$

where c is some constant.

We used Chebyshev's estimate $\psi(x) = O(x)$ in proving the last lemma. The following problem contains a result in the converse direction.

PROBLEM 4.9 Assuming the truth of Lemma 4.8, give another proof that

$$\limsup_{x \to \infty} \frac{\psi(x)}{x} < \infty; \qquad \liminf_{x \to \infty} \frac{\psi(x)}{x} > 0.$$

Hint. Use relation (4.3) with two different values of x.

The last lemma yields estimates of

$$\sum_{p\leq x} p^{-1} \text{ and } \prod_{p\leq x}(1-p^{-1}).$$

These quantities are related, since the log of the product expression is nearly the negative of the sum of the reciprocals of the primes. We have

Lemma 4.10　*There exist constants B_1 and B_2 such that*

$$\sum_{p\leq x} p^{-1} = \log\log x + B_1 + O(\log^{-1} x) \quad (x \geq 2),$$

$$\sum_{n\leq x} \kappa(n)/n = \log\log x + B_2 + O(\log^{-1} x) \quad (x \geq 2).$$

Proof.　Let

$$R(x) := \sum_{p\leq x} p^{-1}\log p.$$

Integration by parts and an application of relation (4.5) yield

$$\sum_{p\leq x} p^{-1} = \int_{2-}^{x} \log^{-1} t\, dR(t) = \frac{R(x)}{\log x} + \int_{2}^{x} \frac{R(t)}{t\log^2 t}\, dt$$

$$= \frac{\log x + O(1)}{\log x} + \int_{2}^{x} \frac{\log t\, dt}{t\log^2 t} + \left(\int_{2}^{\infty} - \int_{x}^{\infty}\right)\left(\frac{R(t)-\log t}{t\log^2 t}\right) dt$$

$$= \log\log x + \left\{1 - \log\log 2 + \int_{2}^{\infty} \frac{R(t)-\log t}{t\log^2 t}\, dt\right\} + O\left(\frac{1}{\log x}\right).$$

We take B_1 to be the quantity in braces.

The estimate of $\sum_{n\leq x}\kappa(n)/n$ is established similarly, but starting with (4.3). We shall show in Lemma 6.18 that B_2 is Euler's constant γ.　□

PROBLEM 4.10　Show that the integers in the interval $[1, x]$ have an average of $\log\log x + B_1$ distinct prime divisors. Hint.

$$\sum_{n\leq x} \omega(n) = \sum_{n\leq x} (1 * 1_{\mathcal{P}})(n),$$

where $\mathcal{P}$ denotes the set of primes.

Lemma 4.11　*There exists a constant c such that*

$$\prod_{p\leq x} \left(1 - \frac{1}{p}\right) \sim e^c/\log x.$$

Proof. It suffices to show that

$$\sum_{p \leq x} \log(1 - p^{-1}) = c - \log\log x + o(1),$$

since $\exp(o(1)) = 1 + o(1)$. If we expand $\log(1 - p^{-1})$ in its Maclaurin series we obtain

$$\sum_{p \leq x} \log(1 - p^{-1}) = -\sum_{p \leq x} \sum_{\alpha=1}^{\infty} p^{-\alpha}/\alpha$$

$$= -\sum_{n \leq x} \frac{\kappa(n)}{n} - \sum_{p \leq x} \sum_{\alpha > \frac{\log x}{\log p}} p^{-\alpha}/\alpha =: I + J.$$

Now,

$$|J| \leq \sum_{p \leq x} \frac{\log p}{\log x} \sum_{\alpha > \frac{\log x}{\log p}} p^{-\alpha}$$

$$\leq \sum_{p \leq x} \frac{\log p}{\log x} \frac{x^{-1}}{1 - p^{-1}} \leq \frac{2\psi(x)}{x \log x} = O\left(\frac{1}{\log x}\right).$$

We evaluate I by Lemma 4.10 and obtain the desired estimate. Note that $c = -B_2$; in §6.4 we shall show that $-c = \gamma$, Euler's constant. $\square$

4.4 Convergent sums over primes

Suppose we want to estimate the sum of a series of positive terms taken over the primes, e.g. the asymptotic difference of the expressions in Lemma 4.10:

$$\lim_{x \to \infty} \left\{ \sum_{n \leq x} \frac{\kappa(n)}{n} - \sum_{p \leq x} \frac{1}{p} \right\} = \sum_p \sum_{\alpha \geq 2} \frac{1}{\alpha p^\alpha} = \sum_p \left\{ \log\left(1 - \frac{1}{p}\right)^{-1} - \frac{1}{p} \right\}.$$

Assuming f to be a positive, decreasing integrable function on $[2, \infty)$, we seek to estimate $S := \sum_p f(p)$. We write

$$S = \sum_{p \leq X} f(p) + \sum_{p > X} f(p) =: S_1(X) + S_2(X),$$

say, for a suitable positive integer X. We calculate $S_1(X)$ exactly and give an upper bound for the nonnegative quantity $S_2(X)$. The simplest estimate is made by extending the latter sum over *all* integers exceeding X and applying the integral estimate, valid for any integer $X \geq 2$:

$$S_2(X) \leq \sum_{n>X} f(n) \leq \int_X^\infty f(x)\, dx.$$

This is clearly somewhat wasteful; techniques to be developed later show that this estimate generally can be reduced by a factor of about $\log X$.

Here we give a simple inequality that reduces the size of the preceding error estimate by a factor of 3 for functions f that are convex on $[a, \infty)$, i.e. have a nonnegative second difference on this ray.

Lemma 4.12 *Let f be a nonnegative convex integrable function on some interval $[a, \infty)$. For N a positive integer for which $6N - 3 \geq a$, we have*

$$\sum_{p>6N-3} f(p) < \frac{1}{3} \int_{6N-3}^\infty f(x)\, dx.$$

Proof. All primes aside from 2 and 3 are congruent to ± 1 (mod 6). Since f is nonnegative, it suffices to show, for each $n \geq N$, that

$$f(6n - 1) + f(6n + 1) \leq \frac{1}{3} \int_{6n-3}^{6n+3} f(x)\, dx.$$

We have

$$f\left(6n + \frac{3}{2}\right) + f\left(6n - \frac{3}{2}\right) - f(6n + 1) - f(6n - 1)$$

$$= \sum_{j=-2}^{2} \left\{ f\left(6n + \frac{j+1}{2}\right) - 2f\left(6n + \frac{j}{2}\right) + f\left(6n + \frac{j-1}{2}\right) \right\} \geq 0$$

by convexity. Also, by convexity, for any $b \geq a + 3/2$,

$$f(b) \leq \frac{1}{3} \int_{b-3/2}^{b+3/2} f(x)\, dx.$$

Combining these inequalities, we find that

$$f(6n - 1) + f(6n + 1) \leq f\left(6n - \frac{3}{2}\right) + f\left(6n + \frac{3}{2}\right)$$

$$\leq \frac{1}{3} \left\{ \int_{6n-3}^{6n} + \int_{6n}^{6n+3} \right\} f(x)\, dx.$$

Since f is positive and there are integers $n \equiv \pm 1 \pmod 6$ that are not prime, the claimed inequality is strict. $\qquad\qquad\qquad\qquad\square$

PROBLEM 4.11   Use the preceding method, with a suitable choice of truncation point, to estimate

$$\lim_{x \to \infty} \left\{ \sum_{n \le x} \frac{\kappa(n)}{n} - \sum_{p \le x} \frac{1}{p} \right\}$$

with an error smaller than 0.0001.

4.5   A lower estimate for Euler's φ function

The definition of φ implies that $\varphi(n) < n$ for all $n \ge 2$ and that $\varphi(p) = p-1$ for each prime p. Thus, the upper estimate $\varphi(n) \le n-1$ holds for all $n \ge 2$ and equality is achieved infinitely often. We shall apply some of the ideas developed in the preceding section to give a lower estimate for φ.

Let $\{p_i\}_{i=1}^\infty$ denote the primes in their natural order and set $n_k = \prod_{i=1}^k p_i$, the product of the first k primes. The numbers n_k are distinguished arguments for $T^{-1}\varphi$ as we show in

Lemma 4.13   *Let k be a positive integer. If $1 \le n < n_{k+1}$, then $\varphi(n)/n \ge \varphi(n_k)/n_k$, with equality holding if and only if $n = jn_k$, where $1 \le j < p_{k+1}$.*

Proof.   Let $n' < n_{k+1}$ be chosen such that

$$\min_{1 \le n < n_{k+1}} \varphi(n)/n = \varphi(n')/n'.$$

Since $n' < n_{k+1}$, there exists at least one prime $p^* \le p_{k+1}$ for which $p^* \nmid n'$. On the other hand, no prime factor of n' can exceed p_{k+1}. To see this, suppose there were a prime $p^+ > p_{k+1}$ for which $(p^+)^\beta \| n'$ with $\beta \ge 1$. If we set $n'' = n'p^*(p^+)^{-\beta}$, then we would have

$$\frac{\varphi(n'')}{n''} = \frac{\varphi(n')}{n'} \left(1 - \frac{1}{p^*} \right) \left(1 - \frac{1}{p^+} \right)^{-1} < \frac{\varphi(n')}{n'}.$$

Since $n'' < n' < n_{k+1}$, the last inequality would violate the minimality of $\varphi(n')/n'$. Hence

$$\frac{\varphi(n')}{n'} = \prod_{p | n'} \left(1 - \frac{1}{p} \right) = \prod_{p \in S} \left(1 - \frac{1}{p} \right)$$

where S is a *proper* subset of $\{2, 3, \ldots, p_{k+1}\}$. But the last product is minimal when $S = \{2, 3 \ldots, p_k\}$. Thus n' has to be a multiple of n_k that is smaller than n_{k+1}. □

It is convenient to introduce here the function $\theta(x) := \sum_{p \le x} \log p$. By the definition of ψ we have

$$\psi(x) = \sum_{p^\alpha \le x} \log p = \theta(x) + \{\theta(x^{1/2}) + \theta(x^{1/3}) + \cdots + \theta(x^{1/r})\},$$

where $r = [\log x / \log 2]$. The sum terminates at r since $x^{1/n} < 2$ for $n > r$. The function θ is clearly nonnegative, increasing, and bounded above by ψ. On the other hand, for any $x \ge 1$ the sum in braces is at most

$$\psi(x^{1/2}) + \frac{\log x}{\log 2} \, \psi(x^{1/3}) = O(x^{1/2})$$

by Chebyshev's upper estimate. Thus we have proved

Lemma 4.14 $\theta(x) = \psi(x) + O(\sqrt{x})$.

PROBLEM 4.12   Prove that

$$\theta(x) = \sum_{n=1}^{\infty} \mu(n) \psi(x^{1/n}).$$

We give the lower estimate for φ in

Theorem 4.15   *There exists a constant $c > 0$ such that the inequality $\varphi(n) \ge cn / \log \log e^2 n$ holds for all $n \ge 1$.*

Proof.   We can assume $n \ge 6$ and adjust the constant c, if necessary, to cover the cases $n = 1, \ldots, 5$. Given $n \ge 6$, choose $k \ge 2$ such that $n_k \le n < n_{k+1}$. We have

$$\frac{\varphi(n)}{n} \ge \frac{\varphi(n_k)}{n_k} = \prod_{i=1}^{k} (1 - p_i^{-1}) > \frac{c'}{\log p_k},$$

by Lemmas 4.13 and 4.11.

We now give an upper estimate for p_k in terms of n_k using the preceding lemma and Chebyshev's lower estimate for ψ. We have

$$\log n_k = \sum_{i=1}^{k} \log p_i = \theta(p_k) \ge \psi(p_k) - K\sqrt{p_k} \ge c'' p_k$$

for suitable constants $K > 0$ and $c'' \in (0,1)$. It now follows that

$$\frac{\varphi(n)}{n} > \frac{c'}{\log p_k} > \frac{c'}{\log\log n_k - \log c''} \geq \frac{c'''}{\log\log n_k} > \frac{c'''}{\log\log e^2 n}$$

by choosing $c''' = c' \log\log 6 / (\log\log 6 - \log c'')$.

Finally, we take

$$c = \min\left\{c''', \frac{\varphi(1)\log\log e^2}{1}, \ldots, \frac{\varphi(5)\log\log e^2 5}{5}\right\}. \qquad \square$$

PROBLEM 4.13 Prove that

$$\liminf_{n\to\infty} \frac{\varphi(n)\log\log n}{n} = e^{-\gamma}.$$

You may assume that the constant c of Lemma 4.11 is known to be $-\gamma$.

4.6 Notes

4.1. Lemma 4.2 goes back to Euclid, Elements, Book 9, Proposition 20. Lemma 4.3 was first stated explicitly by Euler. For detailed references, see [Nark], §1.2.

4.2. Chebyshev's work on the distribution of primes appeared in the following two papers (in French): J. Math. pures appl. (1), vol. 17 (1852), pp. 341–365 and 366–390. Also in Oeuvres, vol. 1, St.-Pétersbourg, 1899 (reprinted by Chelsea Publishing Co., 1962), pp. 27–48 and 49–70. These two papers appeared also in Mémoires présentés à l'Académie impériale des Sciences de St.-Pétersbourg par divers Savants, vol. 6 (1848/1851), pp. 141–157 and vol. 7 (1850/1854), pp. 15–33.

A detailed account of the work of Chebyshev is given in [Mat], which was published in 1892. At that time, the P.N.T. had not yet been proved, and Chebyshev's results and their slight improvement by Sylvester represented the current state of the art.

4.3. See F. Mertens, J. reine angew. Math., vol. 78 (1874), pp. 46–62 and G. H. Hardy, J. London Math. Soc., vol. 2 (1927), pp. 70–72 and vol. 10 (1935), pp. 91–94; also in Collected Papers of G. H. Hardy, vol. 2, Oxford, 1967, pp. 210–212 and 230–233.

The quantity B_1 in Lemma 4.10 was shown by J. B. Rosser, Proc. London Math. Soc. (2), vol. 45 (1938), pp. 21–44, to be .261497 . . . . Further information about calculation of B_1 and related quantities is given in [CrPo].

4.5. Theorem 4.15 is due to E. Landau, Archiv Math. Physik (3), vol. 5 (1903), pp. 86–91; also in [LanC], vol. 1, pp. 378–383.

Chapter 5

An Elementary Proof of the P.N.T.

5.1   Selberg's formula

The results of the preceding sections show the great utility of Chebyshev's identity $L1 = \Lambda * 1$ (Lemma 2.21). It was hoped that a proof of the P.N.T. might be deduced from this relation by arguments akin to our technique of convolving by approximate inverses of dN. All such attempts have proved futile.

Another approach to the problem was discovered by A. Selberg, using a weighted version of Chebyshev's identity. In this section we present Selberg's formula and make some remarks on its features. In the next section we shall apply the formula to give an elementary proof of the P.N.T.

Theorem 5.1   (Selberg's Formula). *For all $x \geq 1$,*

$$\int_1^x (L d\psi + d\psi * d\psi) = 2x \log x + O(x), \tag{5.1}$$

$$\sum_{n \leq x} \Lambda(n) \log n + \sum_{n \leq x} \psi \left(\frac{x}{n} \right) \Lambda(n) = 2x \log x + O(x), \tag{5.2}$$

$$\psi(x) \log x + \int_1^x \psi \left(\frac{x}{t} \right) d\psi(t) = 2x \log x + O(x). \tag{5.3}$$

The second formula follows from the first by replacing the convolution by an iterated integration and rewriting the resulting expression in terms of arithmetic functions. The third formula follows from the first one as well, by integration by parts and using Chebyshev's upper estimate for ψ. It suffices then to prove the first formula.

Proof. We apply the derivation L to Chebyshev's identity to obtain

$$(Ld\psi) * dN + d\psi * (LdN) = L^2 dN.$$

We use the identity again to replace the factor LdN and then convolve each side of the resulting equation by dM, the inverse of dN, to get

$$Ld\psi + d\psi * d\psi = (L^2 dN) * dM. \tag{5.4}$$

We next integrate each side of (5.4) over the interval $[1, x]$. The left side of the integrated equation has the desired form. For the right side, we first estimate $\int_1^x L^2 dN$ via Euler's summation formula (Lemma 3.12):

$$\int_{1-}^x L^2 dN = \int_1^x L^2 dt + (N(x) - x)\log^2 x + 2\int_1^x \frac{t - N(t)}{t} \log t \, dt$$

$$= x \log^2 x - 2x \log x + 2x + O(\log^2 ex).$$

(Alternatively, this estimate follows from the integral test of calculus.)

Earlier, we found that integrals of the form $\int_{1-}^x (dN)^\alpha * (dt)^\beta$ (for α, β nonnegative integers, with $\alpha + \beta \geq 1$) were asymptotically approximated by expressions $xP(\log x)$, with P a polynomial of degree $\alpha + \beta - 1$. This suggests that we might approximate $\int_1^x L^2 dN$ by

$$\int_{1-}^x (cdt * dt + c' dt + c'' \delta_1) * dN$$

for suitable constants c, c', and c''. The main term of this expression has the form $xQ(\log x)$, where Q is a quadratic polynomial. Such an approximation contains a convolution factor dN, which will enable us to handle the dM occurring in (5.4).

We have

$$\int_{1-}^x dt * dt * dN = \int_{1-}^x (Ldt) * dN = \int_{1-}^x \left(\frac{x}{t} \log \frac{x}{t} - \frac{x}{t} + 1\right) dN(t)$$

$$= (x \log x - x) \int_{1-}^x \frac{dN(t)}{t} - x \int_{1-}^x \frac{\log t}{t} dN(t) + N(x)$$

$$= (x \log x - x)\{\log x + \gamma + O(x^{-1})\}$$

$$- x\left\{\frac{\log^2 x}{2} + \gamma_1 + O\left(\frac{\log ex}{x}\right)\right\} + x + O(1),$$

where γ is Euler's constant and $\gamma_1 = \int_1^\infty t^{-1} \log t (dN - dt)$ (cf. Problem 3.14). Thus

$$\int_{1-}^x dt * dt * dN = \frac{x}{2} \log^2 x + (\gamma - 1)x \log x + (1 - \gamma - \gamma_1)x + O(\log ex);$$

$$\int_{1-}^x dt * dN = \int_{1-}^x \left(\frac{x}{t} - 1\right) dN(t) = x \log x + (\gamma - 1)x + O(1);$$

$$\int_{1-}^x \delta_1 * dN = x + O(1).$$

We see from these relations that we should take $c = 2$, i.e. that the main term in the approximation of $\int_1^x L^2 dN$ should be $\int_1^x 2L dt * dN$. The actual value of the constants c' and c'' will not be important. However, it is easy to verify that we should take $c' = -2\gamma$ and $c'' = \gamma^2 + 2\gamma_1$. With these choices we have

$$\int_{1-}^x L^2 dN = \int_{1-}^x \{2dt * dt + c'dt + c''\delta_1\} * dN + S(x), \qquad (5.5)$$

say, where $S(x) = O(\log^2 ex)$.

We can combine (5.4) and (5.5) to obtain

$$\int_1^x (Ld\psi + d\psi * d\psi) = \int_{1-}^x \{2dt * dt + c'dt + c''\delta_1\} * (dN * dM)$$

$$+ \int_{1-}^x S(x/t) dM(t)$$

or

$$\int_1^x (Ld\psi + d\psi * d\psi) = 2x \log x + (c' - 2)x + c'' + 2 - c'$$

$$+ O\left\{ \int_{1-}^x \log^2(ex/t) dN(t) \right\}.$$

We have used dN as an upper estimate of $|dM|$ in the last integral. This integral is $O(x)$, as one can see by two integrations by parts or application of the stability theorem. Cf. Problem 3.23 (iii).

Thus the right side of the last formula is $2x \log x + O(x)$. $\qquad \square$

5.1.1 Features of Selberg's formula

The arithmetic functions $L\Lambda$ and $\Lambda * \Lambda$ occurring in the formula measure the (weighted) influence of one prime or prime power and of two primes or prime powers respectively. Also, we note that if the P.N.T. is valid, then

$$\psi(x)\log x \sim x\log x \sim \int_1^x \psi(x/t)\,d\psi(x),$$

i.e., each term on the left side of Selberg's formula has equal "weight," and the error term clearly has lower "weight" than the main terms. These features of Selberg's formula enable us to give various prime number estimates that were not accessible to Chebyshev.

For example, we can give nontrivial upper bounds for the number of primes lying in relatively short intervals. Using the bounds discovered by Chebyshev we can assert that

$$\limsup_{x\to\infty} \frac{\psi(x+\epsilon x) - \psi(x)}{\epsilon x} \leq \frac{1.106(1+\epsilon) - .921}{\epsilon} = 1.106 + \frac{.185}{\epsilon}.$$

This is a poor estimate for small values of ϵ. On the other hand, Selberg's formula enables us to obtain a good estimate of the difference quotient for arbitrarily small positive ϵ.

PROBLEM 5.1   Show that if $0 < \epsilon < 1$ then

$$0 \leq \psi(x+\epsilon x) - \psi(x) \leq 2\epsilon x + O(x/\log x)$$

and thus for any fixed $\epsilon > 0$,

$$\limsup_{x\to\infty}\{\psi(x+\epsilon x) - \psi(x)\}/(\epsilon x) \leq 2.$$

Hint. Use Selberg's formula and the fact that $\Lambda * \Lambda \geq 0$.

Let a and A denote the limit inferior and limit superior respectively of $\psi(x)/x$. We showed by arguments based on Chebyshev's identity that $0 < a \leq 1 \leq A < \infty$. Selberg's formula enables us to show that a and A are symmetrically located with respect to 1.

PROBLEM 5.2   Prove that $a + A = 2$. Hint.

$$\int_1^x \psi(x/t)\,d\psi(t) \leq \int_1^x (A+o(1))\frac{x}{t}\,d\psi(t).$$

Formula (5.3) can be rewritten as

$$\frac{\psi(x)}{x} + \frac{1}{\log x} \int_1^x \frac{\psi(x/t)}{x/t} \frac{d\psi(t)}{t} = 2 + O\left(\frac{1}{\log x}\right).$$

Recalling that $\int_1^x t^{-1} d\psi(t) \sim \log x$, we see that one of the above terms is a complicated average of the function $u \mapsto \psi(u)/u$. We are going to deduce estimates of ψ from information about this average. Such an "unaveraging" is called a *tauberian* method. In contrast, methods such as the one establishing the stability theorem, which are essentially averaging arguments, are called *abelian*. In §5.3 we shall deduce the P.N.T. from Selberg's formula by a tauberian method.

5.2   Transformation of Selberg's formula

The P.N.T. is equivalent to the assertion $\psi(x) - x = o(x)$. It is convenient to work with a function that is smoother than ψ. To this end we define $R \in \mathcal{V}$ by

$$R(x) := \int_{1-}^x (d\psi - dt - \delta_1) * t^{-1} dt = \int_1^x (\psi(t) - t) t^{-1} dt.$$

It follows from the bound $\psi(x) = O(x)$ that $R(x) = O(x)$. Also,

$$\limsup_{x \to +\infty} \frac{|R(x)|}{x} = \limsup_{x \to +\infty} \frac{1}{x} \left| \int_1^x (\psi(t) - t) t^{-1} dt \right| \leq \limsup_{x \to +\infty} \frac{|\psi(x) - x|}{x}.$$

Thus, if $\psi(x) - x = o(x)$, then $R(x) = o(x)$. Conversely, we have

Lemma 5.2   *If $R(x) = o(x)$, then $\psi(x) - x = o(x)$.*

Proof.   We take a fixed number ϵ in $(0, 1/2)$ and consider $R(x+\epsilon x) - R(x)$. On the one hand we have

$$\int_x^{x+\epsilon x} \frac{\psi(t) - t}{t} dt = R(x + \epsilon x) - R(x) = o(x) - o(x) = o(x).$$

On the other hand, since $(1 + \epsilon)^{-1} > 1 - \epsilon$, we have the inequality

$$\int_x^{x+\epsilon x} \frac{\psi(t) - t}{t} dt \geq \frac{\psi(x)}{x + \epsilon x} \cdot \epsilon x - \epsilon x \geq \epsilon \{\psi(x) - x\} - \epsilon^2 \psi(x)$$

$$\geq \epsilon \{\psi(x) - x\} - K\epsilon^2 x$$

for some absolute constant K. Similarly, using the inequality $(1 - \epsilon)^{-1} < 1+2\epsilon$, we can obtain a lower bound for $\psi(x)-x$ from $R(x)-R(x-\epsilon x) = o(x)$:

$$\int_{x-\epsilon x}^{x} \frac{\psi(t) - t}{t} dt \le \frac{\psi(x)}{x - \epsilon x} \epsilon x - \epsilon x \le \epsilon\{\psi(x) - x\} + 2\epsilon^2 \psi(x)$$

$$\le \epsilon\{\psi(x) - x\} + 2K\epsilon^2 x.$$

Thus $|\psi(x) - x| \le o(x)/\epsilon + 2K\epsilon x$, and so

$$\limsup_{x \to \infty} \frac{|\psi(x) - x|}{x} \le 2K\epsilon.$$

Since ϵ is arbitrary, the lemma is proved. $\qquad\square$

PROBLEM 5.3   Let $f(n) \ge 0$ for $n = 1, 2, \ldots$. For $j = 0, 1, 2, \ldots$ define

$$F_j(x) = \sum_{n \le x} f(n)(\log x/n)^j/j!.$$

For each $j \ge 1$, prove that $F_j(x) \sim x \iff F_0(x) \sim x$. Hint. Express F_j in terms of F_{j-1}.

5.2.1   Calculus for R

The function R is continuous as one can see from the representation $R(x) = \int_1^x \{\psi(t) - t\}t^{-1} dt$. Also, R is differentiable except at prime powers, where ψ is discontinuous. Away from prime powers we have

$$R'(x) = \{\psi(x) - x\}/x, \quad R''(x) = -\psi(x)/x^2.$$

Thus R'' is negative for $x > 2$, x not a prime power. Let $2 \le a < b$ be successive prime powers. Then by Rolle's theorem, R' can have at most one zero in (a, b) and R can have at most two zeros there.

We recast Selberg's formula as a relation involving R in

Lemma 5.3

$$R(x) \log x + \int_1^x dR * d\psi = O(x). \qquad (5.6)$$

Proof.   From (4.3) we have the relation

$$x \log x + \int_1^x \frac{x}{u} d\psi(u) = 2x \log x + O(x).$$

Subtracting this from from Selberg's formula (5.3) we obtain

$$(\psi(x) - x)\log x + \int_{1-}^{x}(d\psi - \delta_1 - du) * d\psi = O(x).$$

We convert this formula into one involving R by changing x to t, dividing by t, and integrating. These operations amount to a convolution by $t^{-1}dt$ (recall the identity $A(t)t^{-1}dt = dA * t^{-1}dt$, valid for any $A \in \mathcal{V}$). We have

$$\int_{1}^{x}(\psi(t) - t)\log t\frac{dt}{t} + \int_{1}^{x}\left\{\int_{1-}^{t}(d\psi - \delta_1 - du) * d\psi\right\}\frac{dt}{t} = \int_{1}^{x}O(t)\frac{dt}{t}.$$

Integrate the first term by parts and rewrite the second term using the preceding convolution identity to obtain

$$R(x)\log x - \int_{1}^{x}\frac{R(t)dt}{t} + \int_{1-}^{x}(d\psi - \delta_1 - dt) * d\psi * \frac{dt}{t} = O(x).$$

If we apply the relations $R(t) = O(t)$ and $(d\psi - \delta_1 - dt) * t^{-1}dt = dR$, the lemma follows. $\qquad\square$

We shall apply an idea of Selberg to replace the $d\psi$ in (5.6) by a more tractable expression and produce an inequality for $|R|$. However, we insert an extra logarithmic factor, thereby avoiding division by the logarithm.

Lemma 5.4

$$|R(x)|\log^2 x \le 2\int_{1}^{x}\left|R\left(\frac{x}{t}\right)\right|\log t\,dt + O(x\log x).$$

Proof. Convolve (5.6) with $d\psi$, using iterated integration and obtaining

$$\int_{1}^{x}R\left(\frac{x}{t}\right)\log\frac{x}{t}\,d\psi(t) + \int_{1}^{x}\left\{\int_{1}^{x/t}dR * d\psi\right\}d\psi(t) = \int_{1}^{x}O\left(\frac{x}{t}\right)d\psi(t)$$

or

$$\log x\int_{1}^{x}dR * d\psi = \int_{1}^{x}dR * Ld\psi - \int_{1}^{x}dR * d\psi * d\psi + O(x\log x).$$

Multiply through (5.6) by $\log x$ and subtract from the last relation to get

$$-R(x)\log^2 x = \int_1^x dR * (Ld\psi - d\psi * d\psi) + O(x\log x)$$

$$= \int_1^x R\left(\frac{x}{t}\right)(Ld\psi - d\psi * d\psi)(t) + O(x\log x).$$

If we now take absolute values, we obtain

$$|R(x)|\log^2 x \le \int_1^x \left|R\left(\frac{x}{t}\right)\right|(Ld\psi + d\psi * d\psi)(t) + O(x\log x).$$

The last relation might appear more complicated than (5.6). However, the integrator $Ld\psi + d\psi * d\psi$ is the one occurring in Selberg's formula, and thus we may hope to replace it by $2Ldt$. To do this we temporarily introduce a new function by setting $U := |R| \in \mathcal{V}$ and call the last integral I. We have

$$I := \int_1^x U\left(\frac{x}{t}\right)(Ld\psi + d\psi * d\psi)(t)$$

$$= \int_1^x dU * (Ld\psi + d\psi * d\psi)$$

$$= \int_1^x \left(\int_1^{x/t} (Ld\psi + d\psi * d\psi)\right) dU(t)$$

$$= \int_1^x \left(2\frac{x}{t}\log\frac{x}{t} - \frac{2x}{t} + 2 + O\left(\frac{x}{t}\right)\right) dU(t),$$

where we have used the first formula in Theorem 5.1. The terms $2x/t$ and 2 are clearly $O(x/t)$. We have introduced them because

$$\int_1^z Ldy = z\log z - z + 1.$$

Thus we have $I := I_1 + I_2$, say, where

$$I_1 := \int_1^x \left(\int_1^{x/t} 2Ldy\right) dU(t) = \int_1^x 2Ldt * dU$$

$$= 2\int_1^x U(x/t)L\,dt = 2\int_1^x |R(x/t)|\log t\,dt,$$

which is the claimed main term, and

$$I_2 := \int_1^x O\left(\frac{x}{t}\right) dU(t) = O\left(\int_1^x \frac{x}{t}|U'(t)|\, dt\right).$$

The last relation is valid because U is a continuous function which has a continuous derivative everywhere except at prime powers and points at which R vanishes. Since R can vanish in at most two places between successive prime powers, these singular points are isolated. Elsewhere,

$$|U'(x)| = |R'(x)| = |\psi(x) - x|/x = O(1),$$

since $\psi(x) = O(x)$. Thus we have

$$I_2 = O\left(\int_1^x \frac{x}{t} dt\right) = O(x \log x). \qquad \square$$

We can rewrite the lemma by making a change of variable in the integral and dividing the resulting expression by $x \log^2 x$:

$$\frac{|R(x)|}{x} \le \frac{2}{\log^2 x} \int_1^x \frac{|R(u)|}{u} \log\left(\frac{x}{u}\right) \frac{du}{u} + O\left(\frac{1}{\log x}\right). \qquad (5.7)$$

Now

$$\frac{2}{\log^2 x} \int_1^x (\log x - \log u) \frac{du}{u} = 1,$$

and thus (5.7) asserts that $T^{-1}|R|$ is, asymptotically, dominated by an average of itself. Such a situation can hold only for functions that are, roughly, constants. The word "roughly" is necessary because of the $O(\log^{-1} x)$ error term. We shall show that $T^{-1}|R|$ tends to (the constant) 0.

Set $\alpha = \limsup_{x \to \infty} |R(x)|/x$. We have $0 \le \alpha \le 1$, because Selberg's formula implies that $\psi(x) \le 2x + o(x)$ and hence

$$|R(x)| \le \int_1^x \frac{|\psi(u) - u|}{u} du \le \int_1^x \{1 + o(1)\} du \le x + o(x).$$

Lemma 5.5

$$\alpha \le \limsup_{x \to \infty} \frac{1}{\log x} \int_1^x \frac{|R(u)|}{u^2} du.$$

Proof.   After an integration by parts, (5.7) becomes

$$\frac{|R(x)|}{x} \le \frac{2}{\log^2 x} \int_1^x \left\{ \frac{1}{\log t} \int_1^t \frac{|R(u)|}{u^2} du \right\} \frac{\log t}{t}\, dt + O\left(\frac{1}{\log x}\right).$$

Also, we have

$$\frac{1}{\log t} \int_1^t \frac{|R(u)|}{u^2}\, du \le \left\{ \limsup_{v \to \infty} \frac{1}{\log v} \int_1^v \frac{|R(u)|}{u^2}\, du \right\} + o(1).$$

We insert this estimate into the double integral and obtain

$$\frac{|R(x)|}{x} \le \limsup_{v \to \infty} \frac{1}{\log v} \int_1^v \frac{|R(u)|}{u^2} du + o(1).$$

The lemma follows from the last inequality by taking the limit superior once more.   □

PROBLEM 5.4   Show that equality holds in the last lemma.

5.3   Deduction of the P.N.T.

Our object, of course, is to show that $\alpha = 0$. We shall do this by using the last lemma and a few general facts about R. However, no further use will be made of Selberg's formula in the proof of the P.N.T.

Let β be any real number satisfying $\alpha < \beta \le 2$. We shall show that

$$\limsup_{x \to \infty} \frac{1}{\log x} \int_1^x \frac{|R(u)|}{u^2} du \le \frac{\beta}{1 + c\beta^2}, \tag{5.8}$$

where c is some positive absolute constant. Then the preceding lemma implies that $\alpha \le \beta/(1 + c\beta^2)$. If α were positive we would obtain a contradiction from the last inequality upon taking β sufficiently close to α.

We establish (5.8) by taking successive zeros y, z of R and estimating

$$I(y, z) := \int_y^z \frac{|R(u)|}{u^2} du = \left| \int_y^z \frac{R(u)}{u^2} du \right|.$$

The last expression is valid because R is continuous and hence of one sign on (y, z). We give separate arguments according to the size of z/y. The following lemma will be useful in estimating $I(y, z)$ for large values of z/y or in case R has no zeros from some point onward.

Lemma 5.6 $\displaystyle\int_1^x \frac{R(u)}{u^2}\,du = O(1).$

Proof. The left hand side equals $-R(x)/x + \int_1^x \{\psi(t)-t\}t^{-2}dt$ by integration by parts and the definition of R. Earlier, we showed that $R(x) = O(x)$, and by relation (4.4) the last integral is $O(1)$. $\qquad\square$

Now we establish (5.8). The preceding lemma implies the existence of a bound M (which we may take to be at least 1) such that $I(a,b) \le M$ whenever R has no sign changes on the interval (a,b). Moreover, if R were of one sign for all $x \ge A$ for some A, then the P.N.T. would follow. Indeed, the preceding two lemmas imply that in this case

$$\alpha \le \limsup_{x\to\infty} \frac{1}{\log x}\Big\{ \int_1^A \frac{|R(u)|}{u^2}\,du + \Big| \int_A^x \frac{R(u)}{u^2}\,du \Big| \Big\}$$

$$\le \limsup_{x\to\infty} O(1/\log x) = 0.$$

We may thus assume that R has arbitrarily large zeros. Let $\alpha < \beta < 2$ and take $x_0 = x_0(\beta)$ large enough so that $|R(x)| \le \beta x$ for all $x \ge x_0$. Also, recall that if x is not a prime power,

$$R'(x) = (\psi(x) - x)/x = O(1).$$

Let k be an upper bound for $|R'(x)|$, and for convenience in estimating, assume that $k \ge 1$. We set $\gamma = 1 + \beta^2/(4Mk) > 1$. Let y and z be successive zeros of R with $x_0 \le y < z$. We treat three cases according to the size of z/y, and in each case we show that $I(y,z) \le \beta\gamma^{-1}\log(z/y)$.

Case 1. (long interval): $\log(z/y) \ge \gamma M/\beta$. Here

$$I(y,z) \le M \le \beta\gamma^{-1}\log(z/y).$$

Case 2. (intermediate interval): $2k/(2k-\beta) \le z/y < \exp(\gamma M/\beta)$. Here we use the facts that $R(y) = 0$ and $|R'| \le k$ to show that R is small in the early part of the interval $[y,z]$. Precisely, we have

$$|R(u)| = |R(u) - R(y)| \le k(u-y) \le \beta u/2$$

if $y \le u \le 2ky/(2k-\beta)$, while $|R(u)| \le \beta u$ for all u in $[y,z]$. Thus

$$I(y,z) = \int_y^z \frac{|R(u)|}{u^2}\,du \le \beta\log\frac{z}{y} - \frac{\beta}{2}\log\frac{2k}{2k-\beta} \le \beta\log\frac{z}{y} - \frac{\beta^2}{4k}.$$

We have used the inequality $\eta \leq -\log(1-\eta)$ for $0 < \eta < 1$. By the definition of γ and the fact that $\log(z/y) < \gamma M/\beta$, we have $I(y,z) < (\beta/\gamma)\log(z/y)$.

Case 3. (short interval): $1 < z/y < 2k/(2k-\beta)$. Here we have $|R(u)| < \frac{1}{2}\beta u$ for $y \leq u \leq z$ by the preceding estimates. Thus

$$I(y,z) \leq \frac{\beta}{2}\int_y^z u^{-1}du \leq \frac{\beta}{\gamma}\log\frac{z}{y},$$

upon using the inequality $M \geq 1$ to obtain $\gamma \leq 2$.

At the end of the proof of Lemma 5.4 we showed that the zeros of R are isolated. Given any $x > x_0$, let $x_1 < x_2 < \cdots < x_N$ be the zeros of R in $[x_0, x]$. We express $[1, x]$ as

$$[1, x_1] \cup (x_1, x_2] \cup \cdots \cup (x_{N-1}, x_N] \cup (x_N, x]$$

and estimate the integral in (5.8) over each section. We have

$$\int_1^x \frac{|R(u)|}{u^2}du \leq \int_1^{x_1} O\left(\frac{1}{u}\right)du + \frac{\beta}{\gamma}\sum_{i=1}^{N-1}\log\frac{x_{i+1}}{x_i} + M$$

$$\leq K\log x_1 + (\beta/\gamma)\log x + M.$$

This inequality implies the truth of (5.8), with $c = 1/(4Mk)$.

Lemma 5.5 and inequality (5.8) imply that $\alpha \leq \beta/(1+c\beta^2)$ for $\alpha < \beta < 2$. Since $\alpha \leq 1$ and $0 < c \leq 1/4$, if α were positive, the choice $\beta = \alpha + c\alpha^3$ would yield $\alpha < \beta \leq 5/4 < 2$ and

$$\alpha \leq \frac{\alpha + c\alpha^3}{1 + c(\alpha + c\alpha^3)^2} < \alpha.$$

This is impossible; hence α must be zero and $R(x) = o(x)$. By Lemma 5.2, $\psi(x) \sim x$, which is equivalent to the P.N.T. by Lemma 4.4. $\qquad\square$

5.4 Propositions "equivalent" to the P.N.T.

Before the discovery of an elementary proof of the P.N.T., a collection of propositions came to be described as "equivalent" to this theorem. This meant that they and the P.N.T. could be derived from one another by arguments that are "elementary," i.e. use perhaps calculus but no complex variable or Fourier theory. This classification lost its logical basis with the discovery of an elementary proof of the P.N.T.

We shall call two true propositions *equivalent* if they can be derived from one another by arguments that are considerably simpler than those used to establish directly the truth of either of them. This notion is obviously subjective and tentative, but it gives rise to a pragmatic classification.

The following general lemma, an integral version of "Axer's theorem," will help us to prove the equivalence of several propositions with the P.N.T.

Lemma 5.7 *Let A and $B \in \mathcal{V}$. Assume that $|A| \leq A_1 \in \mathcal{V}$, where $A_1 \uparrow$ and $\int_1^\infty A_1(u)u^{-2}du < \infty$. Also, assume that $B(x) = o(x)$ and its variation function satisfies $B_v(x) = O(x)$. Then $\int_{1-}^x dA * dB = o(x)$.*

Proof. First, $A_1(x) = o(x)$. Indeed, if $2^{n-1} < x \leq 2^n$ we have

$$\frac{A_1(x)}{x} \leq \frac{A_1(2^n)}{2^{n-1}} \leq 2 \int_{2^n}^\infty A_1(t)t^{-2}dt = o(1),$$

since the tail of a convergent integral tends to zero.

Let $\epsilon > 0$ be given, and let $C > 1$ and $\eta > 0$ be numbers, presently to be specified. We apply the hyperbola method (Th. 3.31), writing

$$\int_{1-}^x dA * dB = \int_{1-}^{x/C} A\left(\frac{x}{t}\right) dB(t) + \int_{1-}^C B\left(\frac{x}{t}\right) dA(t) - A(C)B\left(\frac{x}{C}\right).$$

$$=: I + II - III, \quad \text{say.}$$

$$|I| \leq \int_{1-}^{x/C} A_1\left(\frac{x}{t}\right) dB_v(t) \leq K \int_{1-}^{x/C} A_1\left(\frac{x}{t}\right) (\delta_1 + dt).$$

We have applied Lemma 3.11, noting that $t \mapsto A_1(x/t)$ is decreasing and $B_v(t) \leq Kt$ for all $t \geq 1$. If we change the variable in the last integral and extend the range of integration, we obtain

$$|I| \leq KA_1(x) + Kx \int_C^\infty A_1(u)u^{-2}du \leq \frac{1}{3}\epsilon x$$

provided that $x \geq$ some x_0 and C is chosen sufficiently large.

Now let C be fixed and estimate II. Choose $\eta < \epsilon/\{3A_v(C)\}$. We have $|B(x)| < \eta x$ provided $x \geq$ some x_1. Thus if $x/C \geq x_1$, then

$$|II| \leq \int_{1-}^C \eta \frac{x}{t} dA_v(t) \leq \eta x A_v(C) < \frac{1}{3}\epsilon x.$$

Finally, if $x \geq Cx_1$, then

$$|III| \leq |B(x/C)| A_v(C) < \eta x A_v(C) < \epsilon x/3.$$

Thus $(1/x) \int_{1-}^{x} dA * dB$ is arbitrarily small for x large.   □

Corollary 5.8   *Let $B \in V$ and assume*

$$B(x) = o(x), \quad B_v(x) = O(x), \quad and \quad \int_{1-}^{x} dN * dB = o(x).$$

Then

$$\int_{1-}^{x} t^{-1} dB(t) = o(1).$$

Proof.   It suffices to show that $\int_{1-}^{x} (x/t) \, dB(t) = o(x)$. For this write

$$\int_{1-}^{x} \frac{x}{t} dB(t) = \int_{1-}^{x} dN * dB + \int_{1-}^{x} (\delta_1 + dt - dN) * dB.$$

The last integral is $o(x)$ by Lemma 5.7, since $|x - N(x)| \leq 1$, which is a nondecreasing function and satisfies $\int_{1}^{\infty} 1 \cdot u^{-2} du < \infty$.   □

Theorem 5.9   *The following relations hold:*

$$\int_{1}^{x} d\psi * \frac{dt}{t} = \sum_{n \leq x} \log \frac{x}{n} \Lambda(n) \sim x, \tag{5.9}$$

$$\psi(x) \sim x, \tag{5.10}$$

$$\pi(x) \sim x/\log x \quad (P.N.T.), \tag{5.11}$$

$$M(x) := \sum_{n \leq x} \mu(n) = o(x), \tag{5.12}$$

$$\sum_{n \leq x} \mu(n)/n = o(1), \tag{5.13}$$

$$\int_{1}^{x} \frac{d\psi(t)}{t} = \sum_{n \leq x} \frac{\Lambda(n)}{n} = \log x - \gamma + o(1), \tag{5.14}$$

$$p_n \sim n \log n. \tag{5.15}$$

Here γ is Euler's constant and p_n is the nth prime in the natural order.

Proof. In the last section we showed that

$$R(x) = \int_1^x (d\psi - dt - \delta_1) * \frac{dt}{t} = \int_1^x d\psi * \frac{dt}{t} - (x-1) = o(x).$$

Thus (5.9) holds. We deduced (5.10) from this in Lemma 5.2. Relation (5.11) is a consequence of (5.10) by Lemma 4.4.

(5.10) $\longrightarrow$ (5.12): Chebyshev's identity $d\psi = (LdN) * dM$ gives one relation between ψ and M. This expression is not very convenient here because it is difficult to isolate M. There is a related identity,

$$LdM = -d\psi * dM, \tag{5.16}$$

which is much more useful. We prove (5.16) by noting that

$$0 = L\delta_1 = L(dN * dM) = (LdN) * dM + dN * (LdM).$$

Thus $dN * LdM = -d\psi$ and (5.16) follows by convolving with dM. If we add $dN * dM - \delta_1 = 0$ to the right side of (5.16) and integrate, we get

$$\int_1^x LdM = \int_{1-}^x \left\{ N\left(\frac{x}{t}\right) - \psi\left(\frac{x}{t}\right) \right\} dM(t) - \int_{1-}^x \delta_1.$$

The left side of this equation equals

$$M(x)\log x - \int_1^x M(t)t^{-1}dt = M(x)\log x + O(x),$$

while the right side equals

$$-1 + \int_1^x o(x/t)dM(t) = -1 + o\left(\int_1^x (x/t)dN(t) \right) = o(x\log x).$$

Thus $M(x) = o(x)$.

(5.12) $\longrightarrow$ (5.13): Take $B = M$ in Corollary 5.8. We have $M(x) = o(x)$;

$$\int_{1-}^x |dM| \leq N(x) = O(x); \quad \text{and} \quad \int_{1-}^x dN * dM = \int_{1-}^x \delta_1 = 1 = o(x).$$

It follows that $\int_{1-}^x t^{-1}dM(t) = o(1)$.

(5.12) $\longrightarrow$ (5.14): We write

$$x\int_1^x t^{-1}d\psi(t) = \int_{1-}^x (\delta_1 + dt) * LdN * dM =: I + II$$

say, with

$$I := \int_{1-}^{x} \frac{x}{t}(dt - \gamma\delta_1) = x\log x - \gamma x,$$

$$II := \int_{1-}^{x} (\delta_1 + dt) * (LdN - dt * dN + \gamma dN) * dM.$$

We are going to bound II with the aid of Lemma 5.7. Using estimates from the end of §3.3 we obtain

$$\int_{1-}^{x} (\delta_1 + dt) * (LdN - dt * dN + \gamma dN)$$

$$= x \int_{1-}^{x} \left\{ \frac{\log t \, dN(t)}{t} - \frac{dt}{t} * \frac{dN(t)}{t} + \gamma \frac{dN(t)}{t} \right\}$$

$$= cx + O(\log ex), \quad \text{where} \quad c = 2\gamma_1 + \gamma^2.$$

(The exact value of c will not be needed.) Take

$$A(x) := \int_{1-}^{x} (\delta_1 + dt) * (LdN - dt * dN + \gamma dN) - cN(x) = O(\log ex)$$

and write

$$II = \int_{1-}^{x} dA * dM + c \int_{1-}^{x} dN * dM.$$

The last term equals c. We estimate the first integral using Lemma 5.7 with $M = B$. We have $|A(x)| \le K\log ex$, a monotone function. Also, $\int_{1}^{\infty} (\log ex)x^{-2}dx < \infty$. Thus $\int_{1-}^{x} dA * dM = o(x)$, and

$$\int_{1}^{x} \frac{d\psi(t)}{t} = \{x\log x - \gamma x + c + o(x)\}/x = \log x - \gamma + o(1).$$

(5.11) $\longrightarrow$ (5.15): We remark that the function $F \in \mathcal{A}$ defined by $F(n) = p_n$ is a "generalized inverse" of π. Precisely, we have $\pi(F(n)) = \pi(p_n) = n$ and for $p_n \le x < p_{n+1}$, $F(\pi(x)) = F(\pi(p_n)) = F(n) = p_n$. We establish the desired asymptotic estimate by successively noting that

$$n = \pi(p_n) = (1 + o(1))p_n/\log p_n,$$

$$\log n = o(1) + \log p_n - \log\log p_n = (1 + o(1))\log p_n,$$

$$p_n = (1 + o(1))n\log p_n = (1 + o(1))n\log n. \qquad \square$$

At the outset of this section we defined a notion of equivalence of propositions. We now show the relations in Theorem 5.9 all to be equivalent. This fact is of some utility, since there exist in the literature direct proofs of all of them except (to our knowledge) $p_n \sim n \log n$. Thus we have a wide variety of possible proofs of the P.N.T.

The chains of implications are set forth in the following diagram. Single arrows represent implications of Theorem 5.9; double arrows correspond to those of Theorem 5.11 below.

$$(5.9) \quad \Longleftarrow \atop \longrightarrow \quad (5.10) \quad \Longleftarrow \atop \longrightarrow \quad (5.11) \quad \Longleftarrow \atop \longrightarrow \quad (5.15)$$

$$(5.14) \quad \longleftarrow \quad (5.12) \quad \Longrightarrow \quad (5.13)$$

It is convenient to establish the following result, which was discovered by Kronecker (cf. Problem 1.6).

Lemma 5.10 Let $A \in \mathcal{V}$ and assume that $\int_{1_-}^{\infty} t^{-1} dA(t)$ converges. Then $A(x) = o(x)$.

Proof. Let $\alpha =: \int_{1_-}^{\infty} t^{-1} dA(t)$ and $C(x) := \int_{1_-}^{x} t^{-1} dA(t) = \alpha + o(1)$. Then

$$A(x) = \int_{1_-}^{x} t \, dC(t) = \int_{1_-}^{x} t \, d\{C(t) - \alpha\}$$

$$= x\{C(x) - \alpha\} + \alpha - \int_{1}^{x} \{C(t) - \alpha\} \, dt = o(x). \qquad \square$$

Theorem 5.11 *Relations* (5.9) - (5.15) *are equivalent.*

Proof. We establish the implications shown by double arrows in the above diagram.

(5.14) $\Longrightarrow$ (5.10): We are given that $\int_{1}^{x} t^{-1}(d\psi - dt)$ converges. By Lemma 5.10, $\psi(x) = x + o(x)$.

(5.10) $\implies$ (5.9): $\int_1^x d\psi * t^{-1} dt = \int_1^x \psi(t) t^{-1} dt = x + o(x)$.

(5.13) $\implies$ (5.12): We apply Lemma 5.10.

(5.11) $\implies$ (5.10): We apply Lemma 4.4.

(5.15) $\implies$ (5.11): Let $p_n \leq x < p_{n+1}$. Now $p_{n+1}/p_n \sim 1$ and hence $x \sim p_n \sim n \log n$. Also,

$$\log x \sim \log n + \log \log n \sim \log n.$$

Thus $\pi(x) = n \sim x/\log n \sim x/\log x$. $\square$

PROBLEM 5.5   Give direct proofs of the following implications:

$$(5.12) \longrightarrow (5.10), \quad (5.10) \longrightarrow (5.14), \quad (5.13) \longrightarrow (5.14).$$

Hint for (5.10)$\longrightarrow$(5.14). Write $x \int_1^x t^{-1} d\psi(t)$ as

$$\int_1^x d\psi * dN + \int_1^x (\delta_1 + dt - dN) * dt + \int_1^x (\delta_1 + dt - dN) * (d\psi - dt).$$

PROBLEM 5.6   Let λ denote Liouville's function (see Problem 2.27). Show that

$$\sum_{n \leq x} \lambda(n) = o(x) \Leftrightarrow M(x) = o(x).$$

Hint. Show that $\lambda = \mu * 1_S$, where $S = \{n^2 : n \in \mathbb{Z}^+\}$.

PROBLEM 5.7   Show that

$$\#\{n \leq x : \Omega(n) \text{ is even}\} \sim \frac{x}{2}.$$

PROBLEM 5.8   Prove that

$$M(x) = o(x) \Leftrightarrow \int_1^x M(t) t^{-1} dt = o(x).$$

PROBLEM 5.9   Show that

$$Q(x) := \sum_{n \leq x} |\mu(n)| = 6\pi^{-2} x + o(\sqrt{x}).$$

5.5 Some consequences of the P.N.T.

The law of distribution of prime numbers has several applications. Here we give two number theoretic estimates which follow easily from the P.N.T.

Let us estimate $\pi_2(x)$, the number of positive integers not exceeding x which are the product of two distinct primes. We have $\pi_2(x) \geq cx/\log x$ for some $c > 0$, provided that $x \geq 6$. One can see this by fixing one of the primes to be 2 and letting the other prime range from 3 up to $x/2$. On the other hand, it is reasonable to conjecture that $\pi_2(x) = o(x)$. We give the asymptotic estimate in

Lemma 5.12 $\quad \pi_2(x) \sim x \log \log x / \log x.$

Proof. Let $\mathcal{P}$ denote the set of primes. We have

$$
(1_{\mathcal{P}} * 1_{\mathcal{P}})(n) = \begin{cases} 2 & \text{if } n = pp' \text{ and } p \neq p', \\ 1 & \text{if } n = p^2, \\ 0 & \text{if } \Omega(n) \neq 2. \end{cases}
$$

Thus

$$
\pi_2(x) = \frac{1}{2} \sum_{n \leq x} (1_{\mathcal{P}} * 1_{\mathcal{P}})(n) - \frac{1}{2} \sum_{n^2 \leq x} (1_{\mathcal{P}} * 1_{\mathcal{P}})(n^2).
$$

The last term equals $-\pi(\sqrt{x})/2 = O(\sqrt{x})$. We now estimate the summatory function of $\frac{1}{2}(1_{\mathcal{P}} * 1_{\mathcal{P}})$ by the hyperbola method (Theorem 3.31):

$$
\frac{1}{2} \sum_{n \leq x} (1_{\mathcal{P}} * 1_{\mathcal{P}})(n) = \frac{1}{2} \int_1^x d\pi * d\pi
$$

$$
= \int_1^{\sqrt{x}} \pi\left(\frac{x}{t}\right) d\pi(t) - \frac{1}{2}\pi(\sqrt{x})^2
$$

$$
= \int_1^{\sqrt{x}} \frac{(1 + o(1))x}{t \log(x/t)} d\pi(t) + O\left(\left\{\frac{\sqrt{x}}{\log x}\right\}^2\right).
$$

In the integral we write

$$
\log^{-1}(x/t) = (\log^{-1} x)\left\{1 - \frac{\log t}{\log x}\right\}^{-1} = (\log^{-1} x)\left\{1 + O\left(\frac{\log t}{\log x}\right)\right\},
$$

since $1 \le t \le \sqrt{x}$. The integral equals (by Lemmas 4.8 and 4.10)

$$(1 + o(1)) \frac{x}{\log x} \left\{ \int_1^{\sqrt{x}} \frac{d\pi(t)}{t} + O\left(\int_1^{\sqrt{x}} \frac{\log t \, d\pi(t)}{t \log x} \right) \right\}$$

$$= (1 + o(1)) \frac{x}{\log x} (\log \log \sqrt{x} + O(1)) = (1 + o(1)) \frac{x \log \log x}{\log x}. \quad \square$$

For any $r \in \mathbb{Z}^+$ let $\pi_r(x)$ denote the number of positive integers not exceeding x that are the product of r distinct primes. One can show by induction that

$$\pi_r(x) = (1 + o(1)) x (\log \log x)^{r-1} / ((r-1)! \log x).$$

Another problem we can handle is the estimation of $N_{\mathcal{P}, S}(x)$, the number of positive integers not exceeding x that are the product of a prime and a square. This number clearly exceeds $\pi(x)$, since the square could be 1. We give the asymptotic estimate in

Lemma 5.13 $N_{\mathcal{P}, S}(x) \sim \zeta(2) x / \log x.$

Proof.   Let S denote the squares and $\mathcal{P}$ the primes. $1_{\mathcal{P}} * 1_S$ is the indicator function of positive integers of the form $n = pr^2$. Thus $N_{\mathcal{P}, S}$ is the summatory function of $1_{\mathcal{P}} * 1_S$. Alternatively,

$$N_{\mathcal{P}, S}(x) = \sum_{r \le \sqrt{x}} \#\{n \le x : n = pr^2, \text{ for some } p\}$$

$$= \sum_{r \le \sqrt{x}} \pi(x/r^2) = \sum_{n \le x} \pi(x/n) 1_S(n).$$

Now $\sum_{n \le x} 1_{\mathcal{P}}(n) = \pi(x) \sim x / \log x$ while

$$\sum_{n \le x} 1_S(n) = \sum_{m^2 \le x} 1 = [\sqrt{x}] \sim \sqrt{x}.$$

The disparity in the size of the two summatory functions suggests that the idea of the stability theorem is appropriate. (The negative power of $\log x$ prevents us from simply quoting the theorem.) We have

$$N_{\mathcal{P}, S}(x) = \sum_{r \le x^{1/4}} \pi(x/r^2) + \sum_{r > x^{1/4}} O(x/r^2).$$

The error term is $O(x^{3/4})$. The main term is

$$(1 + o(1)) \sum_{r \leq x^{1/4}} \frac{x}{r^2 \log(x/r^2)}$$

$$= (1 + o(1)) \frac{x}{\log x} \sum_{r \leq x^{1/4}} \left\{ 1 + O\left(\frac{\log r}{\log x}\right) \right\} r^{-2},$$

by the argument we applied to $\log^{-1}(x/t)$ in the estimate of $\pi_2(x)$ just above. The last sum equals

$$\zeta(2) + O(x^{-1/4}) + O(\log^{-1} x).$$

Thus $N_{\mathcal{P},S}(x) \sim \zeta(2)x/\log x$. $\square$

PROBLEM 5.10 If $0 < c < 1$, let $f_c(x)$ denote the number of positive integers $n \leq x$ such that $[n^c]$ is prime. Use the P.N.T. to prove that $f_c(x) \sim x/(c \log x)$. Hint. We have

$$f_c(x) = \sum_{p \leq x^c} \sum_{p^{1/c} \leq n < (p+1)^{1/c}} 1 + O(x^{1-c}).$$

5.6 Notes

5.1. Selberg's formula (Th. 5.1) first appeared in Ann. of Math. (2), vol. 50 (1949), pp. 305–313. For another proof of the P.N.T. based on Selberg's formula, see P. Erdős, Proc. Nat. Acad. Sci., vol. 35 (1949), pp. 374–384.

An interesting mathematical and historical account of tauberian theorems is given in the book of the same name by J. Korevaar, Springer, Berlin, 2004.

5.2–5.3. The proof of the P.N.T. given here is based primarily on the following sources:

- A. Selberg, Ann. of Math. (2), vol. 50 (1949), pp. 305–313
- E. M. Wright, Proc. Roy. Soc. Edinburgh Sect. A, vol. 63 (1952), pp. 257–267
- N. Levinson, Amer. Math. Monthly, vol. 76 (1969), pp. 225–245

Many other proofs of the P.N.T. have been found. We give one based on ideas from Fourier analysis in §§7.2–7.3. Another, using the Cauchy integral theorem was given by D. J. Newman, Amer. Math. Monthly, vol. 87

(1980), pp. 693–696. See also J. Korevaar, Math. Intelligencer, vol. 4, no. 3 (1982), pp. 108–115, and D. Zagier, Amer. Math. Monthly, vol. 104 (1997), pp. 705–708. A proof based on the large sieve inequality was given by A. J. Hildebrand, Mathematika, vol. 33 (1986), pp. 23–30. We present a proof of the P.N.T. with an error term in §§8.6–8.7.

Surveys of elementary methods used in prime number theory were given by H. G. Diamond in Bull. Amer. Math. Soc. (N.S.), vol. 7 (1982), pp. 553–589 and by A. F. Lavrik in Trudy Mat. Inst. Steklov, vol. 163 (1984), pp. 118–142.

5.4. A good discussion of Axer's theorem is presented in [HarD], pp. 378–379. References for Axer's article are given in the Notes for §3.6. Lemma 5.7 can be regarded as a primitive form of the stability theorem (Theorem 3.29).

5.5. Lemma 5.12 was conjectured by Gauss at the same time that he conjectured the P.N.T. It is given in Werke, vol. 10, part 1, Göttingen, 1917, p. 11. It was first proved by Landau, Bull. Soc. Math. France, vol. 28 (1900), pp. 25–38; also in [LanC], vol. 1, pp. 92–105. See also E. M. Wright, Proc. Edinburgh Math. Soc. (2), vol. 9 (1954), pp. 87–90.

The assertion of Problem 5.10, namely that $f_c(x) \sim x/(c \log x)$, holds also for values of c slightly larger than 1, but this is much more difficult to establish. See G. A. Kolesnik, Mat. Zametki, vol. 2 (1967), pp. 117–128. The assertion is clearly false for $c = 2$.

Chapter 6

Dirichlet Series and Mellin Transforms

6.1 The use of transforms

One of the great advances in number theory, dating from the time of Riemann and Dirichlet, is the idea of associating to an arithmetic function a certain infinite series, today called a Dirichlet series (briefly: D.s.). This series defines a function of a complex variable which reflects the algebraic properties of the associated arithmetic function. It is often possible to investigate these properties by applying methods of analysis to the D.s. In this chapter we shall develop a theory based on the notion of Mellin transforms which will enable us to treat integrals as well as series.

We generally let s denote a complex number and follow the curious convention of analytic number theory of writing s as $\sigma + it$, where σ and t are real. For x a positive number we define x^s by $e^{s \log x}$, where the real valued branch of the logarithm is used.

Let $F \in \mathcal{V}$ and form the improper Riemann-Stieltjes integral

$$\lim_{X \to \infty} \int_{1-}^{X} x^{-s} dF(x) := \int x^{-s} dF(x),$$

if it exists. Let S denote the set of points $s \in \mathbb{C}$ for which the integral converges, i.e. for which the preceding limit exists. The integral defines a function on S, which we shall generally denote by $\widehat{F}$ and call the *Mellin transform* (briefly: M.t.) of dF. When discussing a function F, we may refer to $\widehat{F}$ as the M.t. *associated with* F. If we change the variable, setting $x = e^u$, the resulting integral will be a Laplace-Stieltjes transform. For most purposes, the M.t. will be more suitable for us.

There are two special types of M.t.'s that we shall use frequently:
(1) Let $f \in \mathcal{A}$ and have summatory function F. The M.t.

$$s \longmapsto \int_{1-}^{\infty} x^{-s} dF(x) = \sum_{n=1}^{\infty} f(n)n^{-s}$$

is called the *Dirichlet series* or *generating function* of f. Occasionally we shall use subscript notation and express the D.s. as $\sum f_n n^{-s}$.

(2) Now let f denote a complex valued function on $[1, \infty)$ that is Riemann integrable on any bounded interval $[1, X]$. We define $F \in \mathcal{V}$ by setting $F(x) := \int_1^x f(t)dt$ for $x \geq 1$ and $F(x) := 0$ for $x < 1$. By (3.2) we have

$$\int_1^X x^{-s} f(x)dx = \int_1^X x^{-s} dF(x)$$

for any $X > 1$. Consequently, $s \mapsto \int x^{-s} f(x)dx$ is an M.t. provided that the integral converges for some $s \in \mathbb{C}$.

Example 6.1   Here are some Mellin transforms. Points at which the integrals converge are indicated in parentheses.

(1) $\displaystyle \int x^{-s} dN(x) = \sum n^{-s} =: \zeta(s)$ $(\sigma > 1)$,

(2) $\displaystyle \int x^{-s} dx = \frac{1}{s-1}$ $(\sigma > 1)$,

(3) $\displaystyle \int x^{-s} \log x \, dx = \frac{1}{(s-1)^2}$ $(\sigma > 1)$,

(4) $\displaystyle \int x^{-s}(x^{-1}dx) = 1/s$ $(\sigma > 0)$,

(5) $\displaystyle \int x^{-s} \delta_c(x) = c^{-s}, \quad c \geq 1$ $(\mathbb{C})$.

The Riemann zeta function was defined in §1.5 for s real and $s > 1$. Here we have defined $\zeta(s)$ for complex s with $\sigma > 1$ by the same Dirichlet series or equivalently as the Mellin transform of dN.

PROBLEM 6.1   Suppose $F \in \mathcal{V}$ and is real valued. Prove that if $\widehat{F}(s)$ converges, then $\widehat{F}(\bar{s})$ converges and equals $\overline{\widehat{F}(s)}$. The bar denotes complex conjugation.

PROBLEM 6.2   Find $f \in \mathcal{A}$ such that $\zeta(2s) = \sum f(n)n^{-s}$.

Let $F \in \mathcal{V}$ and $s \in \mathbb{C}$. We say that $\widehat{F}$ *converges absolutely at s* if $\int x^{-\sigma} dF_v(x) < \infty$. Here $\sigma = \Re s$ and F_v is the total variation of F. For example, if $dF = dN - dx$, then $\int x^{-s} dF(x)$ converges for any s with positive real part. (Integration by parts!) On the other hand, $dF_v(x) = dN(x) + dx$ and the M.t. converges absolutely at s if and only if $\Re s > 1$. In the special case of a D.s. $\sum f(n) n^{-s}$, we take F as the summatory function of f. Then F_v is the summatory function of $|f|$, and the D.s. converges absolutely at s if and only if $\sum |f(n)| n^{-\sigma}$ converges.

For arithmetic functions or members of the class $\mathcal{V}$ of sufficiently slow growth (cf. §6.3), the map to the associated D.s. or M.t. is linear. Further, it is a homomorphism taking convolutions into pointwise products, as we now show for M.t.'s. In this sense, the transform reflects the algebraic structure of the underlying arithmetic object.

Theorem 6.2   *Let F and $G \in \mathcal{V}$ and suppose s is a point at which one of the M.t.'s $\widehat{F}$ and $\widehat{G}$ converges absolutely and the other converges. Then*

$$\int z^{-s} (dF * dG)(z) = \widehat{F}(s) \cdot \widehat{G}(s).$$

If F and G are summatory functions of arithmetic functions f and g respectively, we can also write the last formula as

$$\sum n^{-s} (f * g)(n) = \left\{ \sum \ell^{-s} f(\ell) \right\} \left\{ \sum m^{-s} g(m) \right\}. \tag{6.1}$$

Proof.   Suppose $\int y^{-s} dG(y)$ converges absolutely. Let Z be any number exceeding 1. By Lemma 3.23,

$$\int_{1-}^{Z} z^{-s} (dF * dG)(z) = \int_{y=1-}^{Z} \left\{ \int_{x=1-}^{Z/y} (xy)^{-s} dF(x) \right\} dG(y)$$

$$= \left\{ \int_{1-}^{\sqrt{Z}} x^{-s} dF(x) \right\} \left\{ \int_{1-}^{\sqrt{Z}} y^{-s} dG(y) \right\}$$

$$+ \int_{y=1-}^{\sqrt{Z}} y^{-s} \left\{ \int_{x=\sqrt{Z}}^{Z/y} x^{-s} dF(x) \right\} dG(y)$$

$$+ \int_{y=\sqrt{Z}}^{Z} y^{-s} \left\{ \int_{x=1-}^{Z/y} x^{-s} dF(x) \right\} dG(y)$$

$$=: I + II + III, \quad \text{say}.$$

As $Z \to \infty$ we have $I \to \widehat{F}(s)\widehat{G}(s)$;

$$|II| \leq \int_{y=1-}^{\sqrt{Z}} y^{-\sigma} o(1) dG_v(y) = o(1),$$

since $\int_{\sqrt{Z}}^{Z/y} x^{-s} dF(x) \to 0$ uniformly for $1 \leq y \leq \sqrt{Z}$ as $Z \to \infty$; and

$$|III| \leq \int_{y=\sqrt{Z}}^{Z} O(1) y^{-\sigma} dG_v(y) = o(1). \qquad \square$$

Example 6.3   For $\sigma > 1$ we have

(1) $\int x^{-s}(dx * dx) = \left(\int x^{-s} dx\right)^2 = (s-1)^{-2}$,

(2) $\sum n^{-s} \tau(n) = \sum n^{-s}(1 * 1)(n) = \left(\sum n^{-s}\right)^2 = \zeta^2(s)$,

(3) $\sum 1 = \sum n^{-s} e(n) = \sum n^{-s}(1 * \mu)(n) = \left(\sum n^{-s}\right)\left(\sum \mu(n) n^{-s}\right)$
    and hence $\sum \mu(n) n^{-s} = 1/\zeta(s)$,

(4) $\sum f(n) n^{-s} = \zeta(s)^{1/2}$ for $f \in \mathcal{A}_1$, $f * f = 1$.

6.2   Euler products

Theorem 6.2 allows us to represent the Dirichlet series of a convolution of a *finite number* of arithmetic functions as a product of Dirichlet series. In Lemma 2.26 we showed how to represent any multiplicative function f as an infinite convolution product $\prod_{j=1}^{\infty} f_j$, where $f_j(p_j^\alpha) = f(p_j^\alpha)$ for all $j \geq 1$ and $\alpha \geq 0$, and $f_j(n) = 0$ otherwise. With a suitable convergence hypothesis we can extend Theorem 6.2 to cover this case also.

Lemma 6.4   *Let $f \in \mathcal{M}$. Then*

$$\sum f(n) = \prod_p \{1 + f(p) + f(p^2) + \cdots\}, \qquad (6.2)$$

provided that the series on the left converges absolutely.

Proof.   For any fixed $x > 1$, let

$$P(x) := \prod_{p \leq x} \{1 + f(p) + f(p^2) + \cdots\}.$$

For each prime p, the sum $\sum_{r=0}^{\infty} f(p^r)$ is a subseries of an absolutely convergent series and thus is absolutely convergent. By multiplying out the finite number of factors of this form and using the fact that f is multiplicative, we obtain

$$P(x) = \sum_{(x)} f(n),$$

where $\sum_{(x)}$ extends over just those positive integers n all of whose prime factors are at most x. (Since $\sum_{(x)} f(n)$ also is an absolutely convergent series, the order of the terms is immaterial.)

Let $\sum'_{(x)}$ indicate that the summation is extended over those positive integers divisible by at least one prime greater than x. Then

$$\left| \sum f(n) - P(x) \right| = \left| \sum f(n) - \sum_{(x)} f(n) \right| = \left| \sum'_{(x)} f(n) \right| \le \sum_{n>x} |f(n)|,$$

and the last series goes to zero as $x \to \infty$. Thus

$$P(x) \to \sum_{n=1}^{\infty} f(n) \quad (x \to \infty).$$

It remains to show that the infinite product in (6.2) does not diverge to zero. Choose $X \ge 1$ such that $\sum_{n>X} |f(n)| < 1$. Then

$$\left| \prod_{p>X} \{1 + f(p) + f(p^2) + \cdots\} - 1 \right| \le \sum''_{(X)} |f(n)|,$$

where $\sum''_{(X)}$ denotes a sum taken over numbers all of whose prime factors exceed X. The last series is at most $\sum_{n>X} |f(n)| < 1$, and thus

$$\prod_{p>X} \{1 + f(p) + f(p^2) + \cdots\} \ne 0.$$

Hence $\lim_{x \to \infty} P(x)$ can equal zero only if a finite number of factors on the right side of (6.2) are equal to zero. $\square$

The above argument is essentially that used to establish Theorem 1.4. The infinite product in (6.2) actually converges absolutely. Indeed $|f|$ is multiplicative, and (6.2) is valid for $|f|$.

Corollary 6.5   *Let f be completely multiplicative and $\sum |f(n)| < \infty$. Then*

$$\sum f(n) = \prod_p \{1 - f(p)\}^{-1}. \tag{6.3}$$

Proof.   We have $f(p^r) = f(p)^r$ for each $r \geq 0$ and each prime p. The right side of (6.2) can now be expressed as a product of (absolutely) convergent geometric series.   □

Corollary 6.6   *Let $g \in \mathcal{M}$ and satisfy $g(n) = O(n^c)$ for some real constant c. If $\Re s > c + 1$ then*

$$\sum g(n)n^{-s} = \prod_p \{1 + g(p)p^{-s} + g(p^2)p^{-2s} + \cdots\}.$$

Proof.   Note that $T^{-s}g \in \mathcal{M}$ and $\sum |n^{-s}g(n)| < \infty$ for $\Re s > c + 1$. We now apply Lemma 6.4.   □

If g is completely multiplicative and $g(n) = O(n^c)$, then we have

$$\sum g(n)n^{-s} = \prod_p \{1 - g(p)p^{-s}\}^{-1} \quad (\sigma > c + 1). \tag{6.4}$$

Example 6.7   Let $\Re s > 1$. Then

(1) $\zeta(s) = \sum n^{-s} = \prod_p (1 - p^{-s})^{-1}$,

(2) $\sum \mu(n)n^{-s} = \prod_p \{1 + \mu(p)p^{-s} + \mu(p^2)p^{-2s} + \cdots\} = \prod_p \{1 - p^{-s}\}$.

In (1) we used the fact that 1 is completely multiplicative. Comparing (1) and (2), we obtain another proof that

$$\sum \mu(n)n^{-s} = \zeta(s)^{-1} \quad (\sigma > 1).$$

PROBLEM 6.3   Find the infinite product representation of $\sum |\mu(n)| n^{-s}$. Determine a region in which the representation is valid. Express the product as a quotient of zeta functions.

Suppose that we wish to numerically estimate an infinite product of the type (6.2). A useful method in some cases – by analogy with the stability theorem (Th. 3.29)—is to seek another product expression that is suitably close to the given one and whose numerical value is known. We multiply and divide the given product by the new one. The resulting quotient will converge more rapidly than the original product, with the degree of improvement depending upon the quality of the approximation. This quotient can be numerically approximated as a finite product times a factor that is near 1.

In case the given product has the form

$$\prod_p \left\{1 + ap^{-r} + O\left(p^{-r-1}\right)\right\}$$

for some $r > 1$, a candidate for an approximating product is $\zeta(r)^a$, where ζ is the Riemann zeta function. The quotient,

$$\prod_p \left\{1 + ap^{-r} + O\left(p^{-r-1}\right)\right\}\left\{1 - p^{-r}\right\}^a$$

has the form $\prod_p\{1 + O(p^{-r-1})\}$ and thus is more rapidly convergent than the original product.

Example 6.8 Estimate

$$I := \prod_p \left(1 - \frac{1}{p^2 - 1}\right).$$

For large p, the factors of I are close to $1 - 1/p^2$, so we multiply and divide I by $\prod_p(1 - p^{-2}) = 1/\zeta(2) = 6/\pi^2$. Thus

$$I = \frac{6}{\pi^2} \prod_p \left(1 - \frac{1}{p^2 - 1}\right)\left(1 - \frac{1}{p^2}\right)^{-1} =: \frac{6}{\pi^2} I_1,$$

where, by a small calculation,

$$I_1 = \prod_p \left\{1 - \frac{1}{(p^2 - 1)^2}\right\}.$$

I_1 clearly converges more swiftly than the original product for I.

We can approximate I_1 by calculating a suitable partial product and estimating its tail

$$\prod_{p>X} \left\{1 - \frac{1}{(p^2 - 1)^2}\right\} = \exp \sum_{p>X} \log\left\{1 - \frac{1}{(p^2 - 1)^2}\right\}.$$

We then apply simple bounds for the logarithm, such as

$$0 > \log\left(1 - \frac{1}{K}\right) = -\log\left(1 + \frac{1}{K-1}\right) > \frac{-1}{K-1}, \quad K > 2,$$

and estimate the sum over large primes using Lemma 4.12.

In the present case,

$$1 - (p^2 - 1)^{-2} \approx 1 - p^{-4},$$

so if we are not satisfied with the rate of convergence of I_1, we can multiply and divide it by the Euler product for $\zeta(4)$, whose value is $\pi^4/90$, and obtain

$$I = \frac{6}{\pi^2}\frac{90}{\pi^4} I_2, \text{ with } I_2 := \prod_p \left\{1 - \frac{2p^2 - 1}{(p^2 - 1)^3(p^2 + 1)}\right\}.$$

We then approximate I_2 as indicated above. Taking the product over the primes of I_2 that are smaller than 23, we find that $I \doteq 0.53071$, with all the given decimal places correct.

PROBLEM 6.4   Show that there is a constant k such that

$$\prod_{2 < p \le x} (1 - 2/p) \sim k/\log^2 x, \qquad x \to \infty.$$

PROBLEM 6.5   Use the result of Example 6.8 to approximate

$$\prod_p \left(1 - \frac{2}{p^2 - 1}\right).$$

PROBLEM 6.6   Let φ denote Euler's function, and for each $j \in \mathbb{Z}^+$ define $a_j = \#\{n \in \mathbb{Z}^+ : \varphi(n) = j\}$. Show that for $\sigma > 1$,

$$\sum_{j=1}^{\infty} a_j\, j^{-s} = \sum_{n=1}^{\infty} \varphi(n)^{-s} = \zeta(s) \prod_p \left\{1 + (p - 1)^{-s} - p^{-s}\right\}.$$

PROBLEM 6.7   Show that

$$\prod_p \left\{1 + (p - 1)^{-1} - p^{-1}\right\} = \zeta(2)\,\zeta(3)/\zeta(6).$$

6.3   Convergence

The first step in an investigation involving an M.t. usually is to discover the points of $\mathbb{C}$ at which the integral converges. Given $F \in \mathcal{V}$, there are three possibilities for the convergence of $\int x^{-s}dF(x)$: it might converge at all $s \in \mathbb{C}$, converge at some points $s \in \mathbb{C}$ and diverge at others, or diverge at all $s \in \mathbb{C}$. Simple examples of these three cases are provided by taking

$$F_1(x) = e^{-x}, \quad F_2(x) = x, \quad \text{and} \quad F_3(x) = \sum_{n \leq x} e^n$$

for $x \geq 1$, and each function equal to zero for $x < 1$.

There is a close relation between the asymptotic growth of F and the set of points of $\mathbb{C}$ at which $\widehat{F}$ converges, as we show in

Theorem 6.9 *Let $F \in \mathcal{V}$. Suppose that $F(x) = c + O(x^\alpha)$ for some $c \in \mathbb{C}$ and some $\alpha < 0$ or $F(x) = O(x^\alpha)$ for some $\alpha \geq 0$. If $\sigma > \alpha$ then $\int x^{-s} dF(x)$ converges. Conversely, suppose $\int x^{-s} dF(x)$ converges for some $s \in \mathbb{C}$. Then*

(1) $F(x) = o(x^\sigma)$, if $\sigma > 0$ (cf. Lemma 5.10),

(2) $F(x) = o(\log ex)$, if $\sigma = 0$,

(3) $F(x) = c' + o(x^\sigma)$, if $\sigma < 0$, where c' is some constant.

Proof. Suppose $F(x) - c = O(x^\alpha)$ and $\sigma = \Re s > \alpha$. Then

$$\int x^{-s} dF(x) = \lim_{X \to \infty} \left(x^{-s} \{ F(x) - c \} \Big|_{1-}^{X} + s \int_1^X x^{-s-1} \{ F(x) - c \} dx \right)$$

$$= c + s \int_1^\infty x^{-s-1} \{ F(x) - c \} dx \tag{6.5}$$

and the last integral converges.

Conversely, assume $\int x^{-s} dF(x)$ converges. Let

$$\psi(y) := - \int_y^\infty x^{-s} dF(x) \quad (= o(1)).$$

Then

$$F(x) = \int_{1-}^x dF = \int_{1-}^x y^s d\psi(y)$$

$$= x^s \psi(x) + \int_{1-}^\infty y^{-s} dF(y) - s \int_1^x y^{s-1} \psi(y) dy.$$

In cases (1) and (2) the required estimate now follows at once.

In case (3), for $1 < x < y < \infty$ write $F(y) - F(x) = \int_x^y u^s d\psi(u)$. Then

$$F(y) - F(x) = y^s \psi(y) - x^s \psi(x) - s \int_x^y u^{s-1} \psi(u) \, du = o(1)$$

as x and $y \to \infty$. By Cauchy's convergence criterion, $\lim_{y \to \infty} F(y) =: c$ exists, and we have

$$F(x) - c = x^s \psi(x) + s \int_x^\infty u^{s-1} \psi(u) du = o(x^\sigma). \qquad \square$$

Example 6.10   1. $\sum (-1)^{n+1} n^{-s}$ converges for $\sigma > 0$ since $\sum_{n \le x} (-1)^{n+1}$ is bounded.

2. $\sum n^{-1-it}$ does not converge for any real t. Indeed, since $\sum_{n \le x} n^{-1} \ne o(\log x)$, the D.s. $\sum n^{-1} n^{-s}$ diverges at all points of the line $\Re s = 0$.

PROBLEM 6.8   Assume that it is known that $\sum \mu(n) n^{-1-it}$ converges for some real t. Deduce from this the validity of the P.N.T.

PROBLEM 6.9   Let $F \in \mathcal{V}$ and assume that $\int x^{-s_0} dF(x)$ converges for some s_0 with $\Re s_0 > 0$. Prove that

$$\int x^{-s_0} dF(x) = s_0 \int x^{-s_0-1} F(x) dx.$$

PROBLEM 6.10   Suppose $F(x) \sim x$. Find the precise set of points $s \in \mathbb{C}$ at which $\widehat{F}$ converges.

Theorem 6.9 gives the following necessary and sufficient conditions for an integral $\int x^{-s} dF(x)$ to converge at some points of the complex plane: there should exist a polynomial P such that $|F(x)| \le P(x)$ for all $x \ge 1$. In this situation we say that F has *polynomial growth*. In case F is the summatory function of an arithmetic function f, note that F has polynomial growth if and only if f has this property.

6.3.1   Abscissa of convergence

A familiar theorem of analytic function theory asserts that a power series converges at all points of a certain open disc and diverges at all points of the complement of the closure of that disc. The power series may converge at all, some, or no points of the circle of convergence. A similar situation prevails for Mellin transforms except that half planes replace discs as the regions of convergence.

Lemma 6.11   *Let $F \in \mathcal{V}$ and $s_0 = \sigma_0 + it_0 \in \mathbb{C}$. If $\int x^{-s_0} dF(x)$ converges, then the integral converges on the open half plane $\{s : \sigma > \sigma_0\}$.*

Proof. We apply both parts of the preceding theorem: First, there exists a constant c such that $F(x) = c + o(x^{\sigma_0 + \epsilon})$ for any fixed positive ϵ. Then, the integral defining $\widehat{F}$ converges on the half plane $\sigma > \sigma_0 + \epsilon$. Since ϵ is arbitrary, the integral converges on $\{s : \sigma > \sigma_0\}$. □

In case an M.t. converges at some points $s \in \mathbb{C}$ and diverges at others, the last lemma tells us that the region of convergence is a half plane of the form $\{s : \sigma > \sigma_0\}$ and possibly some points on the line $\sigma = \sigma_0$. For a given M.t. $\widehat{F}$, we define $\sigma_c = \sigma_c(\widehat{F})$, the *abscissa of convergence* of $\widehat{F}$, to be

$$\sigma_c = \inf\{\sigma : \widehat{F}(s) \text{ converges for some } s \text{ with } \Re s = \sigma\}.$$

If $\int x^{-s} dF(x)$ diverges for all $s \in \mathbb{C}$ we set $\sigma_c = +\infty$; if it converges for all $s \in \mathbb{C}$ we set $\sigma_c = -\infty$. We define the *line of convergence* of an M.t. $\widehat{F}$ to be those points in $\mathbb{C}$ whose real part is $\sigma_c(\widehat{F})$, if this is a finite number.

Example 6.12 Some simple evaluations of σ_c.

$$\zeta(s) : \sigma_c = 1; \qquad \int x^{-s} x^{-1} dx = 1/s : \sigma_c = 0;$$

$$\sum e^{-n} n^{-s} : \sigma_c = -\infty; \qquad \sum e^n n^{-s} : \sigma_c = +\infty.$$

PROBLEM 6.11 Find the abscissa of convergence of the following M.t.'s:

(a) $\sum |\mu(n)| n^{-s}$, (b) $\int x^{-s} d(x^3)$, (c) $\int x^{-s} \cos(x^2) dx$,

(d) $\sum n^{-2s}$, (e) $\sum_{n \geq 10} n^{\log \log n} n^{-s}$, (f) $\sum_{n \geq 10} n^{-\log \log n} n^{-s}$.

PROBLEM 6.12 For each of the following M.t.'s, find the set of points on its line of convergence at which it converges:

(a) $\int x^{-s} dx$, (b) $\int x^{-s} (\log ex)^{-1} dx$, (c) $\int x^{-s} (\log ex)^{-2} dx$.

The determination of the abscissa of convergence of an M.t. can be a difficult or impossible task. By Theorem 6.9 the problem is equivalent to the determination of the growth rate of the associated summatory function. For example, the exact value of σ_c for $\sum \mu(n) n^{-s}$ is presently unknown. It is easy to see that $0 \leq \sigma_c \leq 1$. Later we shall show that $\sigma_c \geq \frac{1}{2}$, by using the existence of nonreal zeros of $\zeta(s)$.

6.3.2   Abscissa of absolute convergence

In the preceding section we defined the notion of absolute convergence. For a given M.t. $\widehat{F}$, define $\sigma_a = \sigma_a(\widehat{F})$, the *abscissa of absolute convergence of* $\widehat{F}$, to be $\sigma_a = \inf\{\sigma : \int x^{-\sigma}dF_v(x) < \infty\}$. If a transform converges at a point $s \in \mathbb{C}$, but not absolutely, we say that the convergence at that point is *conditional*.

PROBLEM 6.13   Let $F \in \mathcal{V}$ and $b > \sigma_a(\widehat{F})$. Show that $\widehat{F}$ is bounded on $\{s : \sigma \geq b\}$.

For power series the (open) disc of convergence coincides with the (open) disc of absolute convergence. One difference between power series and Dirichlet series occurs here, for the half plane of absolute convergence need not coincide with the half plane of convergence. As an example, the D.s. $\sum(-1)^{n+1}n^{-s}$ has $\sigma_a = 1$ and $\sigma_c = 0$ (shown after the proof of Theorem 6.9, or more simply by an alternating series argument for any real positive value of s).

It is clear that $\sigma_a \geq \sigma_c$ for any M.t. On the other hand $\sigma_a \leq \sigma_c + 1$ for Dirichlet series. (See Problem 6.14 below.) For arbitrary Mellin transforms no such estimate exists. Indeed if we define $F \in \mathcal{V}$ by

$$F(x) := \int_1^x e^t \cos(e^{2t}) \, dt$$

for $x \geq 1$, then $\sigma_c(\widehat{F}) = -\infty$ and $\sigma_a(\widehat{F}) = +\infty$ by Theorem 6.9, since $F(x) = c + O(e^{-x})$ and $F_v(x) \gg e^x$. The size of the gap between σ_c and σ_a reflects the oscillation of F.

PROBLEM 6.14   Suppose that $\sum a_n n^{-s}$ converges at $s = s_0$. Prove that $\sum |a_n| n^{-\sigma}$ converges if $\Re s > 1 + \Re s_0$ and conclude that $\sigma_a \leq 1 + \sigma_c$.

6.4   Uniform convergence

It is useful to know some sets on which an M.t. converges uniformly. The simplest result is

Lemma 6.13   *Let $F \in \mathcal{V}$ and let $b > \sigma_a(\widehat{F})$. Then $\widehat{F}$ converges uniformly on the half plane $\{s : \sigma \geq b\}$.*

Proof.   We have $\int x^{-b}dF_v(x) < \infty$. By the integral version of the Weierstrass M-test, $\widehat{F}$ converges uniformly on $\{s : \sigma \geq b\}$. $\square$

Using this simple lemma, we can show the distinguished character of the first coefficient of a D.s.:

Lemma 6.14 *Let $f \in \mathcal{A}$ have polynomial growth. Let $\{s_j\}$ be any sequence with $\Re s_j \to +\infty$. Then*

$$\lim_{j \to \infty} \sum f(n) n^{-s_j} = f(1). \tag{6.6}$$

Proof. Say $f(n) = O(n^\alpha)$ for some real α. On the one hand the D.s. converges uniformly on the half plane $\{s : \sigma \geq \alpha + 2\}$, say, enabling us to take the limit termwise. On the other hand, except for the first term, which is constant, all the other terms tend to zero as $j \to \infty$. $\qquad\square$

If $\sigma_a > \sigma_c$, a result which is usually more useful than Lemma 6.13 is

Theorem 6.15 *Let $F \in \mathcal{V}$. If the integral defining $\widehat{F}$ converges at a point $s_0 \in \mathbb{C}$, then for any fixed positive δ the integral converges uniformly on the sector $S_\delta := \{s : |\arg(s - s_0)| \leq \frac{\pi}{2} - \delta\}$.*

Proof. We shall show that $\int_Y^Z x^{-s} dF(x)$ tends to zero uniformly on S_δ as Y and Z tend to infinity. Let

$$\psi(y) := -\int_y^\infty x^{-s_0} dF(x) \quad (= o(1)).$$

Then

$$\int_Y^Z x^{-s} dF(x) = \int_Y^Z x^{-(s-s_0)} d\psi(x)$$

$$= x^{-(s-s_0)} \psi(x) \Big|_Y^Z + (s - s_0) \int_Y^Z x^{-(s-s_0)-1} \psi(x) dx.$$

Let $s - s_0 = \xi + i\eta$. The conditions that s lie in S_δ and $s \neq s_0$ are equivalent to the conditions that $\xi > 0$ and $|\xi + i\eta|/\xi \leq \csc \delta$. Now if $Z > Y$,

$$\left| \int_Y^Z x^{-s} dF(x) \right| \leq |\psi(Z)| + |\psi(Y)| + |\xi + i\eta| \int_Y^\infty x^{-\xi-1} |\psi(x)| \, dx$$

$$\leq o(1) + \frac{|\xi + i\eta|}{\xi} \sup_{x \geq Y} |\psi(x)| = o(1)$$

as $Y \to +\infty$, uniformly in $S_\delta \setminus \{s_0\}$. Since the integral converges also at s_0, the convergence is uniform on the whole sector. $\qquad\square$

Problem 6.15   Use Theorem 6.15 to give another proof that the region of convergence of an M.t. is a half plane.

Problem 6.16   Let F, δ, and S_δ be as in Theorem 6.15. Show that $\lim \widehat{F}(s) = F(1)$, where $s \to \infty$ in the sector S_δ.

Problem 6.17   Let $F(x) := \int_1^x \cos(\log u) du$ for $x \geq 1$. Show that the integral defining $\widehat{F}$ converges, but not uniformly, in the open half plane $\{s : \sigma > 1\}$.

Problem 6.18   (E. Landau).   If the Dirichlet series $\sum c_n n^{-s}$ converges at $s = \sigma_0 + it_0$, prove that the function defined by the series for $\Re s > \sigma_0$ cannot have a pole on the line $\Re s = \sigma_0$.

Problem 6.19   Using the preceding problem, show that

$$\sum_{n=1}^\infty \mu(n)\, n^{-\frac{1}{2}-it} \quad \text{and} \quad \sum_{n=1}^\infty n^{-1-it}$$

each diverge for all real t. (You may assume that the Riemann zeta function has zeros with real part $1/2$.)

Corollary 6.16   *Suppose that $F \in \mathcal{V}$, $s_0 \in \mathbb{C}$ and that the integral defining $\widehat{F}(s_0)$ converges. Then $\lim_\delta \widehat{F}(s) = \widehat{F}(s_0)$, where the subscript denotes a limit taken as $s \to s_0$ through values in a sector S_δ for any fixed positive δ.*

Proof.   The sequence of functions $\widehat{F}_N$ defined by $\widehat{F}_N(s) = \int_{1-}^N x^{-s} dF(x)$ converges to $\widehat{F}$ uniformly on S_δ. Each $\widehat{F}_N$ is continuous (everywhere in $\mathbb{C}$). Thus $\widehat{F}$ is continuous on the sector S_δ.   □

As an example, we have

$$\lim_{s \to 1}\,_\delta \{\zeta(s) - 1/(s-1)\} = \int x^{-1}(dN - dx) = \gamma.$$

Here γ is Euler's constant and $\lim_\delta$ is interpreted as above. (It will be shown in the next section that the subscript δ is in fact unnecessary in this example, as the function $s \to \zeta(s) - (s-1)^{-1}$ is analytic at 1.)

N.B. We must know that an integral $\int x^{-s_0} dF(x)$ converges before attempting to evaluate it with the aid of Corollary 6.16.

Example 6.17   Let $f(n) := (-1)^k$ if $n = 2^k$ for some nonnegative integer k and $f(n) := 0$ for all other integers. Let F be the summatory function of

f. Then for $\sigma > 0$,

$$\widehat{F}(s) = \sum f(n) n^{-s} = \sum_{k=0}^{\infty} (-1)^k 2^{-ks} = (1 + 2^{-s})^{-1}$$

and $\lim_{\sigma \to 0+} \widehat{F}(\sigma) = 1/2$. On the other hand, $\widehat{F}(0)$ does not exist.

Sometimes, we know that $\int x^{-s_0} dF(x)$ exists, but we don't know its value. If we can evaluate the limit of $\widehat{F}(s)$ as $s \to s_0$ along some path lying in a sector S_δ, then we can apply Corollary 6.16 to evaluate $\int x^{-s_0} dF(x)$. As an example, we identify the constant B_2 of Lemma 4.10 in

Lemma 6.18 *Let γ denote Euler's constant. Then*

$$B_2 = \lim_{x \to \infty} \left\{ \sum_{n \leq x} \frac{\kappa(n)}{n} - \log\log x \right\} = \gamma.$$

Proof. The stated limit exists by Lemma 4.10. We start by comparing $\log\log x$ with $\int_1^x (1 - t^{-1}) \, dt/(t \log t)$, which is better behaved near $x = 1$. This choice is motivated by the relations

$$1 = \exp \kappa \qquad\qquad\qquad\qquad \text{(cf. §2.4)}$$

$$\delta_1 + dt = \exp \left\{ \frac{(1 - t^{-1})}{\log t} dt \right\} \qquad\qquad \text{(cf. §3.4)}$$

and the fact that $[x] = \sum_{n \leq x} 1$ is close to $x = \int_{1-}^x (\delta_1 + dt)$.
Integration by parts yields

$$\lim_{x \to \infty} \left\{ \int_1^x \frac{1 - t^{-1}}{t \log t} dt - \log\log x \right\} = -\int_1^\infty \frac{\log\log t}{t^2} dt = -\int_0^\infty e^{-u} \log u \, du.$$

We can evaluate the last integral by using two representations of the gamma function (cf. Appendix). We differentiate under the integral sign $\Gamma(s) = \int_0^\infty u^{s-1} e^{-u} \, du$, which is valid for $\sigma > 0$, as shown in the next section. We obtain $\int_0^\infty e^{-u} \log u \, du = \Gamma'(1)$. On the other hand, if we logarithmically differentiate the infinite product

$$\Gamma(s) = \frac{e^{-\gamma s}}{s} \prod_{n=1}^{\infty} \left(1 + \frac{s}{n}\right)^{-1} e^{s/n},$$

we obtain $\Gamma'(1)/\Gamma(1) = -\gamma$. Either representation shows that $\Gamma(1) = 1$. Thus we have

$$\lim_{x \to \infty} \left\{ \int_1^x \frac{1 - t^{-1}}{t \log t} dt - \log \log x \right\} = -\Gamma'(1) = \gamma.$$

Differentiation shows that

$$\int_1^x \frac{1 - t^{-1}}{t \log t} dt - \int_x^\infty \frac{dt}{t^2 \log t} - \log \log x$$

is constant, and thus, for $x \to \infty$,

$$\int_1^x \frac{1 - t^{-1}}{t \log t} dt - \log \log x = \gamma + \int_x^\infty \frac{dt}{t^2 \log t} = \gamma + o(1). \qquad (6.7)$$

It follows that

$$B_2 = \lim_{x \to \infty} \left\{ \sum_{n \le x} \frac{\kappa(n)}{n} - \int_1^x \frac{1 - u^{-1}}{u \log u} du \right\} + \gamma.$$

We evaluate this limit using Cor. 6.16. For $\sigma > 1$,

$$\sum \kappa(n) n^{-s} = \log \zeta(s) \quad \text{and} \quad \int_1^\infty \frac{1 - u^{-1}}{u^s \log u} du = \log \frac{s}{s - 1}.$$

The last formula is verified by first noting, by differentiation with respect to s, that the two sides differ by a constant. Then, since each side tends to zero as $s \to +\infty$, that constant is zero. Thus

$$\sum \kappa(n) n^{-s} - \int u^{-s} \frac{(1 - u^{-1})}{\log u} du = \log\{\zeta(s)(s - 1)/s\}. \qquad (6.8)$$

Also, we know that $\zeta(\sigma) = 1/(\sigma - 1) + O(1)$ for $1 < \sigma < 2$. It follows that

$$\log\{\zeta(\sigma)(\sigma - 1)/\sigma\} = \log\left\{ \left(\frac{1}{\sigma - 1} + O(1) \right) \cdot \frac{\sigma - 1}{1 + (\sigma - 1)} \right\}$$

$$= \log(1 + O(\sigma - 1)) \to 0$$

as $\sigma \to 1+$. By Corollary 6.16, we have

$$\lim_{x \to \infty} \left\{ \sum_{n \le x} \frac{\kappa(n)}{n} - \int_1^x \frac{(1 - u^{-1})}{u \log u} du \right\} = 0,$$

and finally $B_2 = \gamma$. $\qquad\qquad\qquad\qquad\qquad\qquad\qquad\qquad\qquad$ $\square$

Corollary 6.19 *Let γ denote Euler's constant. Then, as $x \to \infty$,*

$$\prod_{p \le x} \left(1 - 1/p\right) \sim e^{-\gamma}/\log x.$$

Proof. In Lemma 4.11 we showed that the stated product is asymptotic to $e^c/\log x$, where

$$c = \lim_{x \to \infty} \left\{ \sum_{p \le x} \log(1 - p^{-1}) + \log\log x \right\},$$

and also that

$$\lim_{x \to \infty} \left\{ \sum_{p \le x} \log(1 - p^{-1}) + \sum_{n \le x} \frac{\kappa(n)}{n} \right\} = 0.$$

From these relations and the preceding lemma, we find that $c = -\gamma$. $\square$

PROBLEM 6.20 Let λ denote Liouville's function (cf. Problem 2.27). Find the numerical value of $\sum_{n=1}^{\infty} \lambda(n)/n$, assuming that the series converges.

6.5 Analyticity

A convergent power series defines a function which is analytic at least in the open disc of convergence. We shall establish the corresponding result for M.t.'s on half planes and show that in a number of cases the analytic function can be continued to a larger region.

Theorem 6.20 *Let $F \in \mathcal{V}$ and be of polynomial growth. Then $\widehat{F}$ is analytic on $\{s : \sigma > \sigma_c(\widehat{F})\}$. On this half plane the derivatives of $\widehat{F}$ are represented by the formula*

$$\widehat{F}^{(r)}(s) = \int s^{-s}(-\log x)^r dF(x), \quad r = 1, 2, \ldots.$$

Remarks 6.21 We have noted that L is a derivation on integrators and the M.t. is a homomorphism from integrators to analytic functions. For functions $F \in \mathcal{V}$ of polynomial growth, the sequences of operations

$$dF \longmapsto \widehat{F} \longmapsto \{\widehat{F}\}' \quad \text{and} \quad dF \longmapsto LdF \longmapsto \{\widehat{F}\}'$$

are equivalent.

Proof. For $1 < X < \infty$ set

$$\widehat{F}_X(s) := \int_{1-}^X x^{-s} dF(x).$$

$\widehat{F}_X$ is an entire function whose derivative is obtained by differentiating under the integral sign. This operation can be justified by noting that

$$\frac{x^{-s-\delta} - x^{-s}}{\delta} = -x^{-s} \log x \left(\frac{e^{-\delta \log x} - 1}{-\delta \log x} \right) \to -x^{-s} \log x$$

uniformly for $1 \le x \le X$ (for each fixed value of s).

Given $s_0 = \sigma_0 + it_0$ with $\sigma_0 > \sigma_c$, let $s_1 = \frac{1}{2}(\sigma_c + \sigma_0) + it_0$, and let S be the sector $\{s : |\arg(s - s_1)| \le \pi/4\}$. Let $\{X_n\}_{n=1}^\infty$ be a sequence of real numbers tending to infinity. By Theorem 6.15, $\widehat{F}_{X_n} \to \widehat{F}$ uniformly on S, and so $\widehat{F}$ is analytic at each point s_0 in the open half plane $\{s : \sigma > \sigma_c\}$ by the Weierstrass theorem on the uniform limit of analytic functions.

We justify the formula for $\widehat{F}^{(r)}$ as follows: If $\{f_n\}$ is a set of functions analytic on a set U and if $f_n \to f$ uniformly on U, then for any $r \in \mathbb{Z}^+$, $f_n^{(r)} \to f^{(r)}$ at all interior points of U. As before, we have

$$\widehat{F}_X^{(r)}(s) = \int_{1-}^X x^{-s}(-\log x)^r dF(x)$$

for $r = 1, 2, \ldots$ and $X = X_1, X_2, \ldots$ and furthermore $\widehat{F}_{X_j} \to \widehat{F}$ uniformly on the sector S. Thus

$$\widehat{F}^{(r)}(s) = \lim_{j \to \infty} \widehat{F}_{X_j}^{(r)}(s) = \int x^{-s}(-\log x)^r dF(x). \qquad \square$$

Example 6.22 Let us identify $\widehat{\psi}$, where ψ is Chebyshev's function. We start with Chebyshev's formula $L dN = d\psi * dN$ and form M.t.'s. The integrals defining $\widehat{\psi}$ and $\zeta = \widehat{N}$ converge absolutely for $\sigma > 1$. By the preceding theorem and the homomorphic property of the M.t. (Theorem 6.2), we get

$$-\zeta'(s) = \int x^{-s} \log x \, dN(x) = \int x^{-s}(d\psi * dN)(x) = \widehat{\psi}(s)\,\zeta(s).$$

It follows that

$$\widehat{\psi}(s) := \int x^{-s} d\psi(x) = -\zeta'(s)/\zeta(s) \quad (\sigma > 1). \tag{6.9}$$

6.5.1 Analytic continuation

An analytic function defined by a power series can sometimes be analytically continued outside the disc on which the series converges. Similarly, an M.t. can sometimes be continued beyond its half plane of convergence. If $F \in \mathcal{V}$ and is closely approximated by a smooth, simple "comparison function" $\varphi \in \mathcal{V}$, then we can often analytically continue $\widehat{F}$ as $(F - \varphi)\widehat{} + \widehat{\varphi}$. Some useful conditions for this extension are set out in

Theorem 6.23 *Let $F \in \mathcal{V}$ and let*

$$F(x) = \sum_{j=1}^{r} \sum_{k=1}^{m_j} c_{jk} x^{s_j} (\log x)^{k-1} + E(x) \qquad (6.10)$$

where $s_1, \ldots, s_r \in \mathbb{C}$ and $\sigma_1 \le \sigma_2 \le \cdots \le \sigma_r$; $m_1, \ldots, m_r \in \mathbb{Z}^+$; and the $c_{jk} \in \mathbb{C}$. Further, assume that there exists a real number $\theta < \sigma_1$ such that for any fixed positive ϵ, $E(x) = O(x^{\theta + \epsilon})$.

Then $\widehat{F}$ has a continuation as a meromorphic function on the half plane $\{s : \sigma > \theta\}$, given by the formula

$$\widehat{F}(s) = s \sum_{j=1}^{r} \sum_{k=1}^{m_j} c_{jk} \frac{(k-1)!}{(s - s_j)^k} + s \int_1^{\infty} E(x) x^{-s-1} dx. \qquad (6.11)$$

The singularities of $\widehat{F}$ on this half plane are exhibited by the double sum. For fixed x, the residue of $s \mapsto x^s \widehat{F}(s)/s$ at any point s_j is

$$\sum_{k=1}^{m_j} c_{jk} x^{s_j} (\log x)^{k-1}.$$

Proof. For $\sigma > \sigma_r$ we have

$$\widehat{F}(s) = \int x^{-s} dF(x) = s \int x^{-s-1} F(x) dx$$

since the integrated term $x^{-s} F(x) \to 0$ as $x \to \infty$, by hypothesis (6.10). We replace F, using (6.10) again, and evaluate the main term of the resulting expression by means of the identity

$$\int x^{-s-1+s_j} (\log x)^{k-1} dx = (k-1)!(s - s_j)^{-k} \qquad (\sigma > \sigma_j).$$

This relation can be established by $(k - 1)$ differentiations of the M.t. $\int x^{-s-1+s_j} dx = 1/(s - s_j)$.

Thus formula (6.11) is valid for $\sigma > \sigma_r$. The integral in (6.11) converges and defines an analytic function on $\{s : \sigma > \theta\}$. The double sum in (6.11) is analytic on $\mathbb{C}$ except for poles at the points $\{s_j\}$. By the principle of analytic continuation, $\widehat{F}$ has a unique extension as a meromorphic function to the half plane $\{s : \sigma > \theta\}$.

We can compute the residue at s_j by writing

$$x^s \frac{\widehat{F}(s)}{s} = x^{s_j} e^{(s-s_j)\log x} \sum_{k=1}^{m_j} c_{jk} \frac{(k-1)!}{(s-s_j)^k} + \phi(s),$$

where ϕ is analytic at s_j. Then we expand the exponential function in its Maclaurin series and pick out the coefficient of $(s - s_j)^{-1}$. $\qquad\square$

6.5.2 Continuation of zeta

Here we analytically continue the Riemann zeta function to the half plane $\{s : \sigma > -1\}$ by two applications of the preceding theorem. Since this amounts to an application of the Euler summation formula (Lemma 3.12) and its extension (3.3), we shall be brief. We have

$$\zeta(s) = \frac{1}{s-1} + \frac{1}{2} + s \int_1^\infty x^{-s-1}\left(N(x) - x + \frac{1}{2}\right) dx, \tag{6.12}$$

$$= \frac{1}{s-1} + \frac{1}{2} + s(s+1) \int_1^\infty x^{-s-2}\varphi(x)\, dx, \tag{6.13}$$

where $\varphi(x) = \int_0^x (N(t) - t + 1/2)dt$, a continuous, periodic—and hence bounded—function. The integral in (6.12) converges and defines an analytic function for $\sigma > 0$, and the integral in (6.13) defines an analytic function for $\sigma > -1$. Thus we have the desired continuation. Incidentally, formula (6.12) justifies the definition of $\zeta(\alpha)$ for $0 < \alpha < 1$ given in (3.6).

The process of integrating by parts can be repeated indefinitely for ζ, and one can thereby show that $s \mapsto \zeta(s) - 1/(s-1)$ is an entire function. We shall later continue ζ to $\mathbb{C} \setminus \{1\}$ by use of its functional equation.

In the next chapters, we study where $\zeta(s)$ is nonzero. The following problem takes a small step in this direction.

PROBLEM 6.21   Use the formula

$$\frac{1}{s-1} - \frac{\zeta(s)}{s} = \int_1^\infty \frac{x - [x]}{x^{s+1}}\, dx \qquad (\Re s > 0)$$

(an alternative form of (6.12)) to prove that $\zeta(s) \neq 0$ on the set

$$\{s : \Re s \geq 1/2, \ |\Im s| \leq \sqrt{3}/2\}.$$

Hint. For $\Re s > 0$ show that

$$\left| \frac{1}{s-1} - \frac{\zeta(s)}{s} \right| < \frac{1}{2\sigma}.$$

It is well known that an analytic function defined by a power series must have a singularity at some point of its circle of convergence. The analogous result for Dirichlet series need not hold, as we show in the following example.

6.5.3   Example of analyticity on $\sigma = \sigma_c$

Define $\widehat{F}(s) := \Sigma(-1)^{n+1}n^{-s}$. We saw in Example 6.10 that this series converges for $\sigma > 0$. Also, the series diverges at 0 and hence $\sigma_c = 0$. For $\sigma > 1$ we have the representation

$$(1 - 2^{1-s})\zeta(s) = \sum_{n=1}^{\infty} n^{-s} - 2\sum_{n=1}^{\infty}(2n)^{-s} = \sum_{n=1}^{\infty}(-1)^{n+1}n^{-s} = \widehat{F}(s).$$

(The second equality is most conveniently shown by writing

$$2\sum_{n=1}^{\infty}(2n)^{-s} = 0 + \frac{2}{2^s} + 0 + \frac{2}{4^s} + 0 + \cdots$$

and subtracting convergent series.) $\widehat{F}$ is analytic on $\{s : \sigma > 0\}$ by Theorem 6.20. We have shown ζ to be analytic on $\{s : \sigma > -1\}$ except for a simple pole at $s = 1$. This singularity is cancelled by a zero of the entire function $s \mapsto 1 - 2^{1-s}$. We can continue $\widehat{F}$ as an analytic function on $\{s : \sigma > -1\}$ by setting $\widehat{F}(s) = (1 - 2^{1-s})\zeta(s)$. Thus $\widehat{F}$ is analytic at all points of the imaginary axis. In fact, $\widehat{F}$ is entire as we shall see later.

6.6   Uniqueness

We are going to investigate properties of a function $F \in \mathcal{V}$ by studying the associated M.t. $\widehat{F}$. After performing analysis on $\widehat{F}$, we shall want to return to F. Fortunately, the map from F to $\widehat{F}$ is univalent (one-to-one) as we now show. We are going to give a "moment problem" proof using the Weierstrass theorem on polynomial approximation. The univalence will follow also from the inversion formula for M.t.'s (Theorem 7.10).

To achieve univalence we must make a convention on function values at discontinuities. For example, let $F(x) := 0$ for $x < e$ and $F(x) := \log x$ for $x > e$. Then $\widehat{F}(s) = e^{-s}/s$, regardless of the value of $F(e)$. In the following theorem we assume that the functions are continuous from the right. We shall reconsider this matter when we discuss inversion formulas.

Theorem 6.24 (Uniqueness). *Let F and $G \in \mathcal{V}$ and suppose that the integrals defining $\widehat{F}$ and $\widehat{G}$ converge at some points of $\mathbb{C}$. Further assume that $\widehat{F}(s_j) = \widehat{G}(s_j)$ for a sequence $\{s_j\}_{j=0}^{\infty}$ which either has a limit point in the common open half plane of convergence or else is of the form $s_j = s_0 + j$ for some $s_0 \in \mathbb{C}$. Then $F = G$ identically.*

For Dirichlet series there is a slightly stronger theorem which can be proved more easily.

Theorem 6.25 *Let f and $g \in \mathcal{A}$ and suppose that their D.s. $\widehat{F}$ and $\widehat{G}$ converge at some points of $\mathbb{C}$. Further assume that $\widehat{F}(s_j) = \widehat{G}(s_j)$ for a sequence $\{s_j\}$ which either has a limit point in the common open half plane of convergence or else satisfies $\Re s_j \to +\infty$. Then $f = g$ identically.*

PROBLEM 6.22   Let $\widehat{F}(s) = \sum a_n n^{-s}$ be a D.s. with $\sigma_c(\widehat{F}) < \infty$. Let $N \in \mathbb{Z}^+$ and let $s_1, s_2, \ldots$ be a sequence from $\mathbb{C}$ with $\Re s_j \to +\infty$ as $j \to \infty$. Show that

$$\lim_{j \to \infty} N^{s_j} \sum_{n=N}^{\infty} a_n n^{-s_j} = a_N.$$

Using this relation prove Theorem 6.25.

PROBLEM 6.23   Let $\widehat{F}$ be a nonconstant D.s. with $\sigma_c(\widehat{F}) < \infty$ and let α be any fixed complex number. Show that there exists a half plane $\{s : \sigma > \sigma_0(\widehat{F}, \alpha)\}$ on which $\widehat{F}(s) \neq \alpha$.

Proof of Theorem 6.24.   Let $H := G - F$. We shall show that $H = 0$. Each of $\widehat{F}$ and $\widehat{G}$ has a nonvoid half plane of convergence, and hence $\widehat{H} = \widehat{G} - \widehat{F}$ has one also. Under the first hypothesis $\widehat{H}$ vanishes at a sequence of points having a limit point in the half plane of convergence. Since $\widehat{H}$ is zero there and in particular $\widehat{H}(s_0 + j) = 0$, $j = 0, 1, 2, \ldots$, it suffices to assume the latter hypothesis.

Let $\sigma_0 = \Re s_0$. We can assume that $\sigma_0 > \max(0, \sigma_c(\widehat{H}) + 1)$ by dropping some terms from the beginning of the sequence if necessary. Then we have $H(x) = o(x^{\sigma_0 - 1})$ by Theorem 6.9. It is advantageous to work with a continuous function. To this end we perform two integrations by parts. For $\sigma > \sigma_c(\widehat{H})$ we obtain

$$\widehat{H}(s) = s \int x^{-s-1} H(x)dx = s(s+1) \int x^{-s-2} H_1(x)dx,$$

where $H_1(x) = \int_1^x H(t)dt$. For $j = 0, 1, 2, \ldots$ we have

$$0 = \int x^{-s_j - 2} H_1(x)dx = \int x^{-j-2} x^{-s_0} H_1(x)dx.$$

Now set $f(y) = y^{s_0} H_1(1/y)$ for $0 < y \le 1$. If we define $f(0) = 0$, then f is continuous on $[0, 1]$, since H_1 is continuous on $[1, \infty)$ and

$$\lim_{y \to 0+} f(y) = \lim_{x \to \infty} x^{-s_0} H_1(x) = 0.$$

Replacing x by $1/y$ in the preceding integrals, we obtain

$$\int_0^1 y^j f(y)dy = 0, \quad j = 0, 1, 2, \ldots . \tag{6.14}$$

Equation (6.14) asserts that all the "moments" of f are zero. We shall now show that *a continuous function on $[0, 1]$, all of whose moments are zero, must be identically zero on this interval.*

If f is not real valued, then we can express f in terms of its real and imaginary parts and observe that all the moments of the real part and all those of the imaginary part are zero. Thus it suffices to assume that the function f satisfying (6.14) is real valued.

If P is any polynomial, then by linearity

$$\int_0^1 P(y)f(y)dy = 0. \tag{6.15}$$

The Weierstrass approximation theorem asserts that we can uniformly approximate any continuous real valued function on $[0, 1]$ by a sequence of polynomials. It follows that (6.15) must hold also for P any continuous function. If we choose $P = f$ we obtain $\int_0^1 f(y)^2 dy = 0$, or $f = 0$, which proves the italicized assertion.

It follows that $H_1 = 0$. For any point $x_0 \geq 1$ we then have

$$0 = \frac{1}{\delta}\{H_1(x_0 + \delta) - H_1(x_0)\} = \frac{1}{\delta}\int_{x_0}^{x_0+\delta} H(u)du \to H(x_0)$$

as $\delta \to 0+$, since H is continuous from the right. Thus $H = 0$. $\qquad\square$

6.6.1 Identifying an arithmetic function

We can sometimes identify an arithmetic function by performing manipulations on the associated Dirichlet series. This can generally be justified by formal Dirichlet series methods or, for functions of polynomial growth, by appealing to the preceding uniqueness theorem. Arguments of the latter type are popular with number theorists.

Example 6.26 **Description of $|\mu|^{*-1}$.** Since $|\mu| \in \mathcal{M}$ and $|\mu| \leq 1$, we can apply Corollary 6.6 to write

$$\hat{Q}(s) = \int_{1-}^{\infty} x^{-s}dQ(x) = \sum |\mu|(n)n^{-s} = \prod_p (1 + p^{-s}) \quad (\sigma > 1).$$

Thus we have, for $\sigma > 1$,

$$\hat{Q}(s)^{-1} = \prod_p \frac{1}{1 + p^{-s}} = \prod_p (1 - p^{-s} + p^{-2s} - \cdots) = \sum (-1)^{\Omega(n)} n^{-s},$$

by applying Corollary 6.6 once more. Here $\Omega(p_1^{\alpha_1} \ldots p_r^{\alpha_r}) = \alpha_1 + \cdots + \alpha_r$, and $n \mapsto (-1)^{\Omega(n)}$ is Liouville's function λ, cf. Problem 2.27.

Now, for $\sigma > 1$ we have

$$\sum e_1(n)n^{-s} = 1 = \left(\sum |\mu|(n)n^{-s}\right)\left(\sum \lambda(n)n^{-s}\right) = \sum(|\mu| * \lambda)(n)n^{-s}.$$

By the uniqueness theorem $|\mu| * \lambda = e_1$, and thus $|\mu|^{*-1} = \lambda$. (Of course one also could obtain this formula by applying Lemma 2.25 to factor $|\mu|$ as $\prod_p(e + e_p)$ and then computing the convolution inverse of each factor via the second proof of Theorem 2.7.)

Example 6.27 **Square root of dx.** Does there exist a function $F \in \mathcal{V}$ for which $dF * dF = dx$? If such an integrator dF exists, then $-dF$ also satisfies the convolution equation. As a clue for our problem, recall that

$$(dx)^{*n} = L^{n-1}dx/\Gamma(n) \quad \text{for any } n \in \mathbb{Z}^+,$$

where $\Gamma(n) = (n-1)!$. This suggests that

$$\{L^{-1/2}dx/\Gamma(1/2)\} * \{L^{-1/2}dx/\Gamma(1/2)\} = dx. \qquad (6.16)$$

We verify this formula via the uniqueness theorem for M.t.'s. For $\sigma > 1$,

$$\int_1^\infty x^{-s}(\log x)^{-1/2}dx = \int_0^\infty e^{-(s-1)u}u^{-1/2}du$$

$$= \frac{1}{\sqrt{s-1}}\int_0^\infty e^{-v}v^{-1/2}dv = \frac{\Gamma(1/2)}{\sqrt{s-1}},$$

where the square roots are real and positive for s real and $s > 1$. (The change of variable in the middle equation is made for s real and $s > 1$; the identity remains valid by analytic continuation for complex s with real part exceeding 1.) By Theorem 6.2,

$$\int_1^\infty x^{-s}\left\{\frac{L^{-1/2}dx}{\Gamma(1/2)}\right\}^{*2} = \frac{1}{s-1} = \int_1^\infty x^{-s}dx,$$

and so (6.16) holds. It follows that

$$F(x) := \int_1^x L^{-1/2}dx/\Gamma(1/2)$$

provides a solution in $\mathcal{V}$. Moreover, this solution is unique up to sign in the class of functions of polynomial growth in $\mathcal{V}$. In particular, the solution is unique among monotone increasing functions, since, for all $x \geq 1$,

$$x - 1 = \int_1^x dF * dF \geq \int_1^{\sqrt{x}} F\left(\frac{x}{t}\right)dF(t) \geq F(\sqrt{x})^2;$$

hence $F = F_v$ is of polynomial growth.

6.7 Operational calculus

Suppose $f \in \mathcal{A}$, $f(1) = 0$ and f is of polynomial growth. Associated with f is the D.s. $\widehat{F}(s) = \sum_{n=1}^\infty f(n)n^{-s}$. Let $B : z \mapsto \sum_{k=0}^\infty b_k z^k$ be a function defined by a power series with radius of convergence $\rho > 0$, and let

$$g_r := \sum_{0 \leq k \leq r} b_k f^{*k}.$$

By Lemma 2.12, g_r converges to a limit function, which we call g. In view of the homomorphic property of the M.t. it is reasonable to inquire whether g has a convergent D.s. and if so, what is its relation to B and $\widehat{F}$.

Lemma 6.28   *Let* f, $\widehat{F}$, $\{b_k\}_{k=0}^\infty$, B, *and* g *be as above. The D.s. for* g *converges on some half plane, and there*

$$\widehat{G}(s) := \sum g(n) n^{-s} = \sum_{k=0}^\infty b_k \widehat{F}(s)^k = B(\widehat{F}(s)).$$

Proof.   By (6.6), as $\sigma \to +\infty$,

$$\widehat{F}_v(\sigma) = \sum |f(n)| n^{-\sigma} \to |f(1)| = 0.$$

Thus there exists a number σ_0 such that $|\widehat{F}_v(s)| < \rho$ for $\sigma > \sigma_0$. By the homomorphic property of the M.t., we have for any $f \in \mathbb{Z}^+$ and $\sigma > \sigma_0$

$$\sum g_r(n) n^{-s} = \sum_{0 \leq k \leq r} b_k \widehat{F}(s)^k. \qquad (6.17)$$

As $r \to \infty$, the right side of (6.17) converges to $B(\widehat{F}(s))$ for $\sigma > \sigma_0$. We can show that the left side of (6.17) converges to $\widehat{G}(s)$ on this half plane by noting that

$$\sum_{n=1}^\infty |g(n) - g_r(n)| n^{-\sigma} \leq \sum_{n=1}^\infty \sum_{k=r+1}^\infty |b_k| |f|^{*k}(n) n^{-\sigma}$$

$$= \sum_{k=r+1}^\infty |b_k| \{\widehat{F}_v(\sigma)\}^k \to 0$$

as $r \to \infty$ for any fixed $\sigma > \sigma_0$. It follows that the series for $\widehat{G}$ converges and satisfies $\widehat{G} = B(\widehat{F})$ on $\{s : \sigma > \sigma_0\}$. $\qquad \square$

Example 6.29   The series for $\log \zeta$. In (2.10) we established the representation

$$1 = \sum_{j=0}^\infty \kappa^{*j}/j! =: \exp(\kappa).$$

Now $\kappa(1) = 0$ and $0 \le \kappa \le 1$. Also, the power series $\sum z^k/k!$ has radius of convergence $\rho = \infty$. Thus we have for $\sigma > 1$

$$\zeta(s) = \sum 1 \cdot n^{-s} = \sum_{j=0}^{\infty} \frac{1}{j!} \left\{ \sum \kappa(n)n^{-s} \right\}^j = \exp\left\{ \sum \kappa(n)n^{-s} \right\},$$

and $\sum \kappa(n)n^{-s}$ is a logarithm of $\zeta(s)$ for $\sigma > 1$. Indeed, it is the branch of the logarithm which is real for s real and $s > 1$.

In the last section we computed the convolution inverse of an arithmetic function by using D.s. manipulations and the uniqueness theorem. Here we ask whether the inverse of a function having a D.s. that is convergent in some half plane necessarily has the same property. Equivalently, is the property of polynomial growth shared by a function and its convolution inverse? The next lemma answers these questions in the affirmative.

Lemma 6.30 *Suppose f is an invertible arithmetic function of polynomial growth. Then f^{*-1} also is of polynomial growth.*

We give two proofs, one using Dirichlet series and the other direct estimation. As in previous inversion problems, we may assume that $f(1) = 1$. We set $\varphi = e_1 - f$ and represent f^{*-1} using the second proof of Theorem 2.7:

$$f^{*-1} = e_1 + \varphi + (\varphi * \varphi) + (\varphi * \varphi * \varphi) + \cdots. \tag{6.18}$$

First Proof. The power series $\sum z^n$ has radius of convergence 1. Choose σ_0 such that $\sum |\varphi(n)| n^{-\sigma} < 1$ for $\sigma > \sigma_0$. By Lemma 6.28,

$$\sum f^{*-1}(n) n^{-s} = \sum_{j=0}^{\infty} \widehat{\Phi}(s)^j$$

for $\sigma > \sigma_0$, where $\widehat{\Phi}$ is the D.s. associated with φ. Since f^{*-1} has a convergent D.s., the function is of polynomial growth. $\square$

Second Proof. It suffices to show that $\sum_{n \le x} f^{*-1}(n)$ is of polynomial growth. Let $\Phi(x) := \sum_{n \le x} \varphi(n)$, with φ as in the first proof. There exist constants $K > 0$ and $\alpha \in \mathbb{R}$ such that $|\varphi| \le KT^{\alpha}(1 - e_1)$, i.e. $\varphi(1) = 0$ and $|\varphi(n)| \le Kn^{\alpha}$ for all $n \ge 2$. We may assume without loss of generality that $\alpha \ge 0$. We estimate the summatory function of $|\varphi|$ and its convolution powers by simple integrals. Using the inequality $n^{\alpha} \le n^{\alpha+1} - (n-1)^{\alpha+1}$

for $\alpha \geq 0$ (Show this!), we obtain for any $x \geq 1$,

$$\Phi_v(x) = \sum_{n \leq x} |\varphi(n)| \leq \sum_{2 \leq n \leq x} Kn^\alpha \leq K(x^{\alpha+1} - 1) = K' \int_1^x T^\alpha dt,$$

with $K' = K(\alpha + 1)$. By Corollary 3.26,

$$\sum_{n \leq x} |\varphi|^{*j}(n) \leq (K')^j \int_1^x (T^\alpha dt)^{*j} = (K')^j \int_1^x T^\alpha (dt)^{*j} \qquad (6.19)$$

holds for $1 \leq x < \infty$ and each $j \in \mathbb{Z}^+$.

Combining (6.18) and (6.19) we obtain

$$\left| \sum_{n \leq x} f^{*-1}(n) \right| \leq 1 + \sum_{j=1}^\infty K'^j \int_1^x T^\alpha (dt)^{*j}$$

$$= 1 + K' \int_1^x \sum_{j=1}^\infty \frac{(K' \log t)^{j-1}}{(j-1)!} t^\alpha dt$$

$$= 1 + K' \int_1^x t^{\alpha + K'} dt. \qquad (6.20)$$

The last integral is clearly of polynomial growth.   □

PROBLEM 6.24   Suppose $\varphi \in \mathcal{A}$, $\varphi(1) = 0$ and $f = \exp \varphi$. Show that if one of f, φ is of polynomial growth, then so is the other. Use these relations to give another proof of Lemma 6.30.

PROBLEM 6.25   Suppose $f \in \mathcal{A}$, $f(1) = 1$ and $|f| \leq 1$. Use the equation $f * f^{*-1} = e_1$ to prove that $|f^{*-1}(n)| \leq n^2$ for all n. This result can be used to give still another proof of Lemma 6.30.

PROBLEM 6.26   Let $f = \prod_{n=2}^\infty (e_1 - e_n)^{*-1}$ and let F be the summatory function of f. Show that if $n \geq 2$ then $f(n)$ equals the number of representations of n as a product $k_1 k_2 \cdots k_r$, where $r \geq 1$ and $2 \leq k_1 \leq k_2 \leq \cdots \leq k_r$. (For example $12 = 2 \cdot 6 = 3 \cdot 4 = 2 \cdot 2 \cdot 3$ and $f(12) = 4$.) Show that $\widehat{F}(s) = \prod_{n=2}^\infty (1 - n^{-s})^{-1}$ and $\sigma_c(\widehat{F}) = 1$. Conclude that $F(x) = O(x^{1+\epsilon})$ for any positive number ϵ.

6.8   Landau's oscillation theorem

We have seen that an M.t. $\widehat{F}$ need not have a singularity on its line of convergence $\sigma = \sigma_c$. However, if F is ultimately monotone, then as we show here, $\widehat{F}$ must have a singularity at the real point on the line of convergence. We state the assertion in its contrapositive form, since that is the way we shall most frequently use it.

Theorem 6.31   (Landau's oscillation theorem). *Let $F \in \mathcal{V}$ be real valued and suppose $\widehat{F}$ has abscissa of convergence σ_c. If $\widehat{F}$ has an analytic continuation to a region that includes the point $s = \sigma_c$, then F is not monotone on any infinite interval (x, ∞).*

The main applications of the theorem are to functions that are either the summatory function of an arithmetic function f or the integral of a locally Riemann integrable function φ. Non monotonicity of the summatory function or integral implies in turn that f or φ changes sign infinitely often.

Another use of the theorem, as we noted at the outset of this section, is to determine the abscissa of convergence of a function F that is ultimately monotone. We conclude in this case that $\sigma_c(\widehat{F})$ is the same as the largest real point at which the function has a singularity (cf. Example 6.34).

Proof.   Suppose F is monotone from some point onward and $\widehat{F}$ is analytic at σ_c. We show that the integral defining $\widehat{F}$ would then converge to the left of σ_c, which is impossible, since σ_c is the abscissa of convergence of $\widehat{F}$.

Say that $F \uparrow$ on (x_0, ∞). We expand $\widehat{F}$ in a Taylor series about some real point $\beta > \sigma_c$:

$$\widehat{F}(s) = \sum_{j=0}^{\infty} \widehat{F}^{(j)}(\beta)(s - \beta)^j / j!.$$

For any $X \geq x_0$ we have

$$(-1)^j \widehat{F}^{(j)}(\beta) = \int_{1-}^{\infty} x^{-\beta} \log^j x \, dF(x) \geq \int_{1-}^{X} x^{-\beta} \log^j x \, dF(x).$$

Under the assumption that $\widehat{F}$ is analytic at σ_c, this function is in fact analytic on $\{s : \sigma > \sigma_c\} \cup \{s : |s - \sigma_c| < \delta\}$ for some $\delta > 0$. Let R be a positive number such that the disc of center β and radius R lies in the domain of analyticity but not entirely in $\{s : \sigma \geq \sigma_c\}$. Let s be a real

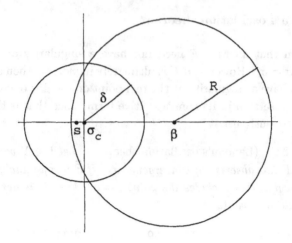

Fig. 6.1   EXTENSION OF CONVERGENCE REGION

number such that $s < \sigma_c$ but $|s - \beta| < R$ (cf. Fig 6.1). Then

$$\widehat{F}(s) \geq \sum_{j=0}^{\infty} \frac{(\beta - s)^j}{j!} \int_{1-}^{X} x^{-\beta} \log^j x \, dF(x)$$

$$= \int_{1-}^{X} \sum_{j=0}^{\infty} \frac{(\beta - s)^j \log^j x}{j!} x^{-\beta} dF(x)$$

since the series converges uniformly for $1 \leq x \leq X$ by the Weierstrass M-test. If we sum the series, we obtain

$$\widehat{F}(s) \geq \int_{1-}^{X} e^{(\beta-s)\log x} x^{-\beta} dF(x) = \int_{1-}^{X} x^{-s} dF(x).$$

We see that the last integral is bounded above for all $X \geq x_0$. Since the integrand is nonnegative and F is monotone on (x_0, ∞),

$$\lim_{X \to \infty} \int_{1-}^{X} x^{-s} dF(x)$$

exists. This is impossible, since $s < \sigma_c(\widehat{F})$, the abscissa of convergence.   □

Here are three examples of the application of Landau's theorem, two trivial and one more substantial. After more properties of the Riemann zeta function have been established, we shall show the oscillation of some other summatory functions.

Example 6.32 Let $F(x) := \sum_{n \le x} (-1)^{n+1}$. Then

$$\widehat{F}(s) = (1 - 2^{1-s}) \zeta(s) = 1 - 2^{-s} + 3^{-s} - 4^{-s} + \cdots.$$

Now $\sigma_c(\widehat{F}) = 0$, $\widehat{F}$ is analytic at 0, and, as predicted by the theorem, F is not ultimately monotonic.

Example 6.33 Let F be as in the preceding example and take $F_1(x) := \int_1^x t^{-1} F(t)\, dt$. Then

$$\widehat{F_1}(s) = \int x^{-s} F(x) x^{-1} dx = s^{-1}(1 - 2^{1-s}) \zeta(s).$$

Now $0 \le F(x) \le 1$ for all $x \ge 1$ and $F(x) = 1$ if $[x]$ is odd, so that $\sigma_c(\widehat{F_1}) = 0$ and, by Landau's theorem, $\widehat{F_1}$ is singular at $s = 0$. It follows that $\zeta(0) \ne 0$. (Actually $\zeta(0) = -1/2$.)

Example 6.34 We showed in the preceding section that if f is an invertible arithmetic function of polynomial growth, then f^{*-1} has the same property. Here we specialize to the case $|f(n)| \le f(1) = 1$ for all n. We apply the convolution comparison method, using (6.20) and the accompanying notation. Here $\alpha = 0$ and $K = K' = 1$, and thus

$$\sum_{n \le x} f^{*-1}(n) = O(x^c) \tag{6.21}$$

with $c = 2$. The D.s. method will enable us to reduce the value of c, as we now show. Moreover, in the case that f^{*-1} is maximal, we shall see that the value of c we obtain is (up to an arbitrary positive ϵ) best possible.

As in the proof of Lemma 6.30, we again set $\varphi := e_1 - f$ and write

$$f^{*-1} = e_1 + \varphi + (\varphi * \varphi) + (\varphi * \varphi * \varphi) + \cdots.$$

$|f^{*-1}|$ is clearly maximized by taking $\varphi = 1 - e_1$, i.e. $\varphi(1) = 0$ and $\varphi(n) = 1$ for all $n \ge 2$. The function f in our class having the extremal inverse is $2e_1 - 1$. Let $g = (2e_1 - 1)^{*-1}$, and note that $g \ge 0$. Thus we have

$$\left| \sum_{n \le x} f^{*-1}(n) \right| \le \sum_{n \le x} g(n) =: G(x)$$

for any $f \in \mathcal{A}$ satisfying $|f(n)| \le f(1) = 1$.

An argument that is now familiar gives

$$\widehat{G}(s) := \sum g(n)n^{-s} = \{2 - \zeta(s)\}^{-1} \qquad (\sigma > 2).$$

By Theorem 6.31, $\sigma_c(\widehat{G})$ is the largest real number at which the function $s \mapsto \{2 - \zeta(s)\}^{-1}$ is singular. Now ζ is analytic on $1 < \sigma < \infty$ and is real and decreasing on this ray. Consequently, the only singularity on the ray occurs at $\sigma = \rho = 1.728647\ldots$, a root of the equation $\zeta(s) = 2$.

Thus $\sigma_c(\widehat{G}) = \rho$ and by Theorem 6.9, $G(x) = O(x^{\rho+\epsilon})$ for any fixed $\epsilon > 0$ and the assertion "$G(x) < x^{\rho-\epsilon}$" must fail for a sequence of x's tending to infinity. In the next chapter we shall show that

$$G(x) \sim x^{\rho}/\{-\rho\zeta'(\rho)\}.$$

PROBLEM 6.27 Let $\widehat{F}(s) = \int_1^{\infty} x^{-s}(\log ex)^{-2}dx$. Observe that $\sigma_c = 1$, $F \uparrow$, and the integral defining $\widehat{F}(s)$ converges at $s = 1$. Is this consistent with Theorem 6.31? Explain.

6.9 Notes

6.1–6.4. Cf. Chapter 9, Dirichlet series, in [TiTF]. Theorem 6.2 was first given by T. J. Stieltjes, Nouvelles Annales de Math. (3), vol. 6 (1887), pp. 210–215.

6.5. Theorem 6.23 is a part of mathematical folklore. It is stated, e.g., in a paper of the first author in Duke Math. J., vol. 25 (1958), pp. 67–72.

6.8. Landau's oscillation theorem (Theorem 6.31) was first proved in Math. Ann., vol. 61 (1905), pp. 527–550; also in [LanC], vol. 2, pp. 206–229. An extensive discussion of this theorem and related matters is presented by R. J. Anderson and H. M. Stark in Springer Lecture Notes in Math., vol. 899 (1981), pp. 79–106. A briefer account is given in an article of P. T. Bateman and H. G. Diamond in Number Theory, R. P. Bambah et al., eds., Hindustan Book Agency and Indian National Science Academy, 2000, pp. 43–54.

Chapter 7

Inversion Formulas

7.1 The use of inversion formulas

An analytic function is completely determined by its value and the value of each its derivatives at a single point of analyticity. We can occasionally extract information about an arithmetic function from knowledge of the behavior of the associated Dirichlet series near a single point. As an example, the divergence of the series for $\zeta(s)$ at $s = 1$ enabled us to deduce the infinitude of primes. Other examples are provided by the uniqueness theorem and Landau's oscillation theorem.

Our knowledge of an analytic function at a point is often incomplete or the data are too complicated for analysis. By examining the analytic function at many points, we are sometimes able to overcome such difficulties. In this chapter we shall present some useful methods of extracting arithmetic information from Mellin transforms by complex integration and related techniques of Fourier analysis.

We have noted (Lemma 6.14) that the sum of a D.s. $\widehat{F}(s) = \sum f(n)n^{-s}$ with $\sigma_c(\widehat{F}) < \infty$ tends to $f(1)$ as $\Re s \to \infty$. Because $\widehat{F}(s)$ is nearly constant for large real values of s, it is difficult to obtain useful information about f by studying $\widehat{F}$ in this region unless one has such simple exact data as that used in the proof of the uniqueness theorem. The region near or to the left of the line of convergence will be most useful for our work.

Passage from an arithmetic function to its D.s. is an abelian process in the sense explained in §5.1.1. As a simple example, if $F(x) \to c$ as $x \to \infty$, then $\widehat{F}(\sigma) = \int x^{-\sigma} dF(x) \to c$ as $\sigma \to 0+$. Passage from a D.s. to the associated arithmetic function involves an "unaveraging" or tauberian process, as was discussed in §5.1.1. We have seen (Example 6.17) that the direct converse of the above abelian assertion is false. However, if F is monotone and $\widehat{F}(\sigma) \to c$ as $\sigma \to 0+$ then $F(x) \to c$ as $x \to \infty$. (Show this!) In general a tauberian process requires some auxiliary condition, such as boundedness or monotonicity, which is unnecessary for the corresponding abelian process. It is not surprising that a tauberian process should be more delicate than the corresponding abelian process.

The inversion theorems we will study involve conditions on an M.t. on its line of convergence $\sigma = \sigma_c$. To see the necessity of such conditions for an asymptotic formula, we give a simple abelian result connecting the relation $F(x) = cx + o(x)$ for a function $F \in \mathcal{V}$ and the behavior of $\widehat{F}$ on σ_c. An integration by parts and the hypothesis on F give

$$\widehat{F}(s) = s \int x^{-s} \{c + o(1)\}\, dx = \frac{cs}{s-1} + |s|\, o\!\left(\frac{1}{\sigma - 1}\right) \quad (\sigma \to 1+) \quad (7.1)$$

uniformly on the half plane $\{s : \sigma > 1\}$. This relation forms the basis of the following

Lemma 7.1 *Suppose that $F \in \mathcal{V}$, $F(x) = cx + o(x)$, and that $\widehat{F}$ can be extended as a meromorphic function to a region containing the closed half plane $\{s : \sigma \geq 1\}$. Then $\widehat{F}$ has no singularities on the closed half plane if $c = 0$, and $\widehat{F}$ has just a simple pole with residue c at $s = 1$ if $c \neq 0$.*

Proof. $\widehat{F}$ is analytic on $\{s : \sigma > 1\}$ by Theorems 6.9 and 6.20. If $\widehat{F}$ had a singularity at $1 + it_0$, it would be a pole by the meromorphy hypothesis. Relation (7.1) implies that $\widehat{F}$ has a simple pole at $s = 1$ unless $c = 0$.

Now assume t_0 is real and nonzero. $\widehat{F}$ has a Laurent expansion about $1 + it_0$ of the form

$$\widehat{F}(s) = \sum_{n=-N}^{\infty} c_n (s - 1 - it_0)^n,$$

where we may assume $c_{-N} \neq 0$. The Laurent coefficient c_{-N} can be determined from the formula

$$c_{-N} = \lim_{\sigma \to 1+} (\sigma - 1)^N \widehat{F}(\sigma + it_0).$$

By (7.1) we have $(\sigma - 1)\widehat{F}(\sigma + it_0) \to 0$ as $\sigma \to 1+$. It follows that $N \leq 0$ and hence $\widehat{F}$ is analytic at $1 + it_0$. $\qquad\square$

As an application we have

Corollary 7.2 *The P.N.T. implies that the Riemann zeta function has no zeros on the line $\Re s = 1$.*

Proof. We have seen that ζ is analytic on $\{s : \sigma > 0\}$ except for a pole at $s = 1$. Thus $-\zeta'/\zeta$ is meromorphic on a region containing $\{s : \sigma \geq 1\}$. Also, we have shown in (6.9) that $-\zeta'/\zeta = \widehat{\psi}$, and in Theorem 5.9 that $\psi(x) \sim x$. It follows from Lemma 7.1 that $-\zeta'/\zeta$ has only the one pole on the line $\Re s = 1$. Thus $\zeta(1 + it) \neq 0$ for all $t \in \mathbb{R}$. $\qquad\square$

Later, we shall show directly that $\zeta(1 + it) \neq 0$ for all real t. This result and the following theorem will lead to another proof of the P.N.T.

7.2 The Wiener-Ikehara theorem

Among the inversion formulas we shall discuss, the theorem of Wiener and Ikehara occurs last historically. We are presenting it first in this chapter because of its ease of application.

Theorem 7.3 (Wiener-Ikehara). *Let F be a real valued monotone non-decreasing function in $\mathcal{V}$. Let $\sigma_c(\widehat{F}) = \alpha > 0$ and suppose that there exist a real number L and a function φ, continuous on the closed half plane $\{s : \sigma \geq \alpha\}$, such that*

$$\widehat{F}(s) = \int x^{-s} dF(x) = L(s - \alpha)^{-1} + \varphi(s)$$

holds on the corresponding open half plane. Then

$$F(x) = Lx^\alpha/\alpha + o(x^\alpha).$$

PROBLEM 7.1 Give an example of a function F for which the hypotheses of the Wiener-Ikehara theorem hold with $\sigma_c(\widehat{F}) = 1$ and $L = 0$. Verify that $F(x) = o(x)$.

The proof of the Wiener-Ikehara theorem rests on two lemmas of Fourier analysis. Let us define, for each positive real number λ, a function K_λ on $\mathbb{R}$, called a *Fejér kernel*, by

$$K_\lambda(x) := \frac{1}{2} \int_{-2\lambda}^{2\lambda} \left(1 - \frac{|t|}{2\lambda}\right) e^{ixt} dt.$$

This function, suitably normalized, will serve as an "approximate identity" and $1-|t|/(2\lambda)$ will serve as a "damping factor" to improve the convergence of a certain integral. We describe some properties of the Fejér kernel in

Lemma 7.4   *Let $x \in \mathbb{R}$, $\lambda > 0$, and $\delta > 0$. Then*

$$K_\lambda(x) = \lambda\Big(\frac{\sin \lambda x}{\lambda x}\Big)^2, \tag{7.2}$$

$$0 \le K_\lambda(x) \le \min(\lambda, \lambda^{-1}x^{-2}), \tag{7.3}$$

$$\int_{-\infty}^{\infty} K_\lambda(u)\,du \text{ exists and is independent of } \lambda, \tag{7.4}$$

$$\int_{|u|>\delta} K_\lambda(u)\,du \le 2/(\lambda\delta). \tag{7.5}$$

Proof. $K_\lambda(0) = \lambda$ by inspection. For $x \ne 0$ we have

$$K_\lambda(x) = \int_0^{2\lambda} \Big(1 - \frac{t}{2\lambda}\Big)\cos(xt)\,dt,$$

and on integrating by parts we obtain

$$K_\lambda(x) = \frac{1}{2\lambda x}\int_0^{2\lambda}\sin(xt)\,dt = \frac{1-\cos 2\lambda x}{2\lambda x^2} = \frac{(\sin \lambda x)^2}{\lambda x^2}.$$

We deduce (7.3) from the inequalities

$$0 \le u^{-2}\sin^2 u \le \min(1, u^{-2}).$$

The integral in (7.4) converges by the estimates of (7.3). A change of variable shows that the integral equals $\int v^{-2}\sin^2 v\,dv$ for any value of λ. One can show (e.g. by contour integration or use of the inversion formula for Fourier transforms) that $\int K_\lambda(x)dx = \pi$. The actual value of the integral will not be important for us here.

Finally, (7.5) follows from the estimate $K_\lambda(x) \le \lambda^{-1}x^{-2}$.   □

PROBLEM 7.2   Show that $\int_{-\infty}^{\infty} K_1(x)dx = \pi$ or obtain upper and lower numerical bounds for the integral.

We shall require a second result from Fourier analysis. It is a special case of the *Riemann-Lebesgue lemma*, which asserts that Fourier transforms "vanish at infinity."

Lemma 7.5 *Let f be a continuous complex valued function on $\mathbb{R}$ which is zero except on a bounded set, and let $y \in \mathbb{R}$. Then*

$$\lim_{y \to \pm\infty} \int_{-\infty}^{\infty} f(t)e^{ity}dt = 0.$$

Proof. Let $I(y) = \int f(t)e^{ity}dt$. By a change of variable,

$$I(y) = \int_{-\infty}^{\infty} f\left(t + \frac{\pi}{y}\right)e^{i\{t+(\pi/y)\}y}dt = -\int_{-\infty}^{\infty} f\left(t + \frac{\pi}{y}\right)e^{ity}dt.$$

Thus

$$2I(y) = \int_{-\infty}^{\infty} \left\{f(t) - f\left(t + \frac{\pi}{y}\right)\right\}e^{ity}dt.$$

Since f is uniformly continuous, the expression in braces tends to zero uniformly as $y \to \pm\infty$. Also, the last integrand vanishes outside a fixed bounded set if $|y| \geq 1$, say. Thus $I(y) \to 0$ as $|y| \to \infty$. $\qquad\square$

Proof of the Wiener-Ikehara theorem. We begin by noting that $F(x) = O(x^{\alpha+\epsilon})$ for any positive ϵ (by Theorem 6.9), since the M.t. converges for $\sigma > \alpha$. Now we change the variable, setting $u = \alpha \log x$ and defining f by $f(u) := F(e^{u/\alpha}) = F(x)$. Then for $\sigma > 1$,

$$\int_{0-}^{\infty} e^{-su}\, df(u) = \widehat{F}(\alpha s) = \frac{L}{\alpha(s-1)} + \varphi(\alpha s).$$

The above estimate of F implies that $f(u) = O(e^{u+\epsilon u})$ for any $\epsilon > 0$, and thus, for $\sigma > 1$,

$$\int_{0}^{\infty} f(u)e^{-su}du = s^{-1}\int_{0-}^{\infty} e^{-su}df(u)$$

$$= \frac{L}{\alpha s(s-1)} + \frac{\varphi(\alpha s)}{s} = \frac{\ell}{s-1} + \varphi_1(s),$$

where $\ell = L/\alpha$ and $\varphi_1(s) = s^{-1}\varphi(\alpha s) - s^{-1}\ell$, so that φ_1 is continuous on $\{s : \sigma \geq 1\}$. If we express $\ell/(s-1)$ as a Laplace integral, we obtain

$$\varphi_1(s) = \int_{0}^{\infty} e^{-su}\{f(u) - \ell e^u\}du \quad (\sigma > 1). \tag{7.6}$$

The proof proceeds in two steps. First we establish the following integral relation, valid for each positive number λ:

$$\lim_{y \to \infty} \int_0^\infty e^{-x} f(x) K_\lambda^*(y-x) dx = \ell, \tag{7.7}$$

where we have set

$$K_\lambda^*(t) = K_\lambda(t) \Big/ \int_{-\infty}^\infty K_1(u) du.$$

In view of (7.4), K_λ^* has total integral 1, independently of λ. When λ is large, K_λ^* is sharply peaked near the origin. Then we give a tauberian argument based on (7.7) and this observation; we obtain $f(x) = \ell e^x + o(e^x)$, which is equivalent to the assertion of the theorem.

We take ϵ and λ to be positive, set $s = 1 + \epsilon + it$, multiply (7.6) by $(1/2)\{1 - |t|/(2\lambda)\}e^{ity}$, and integrate. We obtain

$$\frac{1}{2} \int_{-2\lambda}^{2\lambda} \left(1 - \frac{|t|}{2\lambda}\right) e^{ity} \varphi_1(1 + \epsilon + it) dt$$

$$= \frac{1}{2} \int_{-2\lambda}^{2\lambda} \left(1 - \frac{|t|}{2\lambda}\right) e^{ity} \left\{ \int_0^\infty (f(u) - \ell e^u) e^{-u(1+\epsilon+it)} du \right\} dt$$

$$= \int_0^\infty e^{-u - \epsilon u} (f(u) - \ell e^u) \left\{ \frac{1}{2} \int_{-2\lambda}^{2\lambda} \left(1 - \frac{|t|}{2\lambda}\right) e^{it(y-u)} dt \right\} du.$$

The interchange of integration order is justified by the convergence of $\int_0^\infty e^{-u-\epsilon u} |f(u) - \ell e^u| du$ (cf. [TiTF], §1.85). In terms of K_λ we have

$$\frac{1}{2} \int_{-2\lambda}^{2\lambda} \left(1 - \frac{|t|}{2\lambda}\right) e^{ity} \varphi_1(1 + \epsilon + it) \, dt$$

$$= \int_0^\infty e^{-u - \epsilon u} f(u) K_\lambda(y-u) \, du - \ell \int_0^\infty e^{-\epsilon u} K_\lambda(y-u) \, du. \tag{7.8}$$

Now let $\epsilon \to 0+$ in (7.8). In view of the continuity of φ_1 and the bounded integration range, the limit may be taken inside the integral on the left. Also, since K_λ is continuous, nonnegative, and has a finite integral,

$$\int_0^\infty e^{-\epsilon u} K_\lambda(y-u) \, du \to \int_0^\infty K_\lambda(y-u) \, du$$

as $\epsilon \to 0+$. Thus we have, for any $y \in \mathbb{R}$,

$$\lim_{\epsilon \to 0+} \int_0^\infty e^{-u-\epsilon u} f(u) K_\lambda(y-u)\, du$$

$$= \ell \int_0^\infty K_\lambda(y-u)\, du + \frac{1}{2} \int_{-2\lambda}^{2\lambda} \left(1 - \frac{|t|}{2\lambda}\right) e^{ity} \varphi_1(1+it)\, dt. \quad (7.9)$$

For given y, the limit in the last equation exists and is finite; we want to take the limit inside the integral. We give details, because we are not using Lebesgue theory, where this operation would be immediate. Let $R(y)$ denote the right hand side of (7.9) and let $\eta > 0$ be given. We have $f \geq 0$, since $f \in \mathcal{V}$ and is nondecreasing. Also, $K_\lambda \geq 0$, and hence for $0 < \epsilon \leq \epsilon_0$,

$$0 \leq R(y) - \int_0^\infty e^{-u-\epsilon u} f(u) K_\lambda(y-u)\, du < \eta.$$

Also, in view of the boundedness of $e^{-u-\epsilon_0 u} f(u)$,

$$\int_U^\infty e^{-u-\epsilon_0 u} f(u) K_\lambda(y-u)\, du < \eta$$

holds for all sufficiently large U, and hence

$$0 \leq R(y) - \int_0^U e^{-u-\epsilon_0 u} f(u) K_\lambda(y-u)\, du < 2\eta.$$

Recalling that f and K_λ are nonnegative, we see that the last inequality holds also for $\epsilon_0 = 0$ (for all sufficiently large U); thus we have

$$\int_0^\infty e^{-u} f(u) K_\lambda(y-u)\, du = \ell \int_0^\infty K_\lambda(y-u)\, du$$

$$+ \frac{1}{2} \int_{-2\lambda}^{2\lambda} \left(1 - \frac{|t|}{2\lambda}\right) e^{ity} \varphi_1(1+it)\, dt.$$

If we next let $y \to \infty$ and apply Lemma 7.5, we obtain

$$\lim_{y \to \infty} \frac{1}{2} \int_{-2\lambda}^{2\lambda} \left(1 - \frac{|t|}{2\lambda}\right) e^{iyt} \varphi_1(1+it)\, dt = 0.$$

It follows that

$$\lim_{y \to \infty} \left\{ \int_0^\infty e^{-u} f(u) K_\lambda(y-u)\, du - \ell \int_0^\infty K_\lambda(y-u)\, du \right\} = 0.$$

By monotonicity,

$$\lim_{y\to\infty} \ell \int_{-\infty}^{y} K_\lambda(x)dx = \ell \int_{-\infty}^{\infty} K_1(x)dx.$$

Dividing through by $\int K_1(x)dx$, we obtain (7.7), the key integral formula.

We begin the tauberian argument by showing that $f(y) = O(e^y)$. Let λ and δ be positive numbers and let $J(y,\lambda)$ denote the integral occurring in (7.7). Since the integrand of $J(y,\lambda)$ is nonnegative and both f and the exponential function are monotone, we have for any $y > \delta$

$$J(y,\lambda) \geq \int_{y-\delta}^{y+\delta} e^{-x} f(x) K_\lambda^*(y-x)dx \geq f(y-\delta)e^{-y-\delta} \int_{-\delta}^{\delta} K_\lambda^*(u)du.$$

Combining the last inequality with (7.7) we obtain

$$f(y-\delta)e^{-(y-\delta)} \leq e^{2\delta}\ell \bigg/ \left(\int_{-\delta}^{\delta} K_\lambda^*(u)du \right) + o(1), \qquad (7.10)$$

which implies that $f(y)e^{-y}$ is bounded.

Now choose $\delta = \{(\lambda/2) \int_{-\infty}^{\infty} K_1(u)du\}^{-1/2}$. We have by (7.5)

$$\int_{|u|>\delta} K_\lambda^*(u)du \leq \frac{2}{\lambda\delta \int K_1(u)du} = \delta,$$

and thus

$$\int_{-\delta}^{\delta} K_\lambda^*(u)du \geq 1 - \delta. \qquad (7.11)$$

Let $\epsilon > 0$ be given. We choose λ large enough (and thus δ small enough) to ensure that $e^{2\delta}(1-\delta)^{-1} < 1+\epsilon$. With this choice of λ we deduce from (7.10) and (7.11) that, as $y \to \infty$,

$$f(y)e^{-y} \leq (1+\epsilon)\ell + o(1).$$

This relation holds for arbitrary positive ϵ, and so

$$\limsup_{y\to\infty} e^{-y} f(y) \leq \ell.$$

We now obtain an inequality in the opposite direction. Since $e^{-y}f(y)$ is bounded, there exists a positive number b such that $e^{-y}f(y) \leq b$ for all

$y \geq 0$. For any positive λ and y we have

$$J(y, \lambda) \leq b \int_{|u|>\delta} K_\lambda^*(u)du + f(y+\delta)e^{-y+\delta} \int_{-\delta}^{\delta} K_\lambda^*(u)du$$

$$\leq b\delta + f(y+\delta)e^{-(y+\delta)}e^{2\delta}.$$

By the last inequality and (7.7),

$$f(y)e^{-y} \geq e^{-2\delta}\ell - b\delta e^{-2\delta} + o(1)$$

as $y \to \infty$ for each fixed pair λ, δ satisfying

$$\lambda\delta^2 = 2 \Big/ \int_{-\infty}^{\infty} K_1(u)du.$$

Thus, for each positive δ we have

$$\liminf_{y\to\infty} f(y)e^{-y} \geq e^{-2\delta}\ell - b\delta e^{-2\delta},$$

and so

$$\liminf_{y\to\infty} f(y)e^{-y} \geq \ell.$$

The two inequalities imply that $\lim_{y\to\infty} f(y)e^{-y} = \ell$ or

$$F(x) = Lx^\alpha/\alpha + o(x^\alpha). \qquad \square$$

7.2.1 Example. Counting product representations

In Example 6.34 we estimated the growth of the sum function of f^{*-1} for $f \in \mathcal{A}$ satisfying $|f(n)| \leq f(1) = 1$. We saw that $|f^{*-1}|$ is maximized when $f = 2e_1 - 1$. If we set $g := (2e_1 - 1)^{*-1}$ and $\beta := 1 - e_1$, then

$$g = e_1 + \beta + (\beta * \beta) + (\beta * \beta * \beta) + \cdots. \qquad (7.12)$$

We showed that

$$G(x) := \sum_{n\leq x} g(n) = O(x^{\rho+\epsilon})$$

holds, where $\rho = 1.728647\ldots$ is the positive root of the equation $\zeta(s) = 2$ and ϵ is any positive number.

Here we give a number theoretic interpretation of g and replace the last O-estimate by an asymptotic formula. For $n > 1$, we claim that $g(n)$ equals the number of representations of n as a product of integers exceeding one. Two representations are considered identical if and only if they contain the same factors in the same order. To justify the claim, consider equation (7.12). For $n > 1$,

$\beta(n) = 1 =$ the number of representations of n as a single integer,

$\beta * \beta(n) =$ the number of reps. of n as a product of two integers, etc.

For example,

$$12 = 2 \cdot 6 = 6 \cdot 2 = 3 \cdot 4 = 4 \cdot 3 = 2 \cdot 2 \cdot 3 = 2 \cdot 3 \cdot 2 = 3 \cdot 2 \cdot 2,$$

and thus $g(12) = 8$ (cf. Problem 6.26 in §6.7).

We noted near the end of §6.8 that

$$\widehat{G}(s) = \sum g(n) n^{-s} = \{2 - \zeta(s)\}^{-1}.$$

This relationship is valid for $\sigma > \rho$. At $s = \rho$, $\widehat{G}$ has a simple pole with residue $-1/\zeta'(\rho)$. We now show that $|\zeta(\rho + it)| < 2$ for $t \neq 0$; it follows that $\widehat{G}$ is analytic at all other points of the line $\sigma = \rho$.

We have

$$|\zeta(\rho + it)| = |1 + \sum_{n=2}^{\infty} n^{-\rho} e^{-it \log n}| \leq \sum n^{-\rho} = 2 \qquad (7.13)$$

and strict inequality obtains unless $e^{-it \log n} = +1$ for all positive integers n. Suppose that $e^{-it \log 2} = e^{-it \log 3} = +1$ held for some real nonzero number t. Then we should have $t \log 2 = 2\pi\alpha$ and $t \log 3 = 2\pi\beta$ for some nonzero integers α and β, and hence $2^\beta = 3^\alpha$, violating the unique factorization property of rational integers. It follows that if $t \neq 0$, then

$$|1 + 2^{-\rho-it} + 3^{-\rho-it}| < 1 + 2^{-\rho} + 3^{-\rho}$$

and thus $|\zeta(\rho + it)| < \zeta(\rho)$ for all $t \neq 0$.

We can now express

$$\widehat{G}(s) = -\{(s - \rho)\zeta'(\rho)\}^{-1} + \varphi(s),$$

with φ satisfying the Wiener-Ikehara hypotheses. It follows that

$$G(x) \sim x^\rho / (-\rho \zeta'(\rho)) \doteq .318174 x^{1.728647}.$$

PROBLEM 7.3   By exploiting the behavior of $e^{it \log n}$ for large values of n, show that $|\zeta(\rho + it)| < 2$ for all real $t \neq 0$.

PROBLEM 7.4   Let φ denote Euler's function (no connection with the φ of the Wiener-Ikehara hypotheses above). Prove that as $y \to \infty$,

$$\#\{n : \varphi(n) \leq y\} \sim \frac{\zeta(2)\zeta(3)}{\zeta(6)}\, y.$$

Hint. Represent the generating function $\sum_{n=1}^{\infty} \varphi(n)^{-s}$ as in Problem 6.6 and apply the Wiener-Ikehara theorem.

7.2.2   An O-estimate

Relation (7.1) shows (for the case $\alpha = 1$) how the Wiener-Ikehara theorem can fail: if

$$\limsup_{\sigma \to \alpha+} (\sigma - \alpha)|\widehat{F}(\sigma + 2i\lambda_0)| > 0,$$

then the assertion "$F(x) = cx^{\alpha} + o(x^{\alpha})$" is false. Suppose that F is a monotone real valued function in $\mathcal{V}$ and there exists a positive number λ_0 such that $\widehat{F}$ satisfies the hypotheses of the Wiener-Ikehara theorem in the strip $\{s : \sigma \geq \alpha, |t| < 2\lambda_0\}$, but not in a larger strip. What can one say in this case? Under these conditions, we can assert that *the estimate* $F(x) = O(x^{\alpha})$ *is valid.* Indeed, if we fix λ as a positive number smaller than λ_0, the proof of the Wiener-Ikehara theorem can be repeated verbatim through relation (7.10), which yields the claimed O-estimate.

PROBLEM 7.5   Let $F \in \mathcal{V}$ be defined by

$$F(x) = \int_1^x \{1 - \cos(\lambda_0 \log u)\}\, du \quad (x \geq 1)$$

for some $\lambda_0 > 0$. Find $\widehat{F}$, $\limsup_{x \to \infty} F(x)/x$, and $\liminf_{x \to \infty} F(x)/x$.

7.3   A Wiener-Ikehara proof of the P.N.T.

We shall prove the P.N.T. in the form $\psi(x) \sim x$ by applying the Wiener-Ikehara theorem to $-\zeta'/\zeta$, the M.t. of the nonnegative integrator $d\psi$. We have seen that ζ is analytic on $\{s : \sigma > 0\}$ except for a simple pole at $s = 1$.

Also, the function is nonzero on $\{s : \sigma > 1\}$ since it is represented there by a convergent product of nonzero factors. We can show that

$$s \longmapsto -\zeta'(s)/\zeta(s) - 1/(s-1)$$

is analytic on an open set containing $\{s : \sigma \geq 1\}$ by proving that ζ has no zeros on the line $\Re s = 1$. We established the nonvanishing of zeta in Corollary 7.2 *as a consequence of the P.N.T.* Here we give a direct proof.

Theorem 7.6   *The Riemann zeta function is zerofree on the line $\Re s = 1$.*

Proof.   We show that if zeta had a zero at some point $1 + i\lambda$ with λ real and nonzero, then zeta would have a pole at $1 + 2i\lambda$. For $\sigma > 1$ we have

$$\log|\zeta(s)| = \Re \sum \kappa(n)n^{-s} = \Re \sum_{p} \sum_{k=1}^{\infty} k^{-1}p^{-ks}$$

$$= \sum_{p} p^{-\sigma}\cos(t\log p) + \sum_{p}\sum_{k=2}^{\infty} k^{-1}p^{-k\sigma}\cos(tk\log p).$$

The last double sum is bounded on $\{s : \sigma \geq 3/4\}$, say. We have

$$\log|\zeta(\sigma)\zeta(\sigma + i\lambda)| = \sum_{p}\sum_{k=1}^{\infty} k^{-1}p^{-k\sigma}\{1 + \cos(\lambda k\log p)\} \geq 0.$$

Thus $|\zeta(\sigma)\zeta(\sigma + i\lambda)| \geq 1$ for $\sigma > 1$, and since zeta has a pole of order 1 at $\sigma = 1$, a zero of zeta at $1 + i\lambda$ can be of order at most 1.

Suppose that $1 + i\lambda$ were such a zero. Then $\lim_{\sigma \to 1+}\log|\zeta(\sigma)\zeta(\sigma + i\lambda)|$ would exist, and (since the contributions of the higher prime powers to the sum are bounded) we would have

$$\sum_{p} p^{-\sigma}\{1 + \cos(\lambda\log p)\} = O(1), \qquad (7.14)$$

uniformly for $\sigma > 1$.

Here we give a heuristic argument before completing the proof. Since $\sum_{p} p^{-\sigma} \to +\infty$ as $\sigma \to 1+$, (7.14) would imply that $1 + \cos(\lambda\log p) \approx 0$ holds for most primes p, i.e., $\lambda\log p \approx 2\pi n_p + \pi$ for a suitable integer n_p, and $\cos(2\lambda\log p) \approx +1$. It would follow that

$$\log|\zeta(\sigma + 2i\lambda)| \approx \sum_{p} p^{-\sigma} = \log\zeta(\sigma) + O(1) = \log(\sigma - 1)^{-1} + O(1)$$

as $\sigma \to 1+$, and hence zeta would have a pole at $1 + 2i\lambda$. But zeta is analytic at all points $1 + it$, $t \neq 0$, and so $\zeta(1 + i\lambda) \neq 0$.

It is easy to construct a proof from the above ideas by using (7.14) and the Cauchy-Schwarz inequality. We have by (7.14) (for $\sigma \to 1+$)

$$\{1 + o(1)\}\Big\{\sum_p \frac{1}{p^\sigma}\Big\}^2 = \Big\{\sum_p \frac{-\cos(\lambda \log p)}{p^\sigma}\Big\}^2 \leq \sum_p \frac{\cos^2(\lambda \log p)}{p^\sigma} \sum_p \frac{1}{p^\sigma},$$

$$\{1 + o(1)\}\sum_p \frac{1}{p^\sigma} \leq \sum_p \frac{\cos^2(\lambda \log p)}{p^\sigma} = \sum_p \frac{1 + \cos(2\lambda \log p)}{2p^\sigma},$$

$$\{1 + o(1)\}\sum_p \frac{1}{p^\sigma} \leq \sum_p \frac{\cos(2\lambda \log p)}{p^\sigma}.$$

As before, zeta would have a pole at $1 + 2i\lambda$, which is impossible. $\qquad\square$

Proof of the P.N.T. We have seen that zeta is analytic on $\{s : \sigma > 0\}$ except for a simple pole at $s = 1$ and is nonzero on $\{s : \sigma \geq 1\}$. It follows that we can express

$$-\zeta'(s)/\zeta(s) = \int x^{-s} d\psi(x) = (s - 1)^{-1} + \varphi(s),$$

where φ is analytic on an open set containing $\{s : \sigma \geq 1\}$. If we apply the Wiener-Ikehara theorem to the monotone nondecreasing function ψ, we obtain $\psi(x) \sim x$. $\qquad\square$

PROBLEM 7.6 Show that $M(x) := \sum_{n \leq x} \mu(n) = o(x)$ by applying the Wiener-Ikehara theorem to

$$\zeta(s) + \zeta^{-1}(s) = \int x^{-s}\{dN(x) + dM(x)\}.$$

PROBLEM 7.7 Let $dN^{*1/2}$ denote the positive convolution square root of dN, and recall that $dt^{*1/2} = L^{-1/2}dt/\Gamma(1/2)$ (cf. Examples 6.3 and 6.27). Prove that

$$\int_{1-}^{x} dN^{*1/2} * \frac{L^{-1/2}dt}{\Gamma(1/2)} \sim x.$$

7.4   A generalization of the Wiener-Ikehara theorem

The Wiener-Ikehara theorem applies to an M.t. having a simple pole at the real point on the line of convergence. Here we shall treat more general singularities, including fractional order poles and poles of higher order. This result can be applied for example to estimate the summatory function of $1^{*1/2}$ (the arithmetic function whose D.s. is $\zeta^{1/2}$). Later we shall use this result also to estimate the number of integers in an interval $[1, x]$ which are expressible as a sum of two squares.

Theorem 7.7   *Let F be a real valued nondecreasing function in $\mathcal{V}$ with $\sigma_c(\widehat{F}) = \alpha > 0$. Let φ and ψ be functions which are analytic on the closed half plane $\{s : \sigma \geq \alpha\}$ and assume that $\varphi(\alpha) \neq 0$. For γ a real number distinct from $0, -1, -2, \ldots$, let $s \mapsto (s-\alpha)^{-\gamma}$ be positive valued on the real ray $\{s : s > \alpha\}$. Suppose that*

$$\widehat{F}(s) = \int x^{-s}dF(x) = (s-\alpha)^{-\gamma}\varphi(s) + \psi(s)$$

holds on the open half plane $\{s : \sigma > \alpha\}$. Then

$$F(x) \sim \varphi(\alpha)x^\alpha(\log x)^{\gamma-1}/\{\alpha\Gamma(\gamma)\}.$$

Here Γ denotes the Euler gamma function.

In proving the theorem integrals of the form $\int_0^\infty u^{\gamma-1}K_\lambda(y-u)du$ will arise. We estimate such expressions in

Lemma 7.8   *Let K_λ denote the Fejér kernel and let γ denote a fixed number in $(1, 2)$. There exist a function $g = g_\gamma$ defined on $[2, \infty)$, satisfying $g(\lambda) \to 0$ as $\lambda \to \infty$, and a function $\theta = \theta_\gamma(y, \lambda)$ of modulus at most 1, such that for all $y \geq 2$ and $\lambda \geq 2$ we have*

$$\int_0^\infty u^{\gamma-1}K_\lambda(y-u)du = y^{\gamma-1}\{1 + \theta g(\lambda)\}\int_{-\infty}^\infty K_1(u)\,du. \qquad (7.15)$$

For γ a fixed number in $(0, 1)$ and any fixed $\lambda \geq 2$,

$$\int_0^\infty u^{\gamma-1}K_\lambda(y-u)\,du = o(1) \qquad (y \to \infty).$$

Remark 7.9   One can show that (7.15) is valid also for $0 < \gamma < 1$ by modifying our proof to account for the unboundedness of $u^{\gamma-1}$ at 0 and the fact that $u^{\gamma-1} \downarrow$. However, we shall require only the o-estimate.

Proof of Lemma 7.8.   As we noted earlier,

$$\int K_1 := \int_{-\infty}^{\infty} K_1(u)du = \pi.$$

While we don't need the exact value of $\int K_1$, we do need a positive lower bound. Since $|\sin u| \geq 2|u|/\pi$ for $|u| \leq \pi/2$, we see that $\int K_1 > 4/\pi > 1$.

As in the proof of Theorem 7.3 we define

$$\delta := \left\{ \frac{\lambda}{2} \int K_1 \right\}^{-1/2} \tag{7.16}$$

and note that $0 < \delta < 1$ for $\lambda \geq 2$, and that $\delta \to 0$ as $\lambda \to \infty$.

For $1 < \gamma < 2$, we give a lower estimate of the integral in question by using the fact that $y - \delta > 0$ and applying (7.11). We have

$$\int_0^{\infty} u^{\gamma-1} K_\lambda(y-u)du \geq (y-\delta)^{\gamma-1} \int_{y-\delta}^{y+\delta} K_\lambda(y-u)du$$

$$\geq y^{\gamma-1}\left(1-\frac{\delta}{y}\right)^{\gamma-1}(1-\delta)\int K_1$$

$$\geq y^{\gamma-1}\left(1-\frac{\delta}{2}\right)(1-\delta)\int K_1.$$

We get an upper estimate for this integral by bounding it over four intervals. By Lemma 7.4 and (7.16) (and the condition $y \geq 2$) we have

$$\int_0^{y-\delta} u^{\gamma-1} K_\lambda(y-u)du \leq (y-\delta)^{\gamma-1} \int_0^{y-\delta} K_\lambda(y-u)du$$

$$\leq y^{\gamma-1}\int_\delta^y K_\lambda(v)dv \leq y^{\gamma-1}\frac{\delta}{\lambda\delta^2} = \frac{1}{2}y^{\gamma-1}\delta\int K_1,$$

$$\int_{y+\delta}^{2y} u^{\gamma-1} K_\lambda(y-u)du \leq (2y)^{\gamma-1} \int_{y+\delta}^{2y} K_\lambda(y-u)du \leq y^{\gamma-1}\delta\int K_1,$$

$$\int_{2y}^{\infty} u^{\gamma-1} K_\lambda(y-u)du \leq \int_{2y}^{\infty} u^{\gamma-1}\lambda^{-1}(u-y)^{-2}du$$

$$\leq \frac{4}{\lambda}\int_{2y}^{\infty} u^{\gamma-3}du \leq \frac{4y^{\gamma-2}}{\lambda(2-\gamma)} \leq \frac{2y^{\gamma-1}}{\lambda(2-\gamma)}\int K_1,$$

$$\int_{y-\delta}^{y+\delta} u^{\gamma-1} K_\lambda(y-u)du = \int_0^\delta K_\lambda(v)\{(y-v)^{\gamma-1} + (y+v)^{\gamma-1}\}\, dv$$

$$\leq 2y^{\gamma-1} \int_0^\delta K_\lambda(v)\, dv < y^{\gamma-1} \int K_1.$$

The last integral (the main contribution) was estimated using the symmetry of K_λ about the origin and the concavity of the function $u \mapsto u^{\gamma-1}$.

Altogether, we have

$$\int_0^\infty u^{\gamma-1} K_\lambda(y-u)du \leq y^{\gamma-1}\Big\{1 + \frac{3\delta}{2} + \frac{2}{\lambda(2-\gamma)}\Big\} \int K_1.$$

Combining the two inequalities and recalling (7.16) we obtain (7.15). For fixed $\gamma \in (1,2)$, the error estimate satisfies $g_\gamma(\lambda) = O_\gamma(1/\sqrt{\lambda})$.

We estimate the integral in the case $0 < \gamma < 1$ by separating $[0,\infty)$ into three segments and estimating each part separately. We have

$$\int_0^{y/2} u^{\gamma-1} K_\lambda(y-u)du \leq \frac{4}{\lambda y^2} \int_0^{y/2} u^{\gamma-1}du < \frac{4}{\gamma\lambda} y^{\gamma-2},$$

$$\int_{y/2}^{2y} u^{\gamma-1} K_\lambda(y-u)du \leq 2y^{\gamma-1} \int_{y/2}^{2y} K_\lambda(y-u)du < 2y^{\gamma-1} \int K_1,$$

$$\int_{2y}^\infty u^{\gamma-1} K_\lambda(y-u)du \leq 4y^{\gamma-2}\lambda^{-1}(2-\gamma)^{-1}.$$

The last integral was estimated as in the case $1 < \gamma < 2$. Thus we have for each fixed $\gamma \in (0,1)$

$$\int_0^\infty u^{\gamma-1} K_\lambda(y-u)du = O(y^{\gamma-1}) = o(1) \qquad (y \to \infty). \qquad \square$$

Proof of Theorem 7.7.   The case $1 < \gamma < 2$. We already have treated the case $\gamma = 1$. The proof for $1 < \gamma < 2$ follows closely that of Theorem 7.3. First, let us note that $\varphi(\alpha) > 0$ in this case, because $\widehat{F}(\sigma) > 0$ on the ray $\{\sigma : \sigma > \alpha\}$ and, by hypothesis,

$$\varphi(\alpha) = \lim_{\sigma \to \alpha+} (\sigma - \alpha)^\gamma \widehat{F}(\sigma) \neq 0.$$

Now change the variable, setting $u = \alpha \log x$ and defining $f(u) = F(e^{u/\alpha}) = F(x)$. Then, for $\sigma > 1$,

$$\int_{0-}^{\infty} e^{-su} df(u) = \widehat{F}(\alpha s) = (s-1)^{-\gamma} \alpha^{-\gamma} \varphi(\alpha s) + \psi(\alpha s).$$

Integrating by parts we obtain

$$\int_0^{\infty} e^{-su} f(u) du = \widehat{F}(\alpha s)/s \quad (\sigma > 1).$$

If we expand $\varphi(\alpha s)/s$ in a Taylor series about $s = 1$, we find that

$$\frac{\alpha^{-\gamma}}{(s-1)^{\gamma}} \frac{\varphi(\alpha s)}{s} + \frac{\psi(\alpha s)}{s} = \frac{a}{(s-1)^{\gamma}} + \frac{b}{(s-1)^{\gamma-1}} + \varphi_1(s), \qquad (7.17)$$

where $a := \alpha^{-\gamma} \varphi(\alpha) > 0$, b is some constant, and φ_1 is a continuous function on the closed half plane $\{s : \sigma \geq 1\}$. Now

$$(s-1)^{-\beta} = \int_0^{\infty} e^{-su} e^u u^{\beta-1} du/\Gamma(\beta)$$

for $\sigma > 1$ and $\beta > 0$. This identity can be established for s real and $s > 1$ by changing the variable in the Euler integral representation of the gamma function (cf. Appendix). The result follows for any complex s with $\sigma > 1$ by analytic continuation. Thus we have for $\sigma > 1$,

$$\varphi_1(s) = \int_0^{\infty} e^{-su} \left\{ f(u) - \frac{a}{\Gamma(\gamma)} e^u u^{\gamma-1} - \frac{b}{\Gamma(\gamma-1)} e^u u^{\gamma-2} \right\} du.$$

We are going to show that $f(u) \sim a e^u u^{\gamma-1}/\Gamma(\gamma)$, which is equivalent to the assertion of the theorem. As in the proof of Theorem 7.3, we take $s = 1 + \epsilon + it$ with $\epsilon > 0$, form the integral

$$\frac{1}{2} \int_{-2\lambda}^{2\lambda} \left(1 - \frac{|t|}{2\lambda}\right) e^{ity} \varphi_1(1 + \epsilon + it) dt = \frac{1}{2} \int_{-2\lambda}^{2\lambda} \left(1 - \frac{|t|}{2\lambda}\right) e^{ity} \times$$

$$\int_0^{\infty} \left\{ f(u) - \frac{a}{\Gamma(\gamma)} e^u u^{\gamma-1} - \frac{b e^u}{\Gamma(\gamma-1)} u^{\gamma-2} \right\} \cdot e^{-u(1+\epsilon+it)} du\, dt$$

$$= \int_0^{\infty} e^{-u-\epsilon u} \left\{ f(u) - \frac{a}{\Gamma(\gamma)} e^u u^{\gamma-1} - \frac{b e^u}{\Gamma(\gamma-1)} u^{\gamma-2} \right\} K_\lambda(y-u) du,$$

and let $\epsilon \to 0+$. Using an argument similar to that given near equation (7.9), we obtain

$$\frac{1}{2} \int_{-2\lambda}^{2\lambda} \left(1 - \frac{|t|}{2\lambda}\right) e^{ity} \varphi_1(1 + it) dt$$

$$= \int_0^\infty e^{-u} f(u) K_\lambda(y - u) du - \frac{a}{\Gamma(\gamma)} \int_0^\infty u^{\gamma-1} K_\lambda(y - u) du$$

$$- \frac{b}{\Gamma(\gamma - 1)} \int_0^\infty u^{\gamma-2} K_\lambda(y - u) du.$$

Lemma 7.8 implies that the last integral $= o(1)$ as $y \to \infty$. The Riemann-Lebesgue lemma shows that the integral containing φ_1 also tends to zero as $y \to \infty$. Thus, for each $\lambda \geq 2$, as $y \to \infty$ we have by (7.15)

$$\int_0^\infty e^{-u} f(u) K_\lambda(y - u) du = \frac{ay^{\gamma-1}}{\Gamma(\gamma)} \{1 + \theta g(\lambda)\} \int K_1 + o(1), \qquad (7.18)$$

where $|\theta| \leq 1$ and $g(\lambda) = o(1)$ as $\lambda \to \infty$.

The conclusion of the proof follows in essentially the same way as that of Theorem 7.3. We estimate the left hand side of (7.18) by using only the range $y - \delta \leq u \leq y + \delta$, where $\delta = \delta(\lambda)$ is chosen as in (7.16). Letting $y - \delta = w$ we obtain the estimate

$$e^{-w} f(w) \leq \frac{ae^{2\delta}(w + \delta)^{\gamma-1}}{\Gamma(\gamma)(1 - \delta)} \{1 + \theta g(\lambda)\} + o(1) \qquad (7.19)$$

for fixed $\lambda \geq 2$. Thus (7.19) yields the bound $e^{-w} f(w) \leq B w^{\gamma-1}$ for some $B > 0$ and all $w \geq 2$. We give upper estimates of the integrals $\int_2^{y-\delta}$ and $\int_{y+\delta}^\infty$ in the left hand side of (7.18) by using this bound and some inequalities from the proof of Lemma 7.8. Since the lemma treats only $y \geq 2$, we note that

$$\int_0^2 e^{-u} f(u) K_\lambda(y - u) du < f(2) \int_{y-2}^y K_\lambda(v) dv = o(1).$$

Replacing $y + \delta$ by w we obtain

$$f(w) e^{-w} \geq (w - \delta)^{\gamma-1} e^{-2\delta} \left\{ \frac{a}{\Gamma(\gamma)} - \frac{ag(\lambda)}{\Gamma(\gamma)} - \frac{3B\delta}{2} - \frac{2B}{\lambda(2 - \gamma)} \right\} + o(1).$$

Now λ can be chosen arbitrarily large, and $\delta \to 0+$ as $\lambda \to \infty$. If we combine the last inequality with (7.19) and take λ large, we can conclude

$$f(w)e^{-w} = (1 + o(1))\frac{a}{\Gamma(\gamma)}w^{\gamma-1} + o(1). \tag{7.20}$$

This is equivalent to the assertion of the theorem in case $1 < \gamma < 2$.

The case $\gamma < 1$ and $\gamma \neq 0, -1, -2, \ldots$. If we simply repeated the preceding argument, making some minor changes, e.g. to obtain convergent expressions, we could again deduce (7.20). However, since $\gamma < 1$, we could conclude only that $f(y) = o(e^y)$. We can obtain the desired result by introducing some more "weight" by differentiation.

Let N be the positive integer for which $1 < \gamma + N < 2$. Define $F_1(x) := \int_1^x L^N dF$. We form

$$\widehat{F_1}(s) = \int x^{-s} \log^N x \, dF(x) = (-1)^N \widehat{F}^{(N)}(s)$$

$$= (-1)^N \sum_{j=0}^{N} \binom{N}{j}\{(s-\alpha)^{-\gamma}\}^{(N-j)}\varphi^{(j)}(s) + (-1)^N\psi^{(N)}(s)$$

$$= (s-\alpha)^{-\gamma-N}\frac{\Gamma(\gamma+N)}{\Gamma(\gamma)}\varphi(s) + \cdots + (-1)^N(s-\alpha)^{-\gamma}\varphi^{(N)}(s)$$

$$+ (-1)^N\psi^{(N)}(s)$$

$$= (s-\alpha)^{-\gamma-N}\Phi(s) + (-1)^N\psi^{(N)}(s),$$

where Φ and $\psi^{(N)}$ are analytic functions on $\{s : \sigma \geq \alpha\}$ and

$$\Phi(\alpha) = \Gamma(\gamma+N)\varphi(\alpha)/\Gamma(\gamma).$$

As in the case $1 < \gamma < 2$, we have $\widehat{F_1}(\sigma) > 0$ on $\{\sigma : \sigma > \alpha\}$ and thus

$$\Phi(\alpha) = \lim_{\sigma \to \alpha+}(\sigma-\alpha)^{\gamma+N}\widehat{F_1}(\sigma) > 0.$$

Since $\Gamma(\gamma+N) > 0$, we have $\varphi(\alpha)/\Gamma(\gamma) > 0$ here.

Because $1 < \gamma + N < 2$, we can apply to $F_1(x)$ the form of the theorem that we have already proved to deduce that

$$F_1(x) \sim \frac{\varphi(\alpha)}{\alpha\Gamma(\gamma)}x^\alpha(\log x)^{\gamma+N-1}.$$

Now for $x \geq e$,

$$F(x) = F(e) + \int_e^x L^{-N} dF_1$$

$$= F_1(x) \log^{-N} x + N \int_e^x t^{-1} F_1(t) (\log t)^{-N-1} dt + O(1)$$

$$= \{1 + o(1)\} \frac{\varphi(\alpha)}{\alpha \Gamma(\gamma)} x^\alpha (\log x)^{\gamma - 1} + O\{x^\alpha (\log x)^{\gamma - 2}\}.$$

This establishes the theorem for $\gamma < 1$, $\gamma \neq 0, -1, -2, \ldots$.

The case $\gamma \geq 2$. The proof given for $1 < \gamma < 2$ fails here because $\int_0^\infty u^{\gamma - 1} K_\lambda(y - u) \, du$ diverges for $\gamma \geq 2$. We shall replace K_λ by a power of itself to make the integral convergent. We sketch the argument for $2 \leq \gamma < 4$. The general case follows in the same way.

If we set $f^+ := \max(f, 0)$, we have

$$\lambda^{-1} K_\lambda^2(x) = \lambda \left(\frac{\sin \lambda x}{\lambda x} \right)^4$$

$$= \frac{1}{4\lambda} \int \left(1 - \frac{|t|}{2\lambda} \right)^+ e^{ixt} \, dt \cdot \int \left(1 - \frac{|u|}{2\lambda} \right)^+ e^{ixu} \, du$$

$$= \int_{-4\lambda}^{4\lambda} h(v) e^{ixv} \, dv,$$

where h is the continuous function supported on $[-4\lambda, 4\lambda]$ defined by

$$h(v) = \frac{1}{4\lambda} \int \left(1 - \frac{|t|}{2\lambda} \right)^+ \left(1 - \frac{|v - t|}{2\lambda} \right)^+ dt.$$

An explicit representation of h is not needed, but one can show that h is represented on each of the intervals $[-4\lambda, -2\lambda]$, $[-2\lambda, 0]$, $[0, 2\lambda]$, $[2\lambda, 4\lambda]$ by a cubic polynomial.

In Lemma 7.4 we set out a number of properties of K_λ. The function $\lambda^{-1} K_\lambda^2$ satisfies analogous relations. In particular

$$\int_{-\infty}^\infty \lambda^{-1} K_\lambda^2(u) \, du = \int_{-\infty}^\infty K_1^2(u) \, du$$

for all real λ (which is the *raison d'être* of the factor λ^{-1}). The analogue of Lemma 7.8 holds for $2 \leq \gamma < 4$ if we use $\lambda^{-1} K_\lambda^2$ in place of K_λ.

To prove Theorem 7.7 for $2 \leq \gamma < 4$ we alter (7.17) to exhibit all powers of $s - 1$ occurring with a negative exponent. In (7.18) we use $\lambda^{-1} K_\lambda^2$ in place of K_λ, and the right hand side of (7.18) is altered by the inclusion of terms containing the factor $y^{\gamma-2}$ and $y^{\gamma-3}$ (the last if $\gamma \geq 3$). Of course, these terms are of smaller order than the term containing the factor $y^{\gamma-1}$. The conclusion of the proof is just as before.

For the general case $\gamma \geq 2$, we choose a positive integer N for which $2N > \gamma$ and use the function $\lambda^{1-N} K_\lambda^N$ in place of K_λ. We have

$$\lambda^{1-N} K_\lambda^N = \int e^{ixv} h_N(v) \, dv,$$

where

$$h_N(v) = \lambda^{1-N} 2^{-N} \int \cdots \int \left(1 - \frac{|t_1|}{2\lambda}\right)^+ \times$$

$$\cdots \times \left(1 - \frac{|t_{N-1}|}{2\lambda}\right)^+ \left(1 - \frac{|v - t_1 - \cdots - t_{N-1}|}{2\lambda}\right)^+ dt_1 \cdots dt_{N-1}.$$

The argument proceeds just as we have described for $N = 2$. $\quad\Box$

PROBLEM 7.8   State and prove an O-estimate subordinate to Theorem 7.7 (cf. §7.2.2).

PROBLEM 7.9   Let c be a fixed positive number, and define a completely multiplicative function f by setting $f(p) = 0$ for all primes $p \leq c$ and $f(p) = c$ for all $p > c$. Show that

$$\sum_{n \leq x} f(n) \sim \prod_{p \leq c} (1 - p^{-1})^c \prod_{p > c} \left\{ \frac{(1 - p^{-1})^c}{1 - cp^{-1}} \right\} \frac{x(\log x)^{c-1}}{\Gamma(c)}.$$

PROBLEM 7.10   Let $g \in \mathcal{A}_1$ and $g * g = 1$. Find an asymptotic formula for the summatory function of g (cf. §2.4.2 and Example 6.3(4)).

PROBLEM 7.11   Using the identity

$$\zeta^4(s)/\zeta(2s) = \sum \tau^2(k) k^{-s}$$

and the generalized Wiener-Ikehara theorem, prove that

$$\sum_{k \leq x} \tau^2(k) \sim \pi^{-2} x \log^3 x.$$

(Cf. Problems 2.31 and 3.23, part (4).)

PROBLEM 7.12   (Generalized divisor function.) For c a fixed real number, not 0 or a negative integer, define a multiplicative function τ_c by setting

$$\tau_c(p^j) = c(c+1)\cdots(c+j-1)/j!$$

for primes p and positive integers j. Find an asymptotic formula for the associated summatory function. (Cf. Problem 3.26.)

7.5   The Perron formula

Contour integration offers another method of obtaining inversion formulas. Such an approach requires more information about the generating function than does the Wiener-Ikehara theorem, but it will enable us to obtain an error estimate in addition to an asymptotic formula. The most simply formulated result of this kind is

Theorem 7.10   (Perron inversion formula). *Let $F \in \mathcal{V}$ and assume that $\sigma_a(\widehat{F}) < \infty$. Let $b > \max(\sigma_c(\widehat{F}), 0)$. Then for any $x > 0$,*

$$\lim_{T \to \infty} \frac{1}{2\pi i} \int_{b-iT}^{b+iT} \widehat{F}(s) x^s \frac{ds}{s} = \frac{1}{2}\{F(x+) + F(x-)\}.$$

We noted before proving the uniqueness theorem that the M.t. of an integrator dF is independent of the value of F at any particular point. The Perron inversion formula assigns the value $(1/2)\{F(x+) + F(x-)\}$ at each point x. Consequently, if $F \in \mathcal{V}$ and is not everywhere continuous, its image under the composition of the M.t. and inversion formula will not be everywhere continuous from the right. This point, once noted, will not cause us trouble.

Perron's formula will follow as the limit of an approximate inversion formula involving an integral over a finite range. The approximate formula is generally more practical for applications because of the bounded integration contour. The basis of the inversion formula is

Lemma 7.11   *Let $0 < a < T$. Then*

$$\frac{1}{2\pi i} \int_{a-iT}^{a+iT} x^s \frac{ds}{s} = \begin{cases} 1 + O\{x^a/(T\log x)\} & \text{if } x > 1, \\ \frac{1}{2} + O(T|\log x|) + O(a/T) & \text{if } |\log x| \le 1/T, \\ O\{x^a/(T|\log x|)\} & \text{if } 0 < x < 1. \end{cases}$$

The constants implied by the O's are absolute.

Proof of Lemma 7.11. Suppose first that $x > 1$. For N a positive integer, let R_N denote the rectangle with vertices $a \pm iT$ and $-N \pm iT$, with the usual orientation. By the residue theorem,

$$\frac{1}{2\pi i} \int_{R_N} x^s ds/s = 1.$$

We must estimate the integral along the three unwanted sides. Along the top and bottom sides we have

$$\left| \int x^s \frac{ds}{s} \right| < T^{-1} \int_{-\infty}^{a} x^\sigma d\sigma = \frac{x^a}{T \log x}.$$

Along the left side,

$$\left| \int x^{-s} ds/s \right| < x^{-N} \cdot 2T/N.$$

Letting $N \to +\infty$, the contribution of the left side goes to zero. Thus the first of the three estimates is proved.

Next suppose $0 < x < 1$. Let $N > a$ and let R'_N denote the rectangle with vertices $a \pm iT$ and $N \pm iT$ traversed with a negative orientation. We take the integral over R'_N and again apply the residue theorem. Since the integrand is analytic inside and on R'_N for each $N > a$, the value of the contour integral is zero. We estimate the contribution of the three unwanted sides of R'_N just as we did for R_N and let $N \to +\infty$. The third estimate of the lemma follows.

If x is near 1, the preceding estimates will be poor. In case $|\log x| \leq 1/T$ we can give another estimate. Let $\mathcal{C}$ denote the circular arc centered at the origin which runs from $a - iT$ to $a + iT$ in a counter clockwise direction. By the Cauchy integral theorem,

$$\frac{1}{2\pi i} \int_{a-iT}^{a+iT} x^s \frac{ds}{s} = \frac{1}{2\pi i} \int_{\mathcal{C}} x^s \frac{ds}{s}.$$

On the circular arc we have

$$|s \log x| < (a + T)|\log x| < 2T|\log x| \leq 2,$$

and thus we can estimate the integral over this arc by setting

$$\frac{x^s}{s} = \frac{1}{s} + (\log x) \frac{e^{s \log x} - 1}{s \log x} = \frac{1}{s} + O(|\log x|).$$

Introducing polar coordinates, we have

$$\frac{1}{2\pi i}\int_C \{s^{-1} + O(|\log x|)\}ds = \frac{1}{2\pi}\int_{-\pi/2+\arctan(a/T)}^{\pi/2-\arctan(a/T)} d\theta + O(T|\log x|)$$

$$= \frac{1}{2} - \frac{1}{\pi}\arctan\frac{a}{T} + O(T|\log x|).$$

This yields the stated estimate when $|\log x| \le 1/T$. $\qquad\qquad\qquad\square$

We apply the preceding lemma to establish

Lemma 7.12 *Let* $F \in \mathcal{V}$ *and suppose* $\sigma_a = \sigma_a(\widehat{F}) < \infty$. *Then for any* $x > 0$, $b > \max(\sigma_a, 0)$, *and* $T > b$,

$$\frac{1}{2\pi i}\int_{b-iT}^{b+iT} \widehat{F}(s)\frac{x^s}{s}ds = \frac{1}{2}\{F(xe^{-1/T}) + F(xe^{1/T})\}$$

$$+ O\Big(\frac{x^b}{T}\Big)\Big\{\int_{1-}^{xe^{-1/T}} + \int_{xe^{1/T}}^{\infty}\Big\}\frac{y^{-b}}{|\log x/y|}\,dF_v(y)$$

$$+ O\Big(\int_{xe^{-1/T}}^{xe^{1/T}} (T|\log\frac{x}{y}| + \frac{b}{T})\,dF_v(y)\Big). \qquad (7.21)$$

The constants implied by the O*'s are absolute.*

Proof of Lemma 7.12. The integral defining $\widehat{F}$ converges uniformly on the line $\Re s = b$. Thus we have

$$\frac{1}{2\pi i}\int_{b-iT}^{b+iT} \widehat{F}(s)\frac{x^s}{s}ds = \frac{1}{2\pi i}\int_{b-iT}^{b+iT}\Big\{\int_{y=1-}^{\infty} y^{-s}dF(y)\Big\}\frac{x^s}{s}ds$$

$$= \int_{y=1-}^{\infty}\Big\{\frac{1}{2\pi i}\int_{b-iT}^{b+iT} (x/y)^s\frac{ds}{s}\Big\}dF(y).$$

We represent the outer integral as a sum of three parts taken over the ranges $(0, xe^{-1/T})$, $(xe^{-1/T}, xe^{1/T})$, and $(xe^{1/T}, \infty)$, and apply the preceding lemma. $\qquad\qquad\qquad\square$

7.6   Proof of the Perron formula

We have seen that an M.t. $\widehat{F}$ is bounded on any half plane properly contained in the half plane of absolute convergence. The corresponding result for the half plane of (nonabsolute) convergence is generally false. However, we can give an estimate in this case that will be useful in the proof of Theorem 7.10.

Lemma 7.13 *Let $F \in V$ and assume $\sigma_c(\widehat{F}) < \infty$. Given $\epsilon > 0$, we have $\widehat{F}(\sigma + it) = o(|t|)$ uniformly for $\sigma \geq \sigma_c(\widehat{F}) + \epsilon$ as $t \to \pm\infty$.*

Proof of Lemma 7.13. Let $b = \sigma_c(\widehat{F}) + \epsilon/2$. Then by Theorem 6.9 there exists a number c such that $F(x) = c + O(x^b)$ as $x \to \infty$. (If $b \geq 0$, as happens in the present application, we can take $c = 0$.) For any $X \geq e$ and $\sigma \geq \sigma_c + \epsilon = b + \epsilon/2$, we write

$$\widehat{F}(s) = \int_{1-}^{X} x^{-s} dF(x) + x^{-s}\{F(x) - c\}\Big|_{X}^{\infty} + s \int_{X}^{\infty} x^{-s-1}\{F(x) - c\}\,dx,$$

so that (for K a suitable constant)

$$|\widehat{F}(s)| \leq \int_{1-}^{X} x^{-\sigma} dF_v(x) + KX^{b-\sigma} + (|\sigma| + |t|) \int_{X}^{\infty} Kx^{b-\sigma-1} dx$$

$$\leq \int_{1-}^{X} x^{-b} dF_v(x) + KX^{-\epsilon/2} + K(|\sigma| + |t|)\frac{X^{b-\sigma}}{\sigma - b}.$$

By choosing X sufficiently large we can make $KX^{b-\sigma}/(\sigma - b)$ smaller than any given positive ϵ_1, uniformly for $\{s : \sigma \geq \sigma_c + \epsilon\}$. It follows that

$$\limsup_{t \to \pm\infty} |\widehat{F}(s)/t| \leq \epsilon_1$$

uniformly on $\{s : \sigma \geq \sigma_c + \epsilon\}$. $\square$

PROBLEM 7.13 Suppose $F \in V$ and $\sigma_c(\widehat{F}) < \alpha < \infty$. Let

$$I(T, \alpha) := \frac{1}{2T} \int_{-T}^{T} \widehat{F}(\alpha + it)\,dt.$$

Show that $\lim_{T \to \infty} I(T, \alpha)$ exists and equals $F(1)$. Hint. Show that

$$I(T, \alpha) = F(1) + \int_{1}^{2} x^{-\alpha} \frac{\sin(T \log x)}{T \log x} dF(x) + \frac{\widehat{G}(\alpha + iT) - \widehat{G}(\alpha - iT)}{-2iT},$$

where $\widehat{G}(s) = \int_{2}^{\infty} x^{-s}(\log x)^{-1} dF(x)$.

Proof of Theorem 7.10.   Recall that the functions in $\mathcal{V}$ are locally of bounded variation, and as such, have limits from both the left and right at each point. Thus, for each $x > 0$,

$$\frac{1}{2}\{F(xe^{-1/T}) + F(xe^{1/T})\} \to \frac{1}{2}\{F(x-) + F(x+)\}$$

as $T \to \infty$. We shall first show that if $b > \max\{\sigma_a(\widehat{F}), 0\}$ then the integrals occurring on the right hand side of (7.21) go to zero as $T \to \infty$. We may suppose that x is fixed and $x > 1$. In case $0 < x \le 1$, similar reasoning applies, but some of the integrals are zero since $F = 0$ on $(-\infty, 1)$.

To start, we determine intervals $(x - \delta, x)$ and $(x, x + \delta)$ on which F_v is nearly constant. This is certainly possible, for functions in $\mathcal{V}$ have limits from each side at any point. Given $\epsilon \in (0, \frac{1}{2})$ we can choose $\delta = \delta_x \in (0, \frac{1}{2})$ such that $\delta < x - 1$ and further

$$F_v(x-) - F_v(y) < \epsilon \quad \text{for} \quad x - \delta < y < x,$$

$$F_v(z) - F_v(x+) < \epsilon \quad \text{for} \quad x < z < x + \delta.$$

Assuming that T is large enough that $x(\exp(1/T) - 1) < \delta$ (and hence $x(1 - \exp(-1/T)) < \delta$), we write

$$\left\{ \int_{1-}^{xe^{-1/T}} + \int_{xe^{1/T}}^{\infty} \right\} \frac{(x/y)^b}{T|\log x/y|} \, dF_v(y)$$

$$= \int_{1-}^{x-\delta} + \int_{x-\delta}^{xe^{-1/T}} + \int_{xe^{1/T}}^{x+\delta} + \int_{x+\delta}^{ex} + \int_{ex}^{\infty}$$

and estimate each part separately. We have

$$\int_{1-}^{x-\delta} \le \frac{x^b}{\log\{x/(x-\delta)\}} \frac{F_v(x-\delta)}{T} \to 0 \quad \text{as} \quad T \to \infty,$$

$$\int_{x-\delta}^{xe^{-1/T}} \le \left(\frac{x}{x - \frac{1}{2}}\right)^b \{F_v(xe^{-1/T}) - F_v(x-\delta)\} < 2^b \epsilon,$$

$$\int_{xe^{1/T}}^{x+\delta} \le F_v(x+\delta) - F_v(xe^{1/T}) < \epsilon,$$

$$\int_{x+\delta}^{ex} \le F_v(ex)/\{T\log(1 + \delta x^{-1})\} \to 0 \quad \text{as} \quad T \to \infty,$$

$$\int_{ex}^{\infty} \le \frac{x^b}{T} \int_{1-}^{\infty} y^{-b} dF_v(y) \to 0 \quad \text{as} \quad T \to \infty.$$

In view of Lemma 7.12, it remains to estimate

$$\int_{xe^{-1/T}}^{xe^{1/T}} T|\log \frac{x}{y}| dF_v(y) + \int_{xe^{-1/T}}^{xe^{1/T}} \frac{b}{T} dF_v(y).$$

The second integral clearly tends to zero as $T \to \infty$. We decompose the first integral into three parts (for better legibility, set $\eta := \exp(\epsilon/T)$)

$$\left\{ \int_{xe^{-1/T}}^{x/\eta} + \int_{x/\eta}^{x\eta} + \int_{x\eta}^{xe^{1/T}} \right\} T|\log \frac{x}{y}| dF_v(y)$$

and estimate the three new integrals. The first and third of these each has size at most ϵ, since $T|\log x/y| < 1$ and F_v changes by at most ϵ on each interval. Similar reasoning shows the middle integral to be at most $\epsilon\{F_v(x+) - F_v(x-) + 2\epsilon\}$.

Thus we have shown that the sum of the integrals occurring on the right side of (7.21) has modulus at most

$$\epsilon\{2^b + 4 + F_v(x+) - F_v(x-)\} + o(1)$$

as $T \to \infty$. Since ϵ is arbitrary, that sum has limit zero as $T \to \infty$. This establishes the case where $b > \sigma_a(F)$.

Finally, suppose that $\max(\sigma_c, 0) < b \le \sigma_a$ and let $b' > \sigma_a$. By Cauchy's theorem, the integral $\int x^s \widehat{F}(s) ds/s$ taken over the rectangle with vertices $b' \pm iT$, $b \pm iT$ is zero. The preceding lemma implies that the integrals over the top and bottom sides of the rectangle tend to zero as $T \to \infty$.

Thus we have shown that

$$\lim_{T \to \infty} \frac{1}{2\pi i} \int_{b-iT}^{b+iT} \widehat{F}(s) x^s \frac{ds}{s} = \lim_{T \to \infty} \frac{1}{2\pi i} \int_{b'-iT}^{b'+iT} \widehat{F}(s) x^s \frac{ds}{s}$$

$$= \frac{1}{2}\{F(x+) + F(x-)\}. \qquad \square$$

7.7   Contour deformation in the Perron formula

In this section we first describe a typical way in which the Perron formula is used, which involves a change of integration contour. Then, as an example, we derive a Fourier series expansion that will be used in the next chapter.

The Perron representation is useful because we often can apply the residue theorem to the integral. We obtain a "main term" arising from the contributions of singularities of $\widehat{F}$ and an "error term" whose size will be estimated by the integral over the deformed contour.

In order to obtain a small estimate of the error term, we shall choose the contour such that (1) $\Re s$ is small so that the factor x^s occurring in the Perron formula has a small modulus, (2) the singularities of $\widehat{F}$ within the contour are "manageable," and (3) estimates of $|F(\sigma + it)|$ are of reasonable size on the contour. By taking a contour lying further to the left we may encounter more singularities of $\widehat{F}$ and/or have worse estimates of $|\widehat{F}(\sigma + it)|$ as a function of t. In each application we shall seek a contour to balance the first objective against the second and third.

As an illustration of the preceding remarks we shall treat a class of frequently encountered examples: We are given a real valued monotone increasing function $F \in \mathcal{V}$ whose associated M.t. has a pole of order m at $\alpha = \sigma_c(\widehat{F}) = \sigma_a(\widehat{F}) > 0$. Further, there is an analytic continuation of $\widehat{F}$ having no other singularities in some rectangle

$$\{s : \beta \leq \sigma \leq b, |t| \leq T\}$$

where $0 \neq \beta < \alpha < b$ and $T \geq 2 + 2|\beta| + 2b$. With these conditions we have an estimate for F which we give as

Lemma 7.14 *Suppose that the function $\widehat{F}$ and the parameters T, b, and β satisfy the hypotheses of the preceding paragraph. Then for any $x > 0$,*

$$F(x) = x^\alpha P(\log x) + O\{F(xe^{1/T}) - F(xe^{-1/T})\}$$

$$+ O\left(\left\{\int_{1-}^{xe^{-1/T}} + \int_{xe^{1/T}}^{\infty}\right\} \frac{(x/y)^b \, dF(y)}{T|\log x/y|}\right)$$

$$+ O\left(x^\beta \log(T/|\beta|) \max_{|t| \leq T} |\widehat{F}(\beta + it)|\right)$$

$$+ O\left(T^{-1}(b - \beta) \max_{\beta \leq \sigma \leq b} |\widehat{F}(\sigma + iT)| \, x^\sigma\right) + \delta.$$

Here $\delta = \widehat{F}(0)$ if $\beta < 0$ and $\delta = 0$ if $\beta > 0$, and P is a polynomial of degree $m - 1$ such that $P(\log x)$ equals the residue of $s^{-1}\widehat{F}(s)e^{(s-\alpha)\log x}$ at $s = \alpha$. The constants implied by the O's are absolute.

Proof. We apply Lemma 7.12, and replace

$$\frac{1}{2}\{F(xe^{-1/T}) + F(xe^{1/T})\} + O\Big(\int_{xe^{-1/T}}^{xe^{1/T}} \Big(T\Big|\log\frac{x}{y}\Big| + \frac{b}{T}\Big)dF_v(y)\Big)$$

by

$$F(x) + O\{F(xe^{1/T}) - F(xe^{-1/T})\}.$$

Then we use the residue theorem to replace the contour integral over the line segment $\{b + it : -T \leq t \leq T\}$ by one over the other three sides of the rectangle having vertices $b \pm iT$ and $\beta \pm iT$. The residue of the integrand at $s = \alpha$ gives the stated polynomial by Theorem 6.23. For the integral along the segment $\{\beta + it : -T \leq t \leq T\}$ we use the relation

$$\int_0^T (\beta^2 + t^2)^{-1/2}dt = \log\Big\{\Big(1 + \Big(\frac{T}{|\beta|}\Big)^2\Big)^{1/2} + \frac{T}{|\beta|}\Big\} = O\Big(\log\frac{T}{|\beta|}\Big).$$

The integrals along the top and bottom edges of the rectangle are estimated trivially. $\qquad\square$

In order to use the formula we have just established, information is needed about the analyticity and magnitude of $\widehat{F}$ in a rectangle extending outside the half plane of convergence. In addition we need an estimate of the difference $F(x\delta) - F(x/\delta)$ for a suitable $\delta > 1$. In many cases one can give a satisfactory direct estimate of this difference. For example, in the prime number problem for $y > x$ we have

$$0 \leq \psi(y) - \psi(x) = \sum_{x < n \leq y} \Lambda(n) \leq \sum_{x < n \leq y} \log n \leq (y - x + 1)\log y.$$

7.7.1 The Fourier series of the sawtooth function

We introduced the function $x \mapsto [x] - x + 1/2$ in connection with the Euler–Maclaurin formula. Here we shall use Theorem 7.10 to derive the Fourier series of the related sawtooth function

$$S : y \longmapsto \frac{1}{2}([y] + [y-]) - y + \frac{1}{2},$$

which equals $[y] - y + \frac{1}{2}$ everywhere except at integers n, where $S(n) = 0$.

Lemma 7.15   *For all $y \in \mathbb{R}$,*

$$S(y) = \sum_{n=1}^{\infty} \frac{1}{\pi n} \sin 2\pi ny. \qquad (7.22)$$

The partial sums of the series are uniformly bounded on $\mathbb{R}$, and for any fixed $\delta \in (0, 1/2)$ the series converges uniformly on $\bigcup_{\ell=-\infty}^{\infty} [\ell + \delta, \ell + 1 - \delta]$.

Proof.   We start in a multiplicative setting by using powers of a fixed number, which we choose (arbitrarily) to be 2. After applying the inversion formula, we shall take logarithms to achieve the desired form. We shall assume until the last paragraph of the proof that x is a fixed real number which exceeds $1/2$.

In §2.3 we defined

$$1_2(n) := \begin{cases} 1 & \text{if } n = 2^k, \ k = 0, 1, 2, \ldots, \\ 0 & \text{if } n \text{ is not a power of 2.} \end{cases}$$

For x as above,

$$F(x) := \sum_{n \le x} 1_2(n) = \sum_{k=0}^{[\log x / \log 2]} 1 = \left[1 + \frac{\log x}{\log 2}\right].$$

The D.s. of 1_2 satisfies, for $\sigma > 0$,

$$\widehat{F}(s) = \sum 1_2(n)n^{-s} = \sum_{k=0}^{\infty} 2^{-ks} = (1 - 2^{-s})^{-1}.$$

Let N be a positive integer which is fixed for the moment, and set $T := T(N) := (2N + 1)\pi / \log 2$. To apply Theorem 7.10 we evaluate

$$I(T) := \frac{1}{2\pi i} \int_{1-iT}^{1+iT} (1 - 2^{-s})^{-1} x^s s^{-1} ds$$

by the residue theorem. For $\nu < 0$ take R_ν to be the rectangle with vertices $\nu \pm iT$, $1 \pm iT$. Within R_ν the integrand has the following residues:

$$\frac{1}{2} + \frac{\log x}{\log 2} \quad \text{at} \quad s = 0,$$

$$\frac{1}{2\pi in} \exp\left(2\pi in \frac{\log x}{\log 2}\right) \quad \text{at} \quad s = \frac{2\pi in}{\log 2} \qquad (0 < |n| \le N).$$

It follows that

$$I(T) = \frac{\log x}{\log 2} + \frac{1}{2} + \sum_{n=1}^{N} \frac{1}{\pi n} \sin\left(\frac{2\pi n \log x}{\log 2}\right)$$

$$+ \frac{1}{2\pi i} \int_{R'_\nu} (1 - 2^{-s})^{-1} x^s s^{-1} ds.$$

Here R'_ν denotes the polygonal arc with successive vertices $1 - iT$, $\nu - iT$, $\nu + iT$ and $1 + iT$.

We can estimate the integral along R'_ν by noting that for any $x > 0$ and any $s \in R'_\nu$ we have

$$|x^s (1 - 2^{-s})^{-1}| = O\{(2x)^\sigma\}$$

provided that $\nu \leq -1$, say. Thus for $x > 1/2$, the integral along the left side of R'_ν goes to zero as $\nu \to -\infty$, for each $T > 0$. Also

$$\left| \frac{1}{2\pi i} \left\{ \int_{1-iT}^{-\infty-iT} + \int_{-\infty+iT}^{1+iT} \right\} \{x^s (1 - 2^{-s})^{-1} s^{-1} ds\} \right| = O\left(\frac{x}{N}\right).$$

Now we are going to insert the estimate of $I(T)$ into the formula of Theorem 7.10. Note however that we have used a special sequence of T's. This is immaterial since

$$T(N+1) - T(N) = 2\pi/\log 2 < 10,$$

and for any T' with $|T(N) - T'| < 10$ we have

$$\int_{1+iT}^{1+iT'} x^s (1 - 2^{-s})^{-1} s^{-1} ds = O(x/N).$$

It follows upon letting $N \to \infty$ that

$$\frac{1}{2}\{F(x+) + F(x-)\} = \frac{\log x}{\log 2} + \frac{1}{2} + \sum_{n=1}^{\infty} \frac{1}{\pi n} \sin\left(\frac{2\pi n \log x}{\log 2}\right)$$

for any $x > 1/2$. If we change the variable and set $y = 1 + (\log x)/\log 2$, we obtain formula (7.22) for $y > 0$. Formula (7.22) is trivially valid at $y = 0$ and holds for $y < 0$ since both sides are odd.

7.7.2   Bounded and uniform convergence

It remains to study the convergence of the series (7.22). By periodicity we may assume that $1 \leq y \leq 2$, say, or under the foregoing change of variable, $1 \leq x \leq 2$. We apply Lemma 7.12 to $\widehat{F}(s) = (1 - 2^{-s})^{-1}$, using the special sequence $T = (2N+1)\pi/\log 2$ and the calculation we have made for $I(T)$. We obtain

$$\frac{1}{2}\{F(xe^{-1/T}) + F(xe^{1/T})\}$$

$$= \frac{\log x}{\log 2} + \frac{1}{2} + \sum_{n=1}^{N} \frac{1}{\pi n} \sin\left(\frac{2\pi n \log x}{\log 2}\right)$$

$$+ \frac{1}{2\pi i}\left\{ \int_{1-iT}^{-\infty-iT} + \int_{-\infty+iT}^{1+iT} \right\}(1 - 2^{-s})^{-1}x^s s^{-1}\,ds$$

$$+ O\left(\frac{x}{T}\right)\left\{ \int_0^{xe^{-1/T}} + \int_{xe^{1/T}}^{\infty} \right\}\frac{y^{-1}}{|\log x/y|}\,dF(y)$$

$$+ O\left\{ \int_{xe^{-1/T}}^{xe^{1/T}} \left(T\left|\log\frac{x}{y}\right| + \frac{1}{T}\right)dF(y)\right\}. \tag{7.23}$$

The uniform bound for the partial sums

$$\sum_{n=1}^{\check{N}} \frac{1}{\pi n} \sin\left(\frac{2\pi n \log x}{\log 2}\right) \quad (1 \leq x \leq 2, \quad N = 1, 2, \ldots)$$

holds, since all the other terms in (7.23) are uniformly bounded. (Note that the quantity $T|\log x/y|$ is at most 1 in the last integral and at least 1 in the preceding ones.)

Let $0 < \delta < 1/2$. The uniform convergence of the Fourier series on $1 + \delta \leq y \leq 2 - \delta$ (equivalently, $2^{\delta} \leq x \leq 2^{1-\delta}$) also follows from (7.23). We note that

$$\frac{1}{2}\{F(xe^{-1/T}) + F(xe^{1/T})\} = F(1) = 1$$

provided that $T > (\delta \log 2)^{-1}$. Moreover, all the integrals occurring in (7.23) are either identically zero or tend to zero uniformly for $2^{\delta} \leq x \leq 2^{1-\delta}$ as $T = (2N+1)\pi/\log 2 \to \infty$. $\quad\square$

PROBLEM 7.14   Discuss the reasons for assuming that $x > 1/2$ in most of the preceding proof.

PROBLEM 7.15   Give a real variable proof that the Fourier series for $S(y)$ of Lemma 7.15 converges boundedly on $\mathbb{R}$ and uniformly away from integers. Hint. For y near an integer m, use the inequality

$$\sin|2\pi ny| \le 2\pi n|y - m|.$$

Elsewhere, use summation by parts.

PROBLEM 7.16   Let $S(y)$ denote the sawtooth function and for $N = 1, 2, \ldots$, let $S_N(y)$ denote the Nth partial sum of its Fourier series (7.22).

a. Show that $\int_0^1 \{S^2(y) - S_N^2(y)\}dy \to 0$ as $N \to \infty$.

b. By evaluating $\int_0^1 S^2(y)\,dy$ and $\int_0^1 S_N^2(y)\,dy$, give another proof of Theorem 1.5.

7.8   A "smoothed" Perron formula

Application of the inversion formula of Lemma 7.14 requires an upper bound for the difference $F(x\delta) - F(x/\delta)$ for suitable $\delta > 1$. There are, however, examples for which F is monotone but there exists no satisfactory *a priori* difference estimate for F. In these cases it is usually preferable to use a "smoothed" Perron formula such as

Theorem 7.16   *Let* $F \in \mathcal{V}$ *and* $\sigma_a(\widehat{F}) < \infty$. *Let* $b > \max(\sigma_c(\widehat{F}), 0)$. *Then for any* $x > 0$,

$$\frac{1}{2\pi i}\int_{b-i\infty}^{b+i\infty} x^s\widehat{F}(s)\frac{ds}{s^2} = \int_{1-}^x dF * \frac{dt}{t} = \int_1^x \frac{F(t)}{t}dt.$$

Remarks 7.17   This relation is the formal analogue of the Perron formula, with $\widehat{F}(s)/s = (dF * dt/t)\widehat{}$ in place of $\widehat{F}(s) = (dF)\widehat{}$. The additional factor of s in the denominator can improve integrability in the Perron formula. If we would benefit from additional s factors, we can apply the theorem inductively with $dF * (dt/t)^{*n}$ in place of $dF * (dt/t)^{*n-1}$, $n = 1, 2, \ldots$.

Proof.   Define $G \in \mathcal{V}$ by $G(x) := 0$ $(x < 1)$ and

$$G(x) := \int_{1-}^x dF * t^{-1}dt = \int_1^x F(t)t^{-1}dt \quad (x \ge 1).$$

Then we have $\widehat{G}(s) = \widehat{F}(s)/s$ for $\sigma > \max\{\sigma_c(\widehat{F}), 0\}$. If we apply Theorem 7.10 to G, we obtain

$$\lim_{T \to \infty} \frac{1}{2\pi i} \int_{b-iT}^{b+iT} x^s \frac{\widehat{F}(s)}{s^2} ds = \frac{1}{2}\{G(x+) + G(x-)\}.$$

The right side of the last equation can be replaced by $G(x)$, since G is continuous. The symmetric limit occurring on the left side of the equation is no longer needed, as we now show. This is clear if $b > \sigma_a(\widehat{F})$, for then $\widehat{F}(b+it) = O(1)$ uniformly for $-\infty < t < \infty$.

If $b \le \sigma_a$, we choose $b' > \sigma_a$ and form a rectangle R with vertices

$$b + iT, \quad b + iT', \quad b' + iT', \quad b' + iT,$$

where $T' > T > 0$ but T and T' are otherwise arbitrary. We have

$$\int_R x^s \widehat{F}(s) s^{-2} ds = 0$$

by Cauchy's theorem. Since $\widehat{F}$ is bounded on the right side of R and $o(T)$ and $o(T')$, (by Lemma 7.13) on the bottom and top sides respectively, the last integral taken over these three sides tends to zero as $T, T' \to \infty$. It follows that

$$\lim_{T, T' \to \infty} \int_{b+iT}^{b+iT'} x^s \widehat{F}(s) s^{-2} ds = 0.$$

The same reasoning applies for $T' < T < 0$, and thus we may take a nonsymmetric limit. $\qquad\square$

PROBLEM 7.17   Under the hypotheses of Theorem 7.16, prove that

$$\frac{1}{2\pi i} \int_{b-i\infty}^{b+i\infty} \frac{x^s \widehat{F}(s)}{s(s+1)} ds = \int_1^x \left(1 - \frac{t}{x}\right) dF(t), \quad 1 < x < \infty.$$

Usually we are interested in knowing an approximation of F itself rather than $G = \int dF * t^{-1} dt$. Since G is a kind of average of F, the passage from an estimate of G to one of F is a tauberian process. We can obtain an estimate of F by a differencing argument if F is monotone or if there exists a suitable estimate of differences of F. For G satisfying an asymptotic formula with an explicit error term and F increasing, we shall deduce a corresponding asymptotic formula for F by the methods of Lemma 5.2. We leave as an exercise the case in which estimates are known for differences of F.

Lemma 7.18 *Suppose F is a real valued monotone increasing function in $\mathcal{V}$ and*

$$G(x) = \int_1^x dF * t^{-1}dt = x^\alpha P(\log x) + O\{E(x)\},$$

where α is a positive real number, P is a polynomial with real coefficients not identically zero, and E is an increasing function satisfying $E(2x) \le KE(x)$ for some $K \ge 1$ and all $x \ge 1$ and $E(x) = o\{x^\alpha P(\log x)\}$. Then

$$F(x) = x^\alpha Q(\log x) + O\{\sqrt{x^\alpha P(\log x)E(x)}\},$$

where $Q(u) = \alpha P(u) + P'(u)$.

Proof. For $1 < x < y < ex$ we estimate

$$\int_x^y t^{-1}F(t)dt = G(y) - G(x) = f(\log y) - f(\log x) + O\{E(x)\}. \quad (7.24)$$

Here we have set $f(u) = e^{\alpha u}P(u)$. The difference on the right side of (7.24) can be represented by Taylor's formula centered at $\log x$. Setting $h = \log y - \log x$ we have

$$f(\log y) - f(\log x) = hf'(\log x) + \frac{h^2}{2}f''(\log z), \quad (7.25)$$

where $x < z < y$. We observe that $f'(u) = e^{\alpha u}Q(u)$. Also, for u large enough that $P(u)$ is positive, we have $f''(u) = O\{e^{\alpha u}P(u)\}$.

Since F is monotone increasing,

$$hF(x) \le \int_x^y t^{-1}F(t)dt \le hF(y).$$

By (7.24), (7.25), the first of the above monotonicity inequalities, and the slow growth of E, we obtain

$$F(x) \le x^\alpha Q(\log x) + O\{hx^\alpha P(\log x)\} + O\{h^{-1}E(x)\}.$$

A similar argument, involving a Taylor expansion about $\log y$ in place of $\log x$, leads to the opposite inequality

$$F(y) \ge y^\alpha Q(\log y) + O\{hy^\alpha P(\log y)\} + O\{h^{-1}E(y)\}.$$

The desired formula for F follows from last two inequalities and the choice

$$h = h(x) = (E(x)/\{x^\alpha P(\log x)\})^{1/2}. \qquad \square$$

PROBLEM 7.18   Suppose that the monotonicity condition of Lemma 7.18 is replaced by the difference estimate

$$|F(y) - F(x)| \leq (1 + y - x)x^{\alpha-1}\varphi(x)$$

valid for $1 \leq x < y \leq 2x$. Here φ is a monotone increasing function. Show first that $P(\log x) = O(\varphi(x))$; then show that

$$F(x) = x^\alpha Q(\log x) + O(\sqrt{x^\alpha \varphi(x) E(x)}) + O(x^{\alpha-1}\varphi(x)).$$

In most applications of a Perron inversion formula we shall deform the contour over which the integral is taken. It may happen that on the new integration path $\widehat{F}$ satisfies a bound $\widehat{F}(\sigma + iT) = O(1 + |t|^\alpha)$ with $\alpha \geq 1$. In this case we can repeat k times the technique of Theorem 7.16, where k is an integer greater than α. With the hypotheses of that theorem we have

$$\frac{1}{2\pi i}\int_{b-i\infty}^{b+i\infty} x^s F(s)s^{-k-1}ds = \int_1^x dF * (t^{-1}dt)^{*k}. \tag{7.26}$$

Suppose that the function F is monotone increasing and that we have found an asymptotic estimate for the right side of (7.26). Then we can obtain an estimate of F by k applications of Lemma 7.18. Consideration of the error term suggests that we should take k as small as possible.

7.9   Example. Estimation of $\sum T(1_2 * 1_3)$

Recall that $Tf(n) := nf(n)$ for any $f \in \mathcal{A}$. For p a fixed prime, the arithmetic function 1_p was defined in §2.3 by

$$1_p(n) := \begin{cases} 1 & \text{if } n = p^k, \ k = 0, 1, 2, \ldots, \\ 0 & \text{if } n \text{ is not a power of } p. \end{cases}$$

For $x \geq 1$ the summatory function of $T1_p$ satisfies

$$\sum_{n \leq x} T1_p(n) = \sum_{0 \leq \alpha \leq \frac{\log x}{\log p}} p^\alpha = \frac{p}{p-1}\exp\left\{\left[\frac{\log x}{\log p}\right]\log p\right\} - \frac{1}{p-1}.$$

It follows that

$$x^{-1} \sum_{n \le x} T1_p(n) = O(1), \tag{7.27}$$

but that the ratio has limit inferior $1/(p-1)$ and limit superior $p/(p-1)$ as $x \to \infty$.

A small calculation shows that

$$\sum_{n \le x} T(1_2 * 1_3)(n) = \sum_{\substack{n \le x \\ n=2^\alpha 3^\beta}} n = O(x \log x),$$

since $T(1_2 * 1_3)$ is the convolution of two nonnegative arithmetic functions, each of which satisfies (7.27). Here we establish the mildly surprising result that

$$F(x) := \sum_{\substack{n \le x \\ n=2^\alpha 3^\beta}} n \sim \frac{x \log x}{\log 2 \log 3}, \tag{7.28}$$

in spite of the fact that neither of the two convolution factors $T1_2$, $T1_3$ has a mean value.

Proof of (7.28). For $\sigma > 1$ we have

$$\widehat{F}(s) = \sum n(1_2 * 1_3)(n)n^{-s} = (1 - 2^{1-s})^{-1}(1 - 3^{1-s})^{-1}$$

by (6.4), since $T1_2$ and $T1_3$ are completely multiplicative functions. The singularities of $\widehat{F}$ are located at

$$s = 1 + \frac{2\pi i m}{\log 2} \quad \text{and} \quad s = 1 + \frac{2\pi i n}{\log 3} \quad (m, n \in \mathbb{Z}).$$

By the unique factorization theorem,

$$1 + \frac{2\pi i m}{\log 2} \neq 1 + \frac{2\pi i n}{\log 3} \quad \text{for } (m, n) \neq (0, 0).$$

Consequently $\widehat{F}$ has a double pole at $s = 1$, and all other singularities of $\widehat{F}$ are simple poles.

By Theorem 7.16 we have, for any $b > 1$,

$$\frac{1}{2\pi i} \int_{b-i\infty}^{b+i\infty} x^s (1 - 2^{1-s})^{-1} (1 - 3^{1-s})^{-1} \frac{ds}{s^2} = \int_1^x \frac{F(t)}{t} dt.$$

For suitable positive T we shall evaluate the left hand integral over the line segment $b - iT$, $b + iT$ by deforming the contour to the left and applying the residue theorem. Then we shall show that the integral taken over the remaining part of the path is suitably small.

To carry out this program we must first select T's for which $1 + iT$ is reasonably far from a pole of $\widehat{F}$. Each open segment of length $2\pi/\log 3$ on the line $\sigma = 1$ contains at most two poles of $\widehat{F}$. Thus, each such segment contains an open subsegment of length $2\pi/(3\log 3)$ which is free of poles of $\widehat{F}$. We choose $T > 0$ so that $1 \pm iT$ lies at the midpoint of such a subsegment and hence at a distance at least $\pi/(3\log 3)$ from each pole of $\widehat{F}$.

For $\nu < 0$, $b > 1$, and $T > 0$ as specified above, let R be the rectangle with vertices $\nu \pm iT$ and $b \pm iT$. On the top, bottom and left sides of R we have $x^s \widehat{F}(s) = O(x^\sigma)$. The constant implied by the O is absolute. The sum of the residues of $x^s \widehat{F}(s)s^{-2}$ in R equals $x \log x/(\log 2 \log 3) + x P_T(x)$, where P_T is a bounded function for each fixed T. Applying the residue theorem and letting $\nu \to -\infty$ we obtain

$$\frac{1}{2\pi i} \int_{b-iT}^{b+iT} x^s \widehat{F}(s) \frac{ds}{s^2} = \frac{x \log x}{\log 2 \log 3} + O_T(x) + O(x^b T^{-2}),$$

provided $x \geq 2$, say. The constant implied by the second O-term is absolute.

Setting $b = 1 + \epsilon$, we estimate

$$I_\epsilon(T) := \int_{1+\epsilon+iT}^{1+\epsilon+i\infty} x^s (1 - 2^{1-s})^{-1}(1 - 3^{1-s})^{-1}s^{-2}ds.$$

The unpleasant feature of this integral is that two singularities of $\widehat{F}$ can occur arbitrarily close together. We separate the factors of $\widehat{F}$ by using the Cauchy-Schwarz inequality. Setting $s = 1 + \epsilon + it$, we have

$$|I_\epsilon(T)|^2 \leq x^{2+2\epsilon} \int_T^\infty \left|1 - 2^{-\epsilon-it}\right|^{-2} \frac{dt}{t^2} \int_T^\infty \left|1 - 3^{-\epsilon-it}\right|^{-2} \frac{dt}{t^2}$$

$$=: x^{2+2\epsilon} I_\epsilon'(T)I_\epsilon''(T), \quad \text{say.}$$

We shall estimate $I_\epsilon'(T)$ as a sum over intervals of length $\pi/\log 2$. Suppose first that

$$b(m) := \frac{4\pi m + \pi}{2\log 2} \leq t < \frac{4\pi m + 3\pi}{2\log 2} =: c(m)$$

for some $m \in \mathbb{Z}^+$. Then we have

$$|1 - 2^{-\epsilon - it}| \geq |\Re\{1 - 2^{-\epsilon - it}\}| \geq 1$$

and hence

$$\int_{b(m)}^{c(m)} |1 - 2^{-\epsilon - it}|^{-2} t^{-2} dt < \frac{4 \log 2}{\pi (4m + 1)^2}.$$

Next suppose that

$$a(m) := \frac{4\pi m - \pi}{2 \log 2} \leq t < \frac{4\pi m + \pi}{2 \log 2} = b(m),$$

and let $\tau = t \log 2 - 2\pi m$. Then we have

$$|1 - 2^{-\epsilon - it}|^2 = \{1 - 2^{-\epsilon} \cos(t \log 2)\}^2 + 2^{-2\epsilon} \sin^2(t \log 2)$$

$$\geq (1 - 2^{-\epsilon})^2 + 2^{-2\epsilon} \sin^2 \tau \geq c(\epsilon^2 + \tau^2)$$

for some $c > 0$, uniformly for $0 < \epsilon < 1/2$ and $-\pi/2 \leq \tau < \pi/2$. (We have used the inequality $\sin u > 2u/\pi$ for $0 < u < \pi/2$.) It follows that

$$\int_{a(m)}^{b(m)} |1 - 2^{-\epsilon - it}|^{-2} \frac{dt}{t^2} < \frac{4 \log 2}{c\pi^2 (4m - 1)^2} \int_{-\pi/2}^{\pi/2} \frac{d\tau}{\epsilon^2 + \tau^2}$$

$$< \frac{K}{\pi (4m - 1)^2} \int_{-\infty}^{\infty} \frac{d\tau}{\epsilon^2 + \tau^2} = \frac{K}{\epsilon (4m - 1)^2},$$

for some constant K.

If we set $m_0 = [(2T \log 2 + \pi)/(4\pi)]$, then

$$I'_\epsilon(T) \leq \sum_{m=m_0}^{\infty} \left\{ \frac{4 \log 2}{\pi (4m + 1)^2} + \frac{K}{\epsilon (4m - 1)^2} \right\} = O\left(\frac{1}{\epsilon T} \right),$$

uniformly for $0 < \epsilon < 1/2$ and $T > 10$, say. A similar estimate is valid for $I''_\epsilon(T)$. Thus we have shown that

$$I_\epsilon(T) = O\{x^{1+\epsilon}/(\epsilon T)\} \tag{7.29}$$

uniformly for $0 < \epsilon < 1/2$ and $T > 10$.

Choosing $\epsilon = \epsilon(x) = 1/\log x$, we obtain the estimate

$$\frac{1}{2\pi i} \left\{ \int_{1+\epsilon-i\infty}^{1+\epsilon-iT} + \int_{1+\epsilon+iT}^{1+\epsilon+i\infty} \right\} x^s \widehat{F}(s) s^{-2} ds = O(x T^{-1} \log x).$$

We have used the reflection principle for the estimate on the ray

$$\{s = 1 + \epsilon + it : \; -\infty < t < T\}.$$

The contributions of the horizontal lines at height $\pm T$ are $x^{1+\epsilon}/T^2 \ll x$. Together, we have shown that

$$\int_1^x \frac{F(t)}{t}\,dt = \frac{x \log x}{\log 2 \log 3} + O_T(x) + O(xT^{-1}\log x).$$

The restriction that T lie in a special sequence can be dropped in view of the estimate (7.29). Since T can be taken arbitrarily large, we obtain

$$\int_1^x \frac{F(t)}{t}\,dt = \frac{1 + o(1)}{\log 2 \log 3} x \log x.$$

Finally, we can deduce the desired asymptotic formula (7.28) from the above relation and Lemma 7.18. $\qquad\square$

PROBLEM 7.19 Show that $\sum_{n \le x} T(1_2 * 1_2)(n) = O(x \log x)$ but does not have an asymptotic approximation of the form $cx \log x$. Hint. The sum can be expressed in closed form.

PROBLEM 7.20 (1) Show by a lattice point counting argument that

$$\sum_{n \le x}(1_2 * 1_3)(n) = \frac{\log^2 x}{2 \log 2 \log 3} + O(\log x).$$

Does this estimate lead to another proof of (7.28)?

(2) It is known that actually the sharper estimate

$$\sum_{n \le x}(1_2 * 1_3)(n) = \frac{\log^2 x}{2 \log 2 \log 3} + b \log x + o(\log x)$$

holds for a certain constant b. Does this result imply (7.28)?

7.10 Notes

7.2. The Wiener-Ikehara theorem first appeared (in a rather different form) in J. Math. and Phys. M.I.T., vol. 10 (1931), pp. 1–12, and Annals of Math. (2), vol. 33 (1932), pp. 1–100, particularly pp. 44–50.

The proof given here is based primarily on the following sources:

- S. Bochner, Math. Zeit., vol. 37 (1933), pp. 1–9
- H. Heilbronn and E. Landau, Math. Zeit., vol. 37 (1933), pp. 10–16
- E. Landau, Sitzungsber. Preuss. Akad. Wiss. (Berlin) (1932), pp. 514–521

The last two papers are contained also in volume 9 of [LanC], pp. 207–221. This proof appears also in other works, e.g. [ChanI], pp. 124–128.

There are versions of the Wiener-Ikehara theorem with a weaker assumption on φ than continuity on $\{s : \Re s \geq 1\}$. Two are given in [BaD] and in [Ten], §II.7.5.

The sharpest O-estimate known to hold when the Wiener-Ikehara hypotheses are satisfied only in a strip $\{s : \sigma \geq \alpha, |t| < 2\lambda_0\}$ is that of S. W. Graham and J. D. Vaaler, Trans. Amer. Math. Soc., vol. 265 (1981), pp. 283–302.

The asymptotic formula for $G(x)$ obtained in §7.2.1 is due to L. Kalmár, Acta Szeged, vol. 5 (1931), pp. 95–107.

7.3. The Wiener-Ikehara theorem provided the first proof of the P.N.T. whose only ingredient—other than the meromorphy of the zeta function on a domain including the half plane $\{s : \sigma \geq 1\}$—is its nonvanishing on the line $\{s : \sigma = 1\}$; previous arguments had required, in addition, growth estimates involving ζ on or near this line. The Wiener-Ikehara theorem and Corollary 7.2 are the basis of the assertion that the P.N.T. is "equivalent" to the nonvanishing of zeta on the line $\{s : \sigma = 1\}$.

7.4. A proof of this generalization of the Wiener-Ikehara theorem was given by H. Delange, Ann. sci. Ecole Norm. Sup. (3), vol. 71 (1954), pp. 213–242. For other generalizations, see e.g. [Ten], §7.5.

7.5–7.6. The idea of the Perron inversion formula (Th. 7.10) goes back at least to Riemann, but the first rigorous published proofs of the general result were given by J. Hadamard, Rend. Circ. Mat. Palermo, vol. 25 (1908), pp. 326–330 and 395–396, and by O. Perron, J. reine angew. Math., vol. 134 (1908), pp. 95–143. Perron's hypotheses were slightly weaker than those of Hadamard. Perron refers to the result as the Kronecker-Cahen formula! For further historical remarks on this theme, see §5.2 of [Nark] or p. 729 of the article of H. Bohr and H. Cramér, Die neuere Entwicklung der analytischen Zahlentheorie, Encykl. math. Wiss. II C 8 (1923), pp. 722–849; also in vol. 3 of H. Bohr's Collected Math. Works, Dansk Mat. Forening, København, 1952.

7.8. The smoothed version of the Perron formula given here, in the special case $F = \psi$, the Chebyshev function, was introduced by Hadamard in his proof of the P.N.T.

7.9. Refinements and generalizations of the assertions of Problem 7.20 are treated in Chapter 5 of [HarR].

Chapter 8

The Riemann Zeta Function

8.1 The functional equation

We have seen that zeta is closely related to prime number problems and that it is a factor of many generating functions. Thus ζ ranks among the most important functions in number theory. The main problems we shall consider here are to determine regions of analyticity for ζ and $1/\zeta$ and to estimate the magnitude of each. This information will help to justify such analytic processes as contour deformation in applying the Perron formula (Lemma 7.14) and to obtain qualitative and numerical estimates.

In §6.5.2 we showed that ζ is analytic in $\{s : \sigma > -1\}$ except for a simple pole at $s = 1$. We shall extend ζ as an analytic function on $\mathbb{C} \setminus \{1\}$ by means of the *functional equation*, which connects $\zeta(s)$ with $\zeta(1 - s)$.

Theorem 8.1 (Asymmetric functional equation). *Zeta is an analytic function on $\mathbb{C} \setminus \{1\}$ and satisfies there the functional equation*

$$\zeta(s) = 2^s \pi^{s-1} \sin(\pi s/2) \Gamma(1 - s) \zeta(1 - s).$$

Proof. We integrate by parts the M.t. defining zeta on the half plane $\{s : \sigma > 1\}$ and add and subtract $s(s-1)^{-1} - 1/2$ to obtain

$$\zeta(s) = s \int_1^\infty \frac{N(x) - x + 1/2}{x^{s+1}} dx + \frac{s}{s - 1} - \frac{1}{2}.$$

The integral defines an analytic function on $\{s : \sigma > -1\}$. This can be seen by integration by parts as in (6.13). Now for $\sigma < 0$,

$$s \int_0^1 \frac{N(x) - x + 1/2}{x^{s+1}} dx = -s \int_0^1 x^{-s} dx + \frac{s}{2} \int_0^1 x^{-s-1} dx = \frac{s}{s - 1} - \frac{1}{2}.$$

Thus, for $-1 < \sigma < 0$ we have

$$\zeta(s) = s \int_0^\infty x^{-s-1}(N(x) - x + 1/2)dx.$$

We have shown in (7.22) that for positive $x \notin \mathbb{Z}$,

$$N(x) - x + \frac{1}{2} = \sum_{n=1}^\infty \frac{\sin 2\pi n x}{\pi n}.$$

We combine the last two formulas and interchange the sum and integral. Assuming for the moment the validity of this operation, we have

$$\zeta(s) = s \sum_{n=1}^\infty \int_0^\infty \frac{\sin 2\pi n x}{\pi n} x^{-s-1} dx \quad (-1 < \sigma < 0).$$

We change the variable in the integral, obtaining

$$\zeta(s) = s \sum_{n=1}^\infty \frac{(2\pi n)^s}{\pi n} \int_0^\infty u^{-s-1} \sin u \, du \quad (-1 < \sigma < 0). \tag{8.1}$$

We evaluate the integral for s real and $-1 < s < 0$ by applying Cauchy's theorem to $e^{iz}z^{-s-1}$. Let C be a contour consisting of line segments

$$\{z = u : 0 < \epsilon \le u \le R < \infty\}, \quad \{z = iu : R \ge u \ge \epsilon\}$$

and arcs

$$\{z = Re^{i\theta} : 0 \le \theta \le \pi/2\}, \quad \{z = \epsilon e^{i\theta} : \pi/2 \ge \theta \ge 0\}$$

traversed in the positive sense, and we obtain $\int_C e^{iz}z^{-s-1}dz = 0$. This integral taken over the arc of radius ϵ tends to zero with ϵ (this is easy to see); over the arc of radius R, it is majorized by

$$\int_0^{\pi/2} e^{-R\sin\theta} R^{-s} d\theta < \int_0^\infty e^{-2R\theta/\pi} R^{-s} d\theta = \frac{\pi}{2} R^{-s-1} = o(1), \quad R \to \infty.$$

It follows that

$$\int_0^\infty e^{iu} u^{-s-1} du = \int_0^\infty e^{-u} u^{-s-1} e^{-s\pi i/2} du = \Gamma(-s)e^{-s\pi i/2}$$

or, upon taking imaginary parts (with s still real and $-1 < s < 0$),

$$\int_0^\infty u^{-s-1} \sin u \, du = \sin(-s\pi/2)\Gamma(-s). \tag{8.2}$$

By analytic continuation, (8.2) continues to hold for complex s in the strip $-1 < \sigma < 0$. (The last integral converges by (8.4) below.)

For $-1 < \sigma < 0$, we obtain

$$\zeta(s) = 2^s \pi^{s-1} \zeta(1-s) \sin(s\pi/2) \Gamma(1-s)$$

by combining (8.1), (8.2), and the recurrence formula for the gamma function. The right side of the last equation defines an analytic function on the half plane $\{s : \sigma < 0\}$. This provides a continuation of zeta to $\mathbb{C} \setminus \{1\}$.

It is easy to extend the range of validity of this functional equation. Define F on $\mathbb{C} \setminus \mathbb{Z}^+$ by

$$F(s) := (s-1)\left\{\zeta(s) - 2^s \pi^{s-1} \sin\frac{\pi s}{2}\Gamma(1-s)\zeta(1-s)\right\},$$

and define F by continuity on the positive integers. F is entire and vanishes on the half plane $\{s : \sigma < 0\}$. By the uniqueness theorem for analytic functions, $F = 0$, and hence the functional equation holds on $\mathbb{C} \setminus \{1\}$.

8.1.1 Justification of the interchange of $\sum$ and $\int$

We have seen in Lemma 7.15 that the partial sums of the Fourier series of the sawtooth function S are uniformly bounded on $\mathbb{R}$ and uniformly convergent to S on $\bigcup_{\ell=-\infty}^{\infty} [\ell + \delta, \ell + 1 - \delta]$ for any fixed $\delta \in (0, 1/2)$. We have then ([Apos], Theorem 9.11) for any finite X and $-1 < \sigma < 0$,

$$\int_0^X \sum_{n=1}^\infty \frac{\sin 2\pi n x}{\pi n} x^{-s-1} dx = \sum_{n=1}^\infty \int_0^X x^{-s-1} \frac{\sin 2\pi n x}{\pi n} dx. \qquad (8.3)$$

Also, for $n \geq 1$, $-1 < \sigma < 0$, and $0 < X < Y < \infty$, we have

$$\int_X^Y x^{-s-1} \sin 2\pi n x \, dx = O(X^{-\sigma-1}/n) \qquad (8.4)$$

by integration by parts. For fixed s, the constant is uniform with respect to X, Y, and n. Thus the right side of (8.3) equals

$$\sum_{n=1}^\infty \int_0^\infty x^{-s-1} \frac{\sin 2\pi n x}{\pi n} dx + O\left(\sum_{n=1}^\infty X^{-\sigma-1} n^{-2}\right),$$

and the desired formula follows upon letting $X \to \infty$. $\qquad \square$

The zeta function is well understood on any fixed half plane $\{s : \sigma \geq 1 + \epsilon > 1\}$, where it has an absolutely and uniformly convergent D.s. The functional equation together with Stirling's formula (cf. Appendix) yield estimates of zeta on $\{s : \sigma \leq -\epsilon < 0\}$. There are several important problems that could be solved with sufficient knowledge of zeta in the rest of the complex plane, but here information is most difficult to obtain. The region $\{s : 0 \leq \sigma \leq 1\}$ is called the *critical strip* for zeta.

PROBLEM 8.1 Use Theorem 8.1 to evaluate $\zeta(0)$.

PROBLEM 8.2 The gamma and zeta functions are intimately connected. By substituting the zeta functional equation into itself, establish the gamma function reflection formula, $\Gamma(z)\Gamma(1-z) = \pi / \sin \pi z$, $z \in \mathbb{C} \setminus \mathbb{Z}$.

8.1.2 Symmetric form of the functional equation

The functional equation for zeta has various forms which can be deduced from each other by means of identities for the gamma function. Of particular use is the following symmetric form. Define ξ on $\mathbb{C}$ by

$$\xi(s) := (1/2)s(s-1)\pi^{-s/2}\Gamma(s/2)\zeta(s).$$

Theorem 8.2 ξ *is an entire function and satisfies the functional equation* $\xi(1-s) = \xi(s)$. *Also,* ξ *is real valued on the lines* $\{s : t = 0\}$ *and* $\{s : \sigma = \frac{1}{2}\}$.

Proof. We combine the asymmetric form of the functional equation given in the preceding theorem with two identities for the gamma function:

$$\Gamma\left(\frac{z}{2}\right)\Gamma\left(\frac{z+1}{2}\right) = \sqrt{\pi}\, 2^{1-z}\Gamma(z) \qquad \text{(multiplication)},$$

$$\Gamma(z)\Gamma(1-z) = \pi / \sin \pi z \qquad \text{(reflection)}.$$

We obtain

$$\zeta(s) = 2^s \pi^{s-1} \sin \frac{\pi s}{2} \Gamma(1-s)\zeta(1-s)$$

$$= \pi^{s-3/2} \sin \frac{\pi s}{2} \Gamma\left(\frac{1-s}{2}\right)\Gamma\left(1 - \frac{s}{2}\right)\zeta(1-s)$$

$$= \pi^{s-1/2}\Gamma\left(\frac{1-s}{2}\right)\Gamma\left(\frac{s}{2}\right)^{-1}\zeta(1-s).$$

The last equation yields the relation $\xi(s) = \xi(1-s)$.

The function ξ is entire, as can be seen by examining it on the half plane $\{s : \sigma > 0\}$, say. Its only factor having a singularity here is zeta, which has a simple pole at 1. This pole is cancelled by the zero of the factor $s - 1$.

Each factor of ξ is real on the positive real axis and hence, by the functional equation, ξ is real on $\mathbb{R}$. By the reflection principle we then have $\xi(\bar{s}) = \overline{\xi(s)}$. This fact and the functional equation imply that

$$\xi(1/2 + it) = \xi(1 - \{1/2 + it\}) = \xi(1/2 - it) = \overline{\xi(1/2 + it)},$$

and hence ξ is real for $\sigma = 1/2$. $\qquad\qquad\qquad\qquad\qquad\qquad\qquad\square$

An elegant direct proof of the symmetric functional equation for ζ can be based on the functional equation of the theta function (cf. [TiHB]). The theta functional equation is presented in §A3 of the Appendix.

8.2 *O*-estimates for zeta

It is easy to see that $|\zeta(s)| \leq \zeta(1 + \epsilon) < \infty$ for $\sigma \geq 1 + \epsilon > 1$ and that $\zeta(s) = 1 + O(2^{-\sigma})$ as $\sigma \to +\infty$. We shall estimate $|\zeta(s)|$ for $\sigma < 0$ by using the functional equation and the following special form of Stirling's formula:

Lemma 8.3 *There exists a constant K such that*

$$|\Gamma(\sigma + it)| \leq K e^{K|\sigma|} |t|^{\sigma - 1/2} e^{-\pi|t|/2}$$

on the set $\{s : |\sigma| + 2 \leq |t| < \infty\}$.

The proof follows immediately from the general version of Stirling's formula (cf. Appendix) and the fact that

$$t \arg(\sigma + it) - t\pi/2 = -\sigma \frac{\arctan(\sigma/t)}{\sigma/t} = O(|\sigma|)$$

uniformly on the set.

By Theorem 8.1 and Lemma 8.3 we have for any fixed $a > b > 0$

$$\zeta(\sigma + it) = O\big(|t|^{1/2 - \sigma}\big) \qquad\qquad (8.5)$$

on $\{s : -a \leq \sigma \leq -b < 0, \ |t| \geq 1\}$. The O constant depends on a and b.

As was noted earlier, the region in which we know the least about zeta is in and near the critical strip. The following lemma gives some simple estimates here which will be useful.

Lemma 8.4 *Let $0 < \delta < 1$. Then*

$$\zeta(s) = O(\log t) \qquad (\sigma \geq 1, \ t \geq 2),$$

$$\zeta(s) = O_\delta(t^{1-\delta}) \qquad (\sigma \geq \delta, \ t \geq 2),$$

$$\zeta'(s) = O(\log^2 t) \qquad (\sigma \geq 1, \ t \geq 2).$$

Proof. We estimate trivially the initial terms of the D.s. for zeta, because, for some large t, many of the initial terms n^{-s} might have nearly the same argument (mod 2π). However, the later terms of the series are distributed rather regularly and can be estimated better by partial summation.

Let $t \geq 2$ and $X \in \mathbb{Z}^+$. By manipulation and analytic continuation

$$\zeta(s) = \sum_1^X n^{-s} + \int_X^\infty x^{-s}(dN - dx) + \int_X^\infty x^{-s} dx \qquad (\sigma > 1)$$

$$= \sum_1^X n^{-s} + s \int_X^\infty \frac{N(x) - x}{x^{s+1}} dx + \frac{X^{1-s}}{s-1} \qquad (\sigma > 0).$$

Thus

$$|\zeta(s)| \leq \sum_1^X n^{-\sigma} + \frac{\sigma + t}{\sigma} X^{-\sigma} + t^{-1} X^{1-\sigma} \qquad (\sigma > 0).$$

If we now choose $X = [t]$, we obtain

$$|\zeta(s)| \leq \log t + 4 \quad (\sigma \geq 1, \ t \geq 2),$$

$$|\zeta(s)| \leq \int_0^X u^{-\delta} du + 2X^{-\delta} + (t/\delta) X^{-\delta}$$

$$\leq \frac{t^{1-\delta}}{1-\delta} + 2 + \frac{2t^{1-\delta}}{\delta} \qquad (\sigma \geq \delta, \ t \geq 2).$$

We can estimate $|\zeta'|$ in a similar fashion, by starting with the D.s. $\zeta'(s) = -\sum n^{-s} \log n$. $\qquad\square$

PROBLEM 8.3 Show that

$$\zeta(s) = O(\log t) \quad (t \geq 2, \ \sigma \geq 1 - \log^{-1} t).$$

Use this estimate and Cauchy's inequality for the coefficients of a power series to provide an alternative estimate of $|\zeta'(s)|$ on $\{s : \sigma \geq 1, \ t \geq 2\}$.

PROBLEM 8.4   For $1/2 \leq \sigma \leq 1$ and $t \geq 2$, show that

$$\zeta(s) = O(t^{1-\sigma} \log t).$$

PROBLEM 8.5   Let $0 \leq \sigma \leq 1/2$, $t \geq 2$. By using the functional equation for zeta, show that

$$\zeta(s) = O(t^{1/2} \log t).$$

It is convenient to discuss O-estimates for zeta in terms of the so-called Lindelöf μ function, which is defined for real σ by

$$\mu(\sigma) := \inf\{b \in \mathbb{R} : \zeta(\sigma + it) = O(t^b) \text{ as } t \to \infty\}.$$

We have $\mu(\sigma) = 0$ for $\sigma > 1$ and $\mu(\sigma) = 1/2 - \sigma$ for $\sigma < 0$ by (8.5). Lemma 8.4 implies that $\mu(\sigma) \leq 1 - \sigma$ for $0 \leq \sigma \leq 1$, and the last problem yields the improved estimate $\mu(\sigma) \leq 1/2$ for $0 \leq \sigma \leq 1/2$.

There are better estimates of μ in the interval $(0, 1)$ (cf. [TiHB], [Ivic], [KarVo]). It is known that μ is a continuous, nonnegative, nonincreasing function with a nonnegative second difference (cf. [TiTF], §§6.65, 9.41), but its precise nature is unknown. The *Lindelöf hypothesis* asserts that the graph $y = \mu(\sigma)$ consists of the straight lines

$$y = 1/2 - \sigma \ \ (\sigma \leq 1/2) \ \text{ and } \ y = 0 \ \ (\sigma \geq 1/2).$$

8.3   Zeros of zeta

The M.t. associated with the Chebyshev ψ function is

$$\int x^{-s} d\psi(x) = -\zeta'(s)/\zeta(s).$$

We shall estimate $\psi(x) - x$ for $x \to \infty$ by using the Perron formula and contour deformation. To this end we show that there is a region to the left of the line $\sigma = 1$ in which zeta has no zeros and hence ζ'/ζ is analytic.

We have seen that $\zeta(s) \neq 0$ for $\Re s \geq 1$. However there *are* zeros of zeta elsewhere in $\mathbb{C}$, as we now show. If we apply the functional equation

$$\zeta(s) = 2^s \pi^{s-1} \sin(\pi s/2) \Gamma(1 - s) \zeta(1 - s)$$

for $\sigma < 0$, we see that all the factors on the right side are zerofree except the sine, which vanishes at the negative even integers (where the other factors have no poles). Thus, the only zeros of zeta in $\{s : \sigma < 0\}$ occur at the negative even integers. These are called the *trivial zeros* of zeta.

This definition suggests that there are some other zeros of zeta. We fulfill this expectation in

Theorem 8.5 *There are infinitely many zeros of zeta in the critical strip.*

(This result clearly follows also from the quantitative estimates of §8.9.) It is convenient for the proof to use the function ξ in place of ζ, for ξ is entire and has a simpler functional equation. This is justified, since by the definition of ξ, it and ζ have the same zeros (if any) in the critical strip and ξ has no zeros outside the critical strip. We require an estimate of $|\xi|$ to prove the theorem. Let $M(r) := \max\{|\xi(s)| : |s| = r\}$.

Lemma 8.6 $\log M(r) \sim (1/2)r \log r$.

Proof of Lemma 8.6. Lemma 8.4 implies that $|\zeta(s)| \leq K|s|^{1/2}$ on

$$S := \{s : \sigma \geq 1/2, \, |s| \geq 3\}.$$

Also, Stirling's formula is valid on S. It follows that

$$|\xi(s)| \leq \exp\{(1/2)|s \log(s/2)| + A|s|\} \leq \exp\{(1/2)r \log r + A'r\}$$

holds on S (where $r = |s|$), for here $|\log(s/2)| \leq \log|s| - \log 2 + \pi/2$. We can extend the inequality for $\xi(s)$ to all s with $|s| \geq 4$, say, since by the functional equation, for $\sigma < 1/2$,

$$|\xi(s)| = |\xi(1-s)| \leq \exp\{(1/2)|1-s| \log|1-s| + A'|1-s|\}$$

$$\leq \exp\{(1/2)r \log r + A''r\}.$$

On the other hand, for $r \geq 2$,

$$M(r) \geq \xi(r) \geq \pi^{-r/2}\Gamma(r/2) > \exp\{(1/2)r \log r - Kr\}.$$

The asymptotic formula for $M(r)$ follows from these bounds. $\quad\square$

Proof of Theorem 8.5. By the Hadamard factorization theorem (cf. [TiTF], §8.24, [Con], pp. 287–290, or [Ing], pp. 51–56) we can write

$$\xi(s) = e^{\varphi(s)} \prod \{(1 - s/\rho)e^{s/\rho}\},$$

where φ is a polynomial of degree at most 1 and ρ ranges over the zeros of ξ, with appropriate repetition of any multiple zeros (none are known to exist!). If ξ had only a finite number of zeros, then its growth order would equal that of e^φ. However,

$$\max_{|z|=r} |e^{\varphi(z)}| \le e^{Ar} = o(e^{(1/2)r \log r}).$$

It follows that ξ has an infinite number of zeros, which must all be located in the critical strip. □

The preceding argument does not use the full strength of the factorization theorem. It is reasonable to ask whether we could have proved Theorem 8.5 by using a weaker result. In fact we could have done so by appealing to Lemma 8.9 of Borel-Carathéodory, which is itself an ingredient in the usual proof of the Hadamard theorem.

If $\rho = \beta + i\gamma$ denotes a zero of $\zeta(s)$ with $\beta \in (0,1)$ and some real γ, then by the functional equation $\zeta(1 - \beta - i\gamma) = 0$ also. Clearly, one of β, $1 - \beta$ is $\ge 1/2$, so

$$\sup\{\beta \in \mathbb{R} : \zeta(\beta + i\gamma) = 0\} \ge 1/2.$$

The *Riemann hypothesis*, briefly: R.H., is the assertion that all nonreal zeros of zeta lie on $\{s : \sigma = 1/2\}$, the so-called *critical line*. This conjecture, to date still unproved, is among the outstanding problems in mathematics.

The presence of zeros of zeta in the critical strip is reflected in theoretical limits for the quality of estimation of many arithmetic functions. This is arithmetic information which is generally unobtainable except by analytic methods. We shall study this theme in Chapter 11. Here we give a simple oscillation result for the Chebyshev ψ function.

Theorem 8.7 $\quad \psi(x) - x \ne o(x^{1/2})$.

Proof. Define $F \in \mathcal{V}$ for $x \ge 1$ by $F(x) = \psi(x^2) - x^2 + 1$. We have

$$\widehat{F}(s) = \int x^{-s} \{d\psi(x^2) - d(x^2)\} = -\frac{\zeta'}{\zeta}\left(\frac{s}{2}\right) - \frac{2}{s-2},$$

and $\widehat{F}$ is meromorphic on $\mathbb{C}$. If $F(x) = o(x)$, then it would follow from Lemma 7.1 that $\widehat{F}$ is analytic on the closed half plane $\{s : \sigma \ge 1\}$, i.e. that zeta has no zeros in $\{s : \sigma \ge 1/2\}$. This is false. □

PROBLEM 8.6 Use Landau's oscillation theorem (Th. 6.31) to prove the weaker result that $\psi(x) - x \ne O(x^\alpha)$ for any (fixed) $\alpha < 1/2$. Hint. Consider expressions $\psi(x) - x + Cx^\alpha$.

PROBLEM 8.7   Show that $M(x) := \sum_{n \leq x} \mu(n) \neq o(x^{1/2})$.

8.4   A zerofree region for zeta

We have seen so far that $\zeta(s) \neq 0$ in the closed half plane $\{s : \sigma \geq 1\}$. Here we exhibit a specific (but thin) region to the left of the line $\sigma = 1$ in which zeta is zerofree. Precisely, we prove

Theorem 8.8   *There is a constant $K > 0$ such that $\zeta(s) \neq 0$ in the region*

$$\{\sigma + it : \sigma > 1 - K/\log(|t| + 2)\}.$$

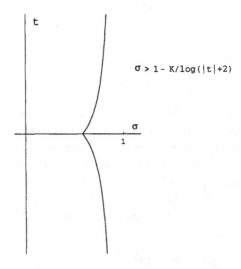

Fig. 8.1   ZEROFREE REGION OF TH. 8.8   (Not to scale!)

If the R.H. is true, then the optimal zerofree region is the half plane $\{s : \sigma > 1/2\}$. What is presently known is that zeta has no zeros in

$$\left\{s : \sigma > 1/2, \ |t| \leq 70 \cdot 10^9\right\} \cup \left\{s : \sigma > 1 - \frac{.01737}{\log^{2/3} |t| \log \log^{1/3} |t|}\right\}.$$

The rectangular region is obtained by an explicit count of zeros on the line $\Re s = 1/2$ combined with a theoretical counting method which we shall describe in §8.9; the other region is found by estimates of zeta using the method of trigonometric sums (cf. Notes).

In §§8.4-8.8 we present a proof of the P.N.T. with a remainder term. We shall use the preceding zerofree region for zeta to justify contour deformation in the Perron formula. Theorem 8.7 shows that the nontrivial zeros of zeta impose limitations on possible O-estimates for $|\psi(x) - x|$; we give upper estimates of this quantity based on the width of the zerofree region.

Our arguments do not depend on the functional equation or on global function theoretic properties of zeta, but rather are based on a local method of Landau. This result, in turn (and some related material below), depends on the following lemma, which estimates the modulus of an analytic function in terms of its real part.

Lemma 8.9 (Borel-Carathéodory). *Suppose that f is analytic in the disc $\{z : |z - z_0| < R\}$ and satisfies there the one sided inequality $\Re f(z) \leq U$. Then, for $|z - z_0| \leq r < R$ we have*

$$|f(z) - f(z_0)| \leq \frac{2r}{R - r}\{U - \Re f(z_0)\}$$

and

$$|f^{(\nu)}(z)|/\nu! \leq \frac{2R}{(R - r)^{\nu+1}}\{U - \Re f(z_0)\} \qquad (\nu \in \mathbb{Z}^+).$$

Proof of Lemma 8.9. Without loss of generality, say $z_0 = 0$. Let

$$g(z) := U - f(z) =: \sum c_n z^n =: P(z) + iQ(z)$$

for real valued functions P and Q, and let $0 < r < R$. Two applications of Cauchy's formula yield

$$c_n = \frac{1}{2\pi i} \int_{|z|=r} g(z)z^{-n-1}dz, \quad n \geq 0, \tag{8.6}$$

$$0 = \frac{1}{2\pi i} \int_{|z|=r} g(z)z^{n-1}dz, \quad n \geq 1. \tag{8.7}$$

Adding the conjugate of (8.7) to (8.6) we obtain (in polar form)

$$r^n c_n = \frac{1}{2\pi} \int_0^{2\pi} \{g(re^{i\theta}) + \overline{g(re^{i\theta})}\}e^{-in\theta}d\theta, \quad n \geq 1.$$

By hypothesis, $g(z) + \overline{g(z)} = 2P(z) \geq 0$. Thus

$$r^n|c_n| \leq \frac{2}{2\pi} \int_0^{2\pi} P(re^{i\theta})d\theta = \Re\frac{2}{2\pi i} \int_{|z|=r} g(z)\frac{dz}{z} = 2\Re c_0.$$

Letting $r \to R-$, we see that $R^n |c_n| \leq 2\Re c_0$ for $n \geq 1$. Thus

$$|f(z) - f(0)| \leq \sum_{n=1}^{\infty} |c_n| r^n \leq \frac{2\Re c_0 \, r/R}{1 - r/R} = \frac{2\{U - \Re f(0)\} r}{R - r}.$$

Also, for $\nu \geq 1$,

$$\frac{|f^{(\nu)}(z)|}{\nu!} \leq \sum_{n=\nu}^{\infty} |c_n| \binom{n}{\nu} r^{n-\nu} \leq \frac{2\Re c_0}{R^\nu} \sum_{n=\nu}^{\infty} \binom{n}{\nu} \frac{r^{n-\nu}}{R^{n-\nu}}$$

$$= \frac{2\{U - \Re f(0)\}}{R^\nu (1 - r/R)^{\nu+1}} = \frac{2R\{U - \Re f(0)\}}{(R - r)^{\nu+1}}. \qquad \square$$

The following lemma of Landau approximates the logarithmic derivative of an analytic function in terms of the nearby zeros of the function.

Lemma 8.10   *Let f be analytic on $D = \{s : |s - s_0| \leq r\}$ for some $r > 0$ and $s_0 \in \mathbb{C}$. Suppose that $f(s_0) \neq 0$ and that for some real number $M > 0$,*

$$|f(s)/f(s_0)| < e^M$$

holds in D. If ρ runs through the zeros of f (with k repetitions for a zero of order k) satisfying $|\rho - s_0| \leq r/2$, and if $|s - s_0| \leq r/3$, then

$$\left| \frac{f'}{f}(s) - \sum_\rho \frac{1}{s - \rho} \right| < 36M/r$$

and

$$-\Re \frac{f'}{f}(s) < \frac{36M}{r} - \Re \sum_\rho \frac{1}{s - \rho}.$$

Proof of Lemma 8.10.   The function defined by $g(s) := f(s) \prod_\rho (s - \rho)^{-1}$ is analytic on D and nonvanishing for $|s - s_0| \leq r/2$. For s on the boundary of D we have $|s - \rho| \geq r/2 \geq |s_0 - \rho|$ and hence

$$\left| \frac{g(s)}{g(s_0)} \right| = \left| \frac{f(s)}{f(s_0)} \prod_\rho \left(\frac{s_0 - \rho}{s - \rho} \right) \right| \leq \left| \frac{f(s)}{f(s_0)} \right| < e^M.$$

By the maximum modulus theorem the estimate holds inside D also. Define $h(s) := \log\{g(s)/g(s_0)\}$ by taking $h(s_0) = 0$. This function is analytic for

$|s - s_0| \leq r/2$ and satisfies $\Re\{h(s)\} < M$ there. The preceding lemma applied with discs of radius $r/3$ and $r/2$ yields

$$\left|\frac{f'}{f}(s) - \sum_\rho \frac{1}{s - \rho}\right| = \left|\frac{g'}{g}(s)\right| = |h'(s)| < 36M/r$$

for $|s - s_0| \leq r/3$. The second assertion of the lemma follows from the last inequality and the relation $\Re w \leq |w|$. $\qquad\qquad\square$

Proof of Theorem 8.8. The main argument is based on the preceding lemma and on the inequality

$$-\Re\{3\frac{\zeta'}{\zeta}(\sigma) + 4\frac{\zeta'}{\zeta}(\sigma + it) + \frac{\zeta'}{\zeta}(\sigma + 2it)\} \geq 0 \quad (\sigma > 1). \qquad (8.8)$$

The inequality, in turn, depends on (a) the trigonometric relation

$$3 + 4\cos\theta + \cos 2\theta = 2(1 + \cos\theta)^2 \geq 0 \quad (\theta \in \mathbb{R}),$$

(b) formula (6.9) in the form

$$-\Re\zeta'(s)/\zeta(s) = \sum \Lambda(n)n^{-\sigma}\cos(t\log n) \quad (\sigma > 1),$$

and (c) the fact that $\Lambda \geq 0$.

Since ζ has a simple pole at $s = 1$, we have

$$-\frac{\zeta'}{\zeta}(\sigma) = \frac{1}{\sigma - 1} + O(1) \quad (\sigma \to 1).$$

The two remaining terms in (8.8) should reflect the influence of a zero of zeta with ordinate near t or $2t$. We exploit this idea using Lemma 8.10. Let $\rho = \beta + i\gamma$ denote a zero of zeta with $\gamma \geq 10$, say, and $\beta < 1$ (of course). Let σ_0 be a number satisfying

$$1 + (2 + \gamma)^{-1} \leq \sigma_0 \leq 9/8, \qquad (8.9)$$

later to be specified. Let $s_0 = \sigma_0 + i\gamma$, $s_0' = \sigma_0 + 2i\gamma$, $r = 3/2$.

We have the estimates

$$|\zeta(s_0)^{-1}| = \left|\sum \mu(n)n^{-s_0}\right| \leq \sum n^{-\sigma_0} < 1 + \frac{1}{\sigma_0 - 1} \leq \gamma + 3$$

and similarly for $|\zeta(s_0')^{-1}|$. Also, by (6.13), $\zeta(s) = O(t^2)$ for $t \geq 2$ and $\sigma \geq -1/2$. Consequently, there exists a constant $A > 0$ such that

$$|\zeta(s)/\zeta(s_0)| < \gamma^A, \quad |\zeta(s)/\zeta(s_0')| < \gamma^A \qquad (8.10)$$

in the discs $\{s : |s - s_0| \le 3/2\}$ and $\{s : |s - s_0'| \le 3/2\}$ respectively.

We now apply Lemma 8.10 to estimate $\Re(\zeta'/\zeta)$ at s_0 and s_0'. We have

$$-\Re\frac{\zeta'}{\zeta}(s_0') < \frac{36A\log\gamma}{3/2} - \Re\sum_{|\rho-s_0'|\le 3/4}\frac{1}{s_0'-\rho} < A'\log\gamma,$$

since

$$\Re(s_0' - \rho)^{-1} = (\sigma_0 - \beta)/|s_0' - \rho|^2 > 0,$$

and analogously,

$$-\Re\frac{\zeta'}{\zeta}(s_0) < \frac{36A\log\gamma}{3/2} - \Re\sum_{|\rho-s_0|\le 3/4}\frac{1}{s_0-\rho} < A'\log\gamma - \frac{1}{s_0-\beta}.$$

We insert these estimates into (8.8) and obtain

$$\frac{3}{\sigma_0-1} + A'' + 4\left(A'\log\gamma - \frac{1}{\sigma_0-\beta}\right) + A'\log\gamma > 0$$

or

$$\frac{4}{\sigma_0-\beta} - \frac{3}{\sigma_0-1} < A'''\log\gamma. \qquad (8.11)$$

If we write $\sigma_0 - \beta = \sigma_0 - 1 + 1 - \beta$ and isolate $1 - \beta$, we obtain

$$1 - \beta > \frac{4(\sigma_0-1)}{A'''(\sigma_0-1)\log\gamma + 3} - (\sigma_0 - 1).$$

We next choose σ_0 such that $A'''(\sigma_0 - 1)\log\gamma = 1/2$, say. Then σ_0 satisfies (8.9) if $\gamma \ge$ some t_1, and we obtain the main estimate

$$1 - \beta > \left(\frac{4}{7/2} - 1\right)(\sigma_0 - 1) = \frac{1}{14A'''\log\gamma}.$$

Finally, we consider the strip $\{s : |t| \le \max(10, t_1)\}$. We know that $\zeta(1+it) \ne 0$ for all real t and that ζ is analytic except for its pole at $s = 1$. It follows that there exists a real $\alpha < 1$ such that $\zeta(s) \ne 0$ on

$$\{s : \alpha \le \sigma \le 1, \ |t| \le \max(10, t_1)\}.$$

The theorem now holds if we choose

$$K = \min\left(\frac{1}{14A'''}, \ (1-\alpha)\log 2\right). \qquad \square$$

PROBLEM 8.8   Show that

$$\zeta^3(\sigma)|\zeta^4(\sigma+it)\zeta(\sigma+2it)| \geq 1 \quad (\sigma > 1,\ t \in \mathbb{R}).$$

Give another proof of the nonvanishing of zeta on the line $\Re s = 1$.

PROBLEM 8.9   Use the inequality of the preceding problem to show that

$$\left|\zeta\left(1 + \frac{K}{\log^9 t} + it\right)\right| > \frac{c\,K^{3/4}}{\log^7 t} \quad (t \geq 2)$$

for any positive constant K. Also, show that the estimate $\zeta'(s) = O(\log^2 t)$ holds on $\{s : \sigma > 1 - \log^{-1} t,\ t \geq 2\}$ (cf. Lemma 8.4). Combine these relations to prove that there exists a $K > 0$ for which

$$1/\zeta(s) = O\{\log^7(|t|+2)\} \quad (\sigma > 1 - K\log^{-9}\{|t|+2\}).$$

8.5   An estimate of ζ'/ζ

We give an upper estimate of $|\zeta'/\zeta|$ for use in the Perron inversion formula. The integration path will be taken in a zerofree region for zeta lying to the left of the line $\sigma = 1$. In the last section we found such a region, but here we assume a rather general zerofree region. This shows how improvements in the results of §8.4 would reflect in better estimates of $|\psi(x) - x|$.

Theorem 8.11   *Let η be a continuous real valued decreasing function on $[0,\infty)$ satisfying*

$$0 < \eta(t) \leq 1/2 \quad (t \geq 0) \tag{8.12}$$

$$\eta(t+1) = \eta(t) + o(\eta(t)) \quad (t \to \infty) \tag{8.13}$$

$$1/\eta(t) = O(\log(t+2)) \quad (t \geq 0). \tag{8.14}$$

Suppose that ζ has no zeros in the region $\{s : \sigma > 1 - \eta(|t|)\}$. Let α be a fixed number satisfying $0 < \alpha < 1$. Then

$$-\frac{\zeta'}{\zeta}(s) - \frac{1}{s-1} = O(\log(|t|+2))$$

uniformly in the region

$$\{s : \sigma \geq 1 - \alpha\eta(|t|)\}.$$

Proof. We may assume that $t > 0$, and that s lies in the region

$$\{s : 1 - \alpha\eta(t) \le \sigma < 1 + 1/\log(t+2)\},$$

as we now show. If $\sigma \ge 1 + 1/\log(t+2)$ then

$$\left|\frac{\zeta'}{\zeta}(s)\right| \le \sum_{n=1}^{\infty} \frac{\Lambda(n)}{n^\sigma} = -\frac{\zeta'}{\zeta}(\sigma) = O\left(\frac{1}{\sigma-1}\right) = O(\log(t+2)),$$

since ζ'/ζ has a simple pole at 1. The rest of the argument proceeds in two steps, first with an estimate of $-\Re(\zeta'/\zeta)$ and then of $|\zeta'/\zeta|$.

Let $s_0 = 1 + 1/L_0 + it_0$, where $L_0 = \log(t_0 + 2)$ and $t_0 \ge e^{10}$ (and is subject to one subsequent restriction). We have

$$|\zeta(s)/\zeta(s_0)| \le e^{A\log t_0} \qquad (|s - s_0| \le 9/5)$$

for some constant A, by the argument used to establish (8.10). We apply Lemma 8.10 to zeta in a disc of radius $9/5$ centered at s_0. We have

$$-\Re\frac{\zeta'}{\zeta}(s) < 20A\log t_0 - \Re\sum_\rho \frac{1}{s-\rho} \qquad (|s - s_0| \le 3/5). \qquad (8.15)$$

Here ρ extends over the zeros of zeta satisfying $|\rho - s_0| \le 9/10$.

Let $\eta_0 = \eta(t_0)$ and $\alpha' = (1 + \alpha)/2$. Suppose that s lies in the disc $\{s : |s - s_0| \le 1/L_0 + \alpha'\eta_0\}$, whose radius is smaller than $3/5$. With this restriction, $\Re(s - \rho)^{-1} > 0$ for all zeros $\rho = \beta + i\gamma$ satisfying $|\rho - s_0| \le 9/10$. Indeed, we have on the one hand

$$\sigma \ge \sigma_0 - (1/L_0 + \alpha'\eta_0) = 1 - \alpha'\eta_0,$$

and on the other hand, by the zerofree condition, $\beta \le 1 - \eta(t_0 + 1)$. Now

$$\eta(t_0 + 1) = \eta(t_0) + o(\eta(t_0)) > \alpha'\eta_0,$$

holds by (8.13), provided t_0 exceeds a sufficiently large number t_1. Thus $\beta < 1 - \alpha'\eta_0$ and so

$$\Re(s - \rho)^{-1} = (\sigma - \beta)/|s - \rho|^2 > 0 \quad (|\rho - s_0| \le 9/10);$$

therefore (8.15) remains true with the sum omitted.

Finally, we apply the Borel-Carathéodory lemma (Lemma 8.9) to the function ζ'/ζ using the discs

$$\{s : |s - s_0| \le 1/L_0 + \alpha'\eta_0\} \supset \{s : |s - s_0| \le 1/L_0 + \alpha\eta_0\}.$$

For points s in the smaller disc we have

$$\left|\frac{\zeta'}{\zeta}(s) - \frac{\zeta'}{\zeta}(s_0)\right| \leq \frac{4/(L_0\eta_0) + 4\alpha}{1 - \alpha}\left\{20A\log t_0 - \Re\frac{\zeta'}{\zeta}(s_0)\right\}.$$

Recalling that $\zeta'(s_0)/\zeta(s_0) = O(\log t_0)$, we have for fixed $\alpha < 1$

$$\frac{\zeta'}{\zeta}(\sigma + it_0) = O(\log t_0) \quad (\sigma \geq 1 - \alpha\eta(t_0), \ t \geq t_1).$$

Since $-\zeta'/\zeta(s) - 1/(s-1)$ is analytic on the compact set

$$\{s = \sigma + it : 1 - \alpha\eta(t) \leq \sigma \leq 1 + 1/\log(t+2), \ 0 \leq t \leq t_1\},$$

the claimed bound holds here too, for a suitable O-constant. $\quad\square$

8.6 Estimation of ψ

In Chapters 5 and 7 we have given proofs that $\psi(x) = x + o(x)$, i.e. that the P.N.T. holds. Here we obtain estimates of the error term in the P.N.T. in terms of the size of the zerofree region of zeta.

Theorem 8.12 *Let η be as in Theorem 8.11 and assume that $\zeta(s) \neq 0$ for all s satisfying $\sigma > 1 - \eta(|t|)$. Let α be a fixed number in $(0,1)$. For each $x > 1$, first define $T = T(x)$ to be the solution of the equation*

$$(\log T)/\eta(T) = \alpha \log x, \tag{8.16}$$

and next define a function ω on $(1, \infty)$ by setting $\omega(x) = \log T$. Then as $x \to \infty$,

$$\psi(x) = x + O\left(x\log^2 x e^{-\omega(x)}\right).$$

With $\eta(t) = K/\log(t+2)$ (cf. Theorem 8.8),

$$\psi(x) = x + O\left(x e^{-c\sqrt{\log x}}\right)$$

is valid for some $c > 0$. If the R.H. is true, then, for any $\epsilon > 0$,

$$\psi(x) = x + O(x^{(1/2)+\epsilon}).$$

We note that $x\exp(-a\sqrt{\log x})$ tends to infinity more swiftly than $x^{1-\epsilon}$ for any given $\epsilon > 0$ but grows more slowly than $x\log^{-m} x$ for any given real number m.

Proof. The function ω is well defined since the left side of (8.16) goes to ∞ monotonically with T, and hence a unique solution of (8.16) exists. The conditions $\alpha < 1$ and $\eta \leq 1/2$ imply that $T < \sqrt{x}$.

We estimate ψ by applying the Mellin inversion formula as given in Lemma 7.14. The condition that $-\zeta'/\zeta$ must satisfy is that it be analytic in a rectangle $\{s : a \leq \sigma \leq b, |t| \leq T\}$ except for the simple pole with residue 1 at $s = 1$. We have

$$\psi(x) = x + O\{\psi(xe^{1/T}) - \psi(xe^{-1/T})\}$$

$$+ O\left\{ \left(\sum_{n \leq xe^{-1/T}} + \sum_{n > xe^{1/T}} \right) \frac{(x/n)^b \Lambda(n)}{T |\log(x/n)|} \right\}$$

$$+ O\{x^a \log(T/a) \max_{|t| \leq T} |(\zeta'/\zeta)(a + it)|\}$$

$$+ O\{T^{-1}(b - a) \max_{a \leq \sigma \leq b} |(\zeta'/\zeta)(\sigma + iT)| x^\sigma\}. \qquad (8.17)$$

Choose $b = 1 + (\log x)^{-1}$, which makes the series involving Λ convergent, and choose $a = 1 - \alpha\eta(T) > 1/2$. We may assume $T \in [2, \sqrt{x}]$ by requiring that $x \geq$ some x_0 (since $\omega(x) \to \infty$ as $x \to \infty$).

We begin making estimates by noting that $\Lambda(n) \leq \log n$ for all n. Thus

$$\psi(xe^{1/T}) - \psi(xe^{-1/T}) = O(xT^{-1} \log x).$$

Also, we estimate

$$\sum_{x/2 < n \leq xe^{-1/T}} \frac{(x/n)^b \Lambda(n)}{T |\log x/n|} \leq \frac{2^b \log x}{T} \int_{x/2}^{xe^{-1/T}} \frac{du}{\log(x/u)} + 2^b \log x$$

$$\leq \frac{2^b x \log x}{T} \int_{x/2}^{xe^{-1/T}} \frac{du}{u \log(x/u)} + 2^b \log x$$

$$= 2^b x T^{-1} \log x \log(T \log 2) + 2^b \log x.$$

The same sum over the range $xe^{1/T} < n \leq 2x$ is estimated similarly. In the remaining range of n's we have $|\log x/n| \geq \log 2$ and hence

$$\left\{ \sum_{n \leq x/2} + \sum_{n > 2x} \right\} \frac{(x/n)^b \Lambda(n)}{T |\log x/n|} < \frac{-x^b}{T \log 2} \frac{\zeta'}{\zeta}(b) = O(xT^{-1} \log x).$$

We use Theorem 8.11 to estimate ζ'/ζ in the remaining O-terms of (8.17):

$$x^a \log \frac{T}{a} \max_{|t| \leq T} \left| \frac{\zeta'}{\zeta}(a+it) \right| = O(x^{1-a\eta(T)} \log^2 T),$$

$$\frac{(b-a)}{T} \max_{a \leq \sigma \leq b} \left| \frac{\zeta'}{\zeta}(\sigma + iT) \right| x^\sigma = O(\frac{x}{T} \log T).$$

When we combine the estimates of the quantities in (8.17) and use the condition $T < \sqrt{x}$, we find that

$$\psi(x) = x + O(xT^{-1} \log x \log T) + O(x^{1-a\eta(T)} \log^2 T).$$

We now choose T satisfying (8.16) and obtain the estimate

$$\psi(x) = x + O\{x \log^2 x \, e^{-\omega(x)}\}.$$

We can take $\eta(t) = (K/2) \log(|t|)$ for $|t| \geq 2$ and $K/(2 \log 2)$ for $|t| < 2$ in accordance with Theorem 8.8. (The factors 2 here enable us to replace the $|t| + 2$ in that theorem by $|t|$.) Then equation (8.16) becomes $\log^2 T = (\alpha K/2) \log x$ and the solution gives $\omega(x) = \sqrt{(\alpha K/2) \log x}$. If we take $0 < c < \sqrt{\alpha K/2}$, we obtain

$$\psi(x) = x + O(xe^{-c\sqrt{\log x}}).$$

(We have "absorbed" the logarithmic factors into c.)

Suppose there is a number $c \in [1/2, 1)$ such that $\zeta(s) \neq 0$ for $\Re s > c$. Then we could take η identically equal to $1 - c$ and have

$$\omega(x) = \log T = (1-c)\alpha \log x$$

for any $\alpha \in (0,1)$. Given $\epsilon > 0$ we could take $\alpha = 1 - \epsilon$ and obtain

$$\psi(x) = x + O(x \log^2 x \exp\{-(1-\epsilon)(1-c) \log x\})$$

$$= x + O(x^{c+\epsilon}).$$

In particular, if the Riemann hypothesis is true, then

$$\psi(x) = x + O(x^{(1/2)+\epsilon}). \qquad \square$$

PROBLEM 8.10 Show that if $\psi(x) - x = O(x^a)$ for some $a \in (1/2, 1)$, then $-\zeta'(s)/\zeta(s) - 1/(s-1)$ is analytic on $\{s : \sigma > a\}$.

8.7   The P.N.T. with a remainder term

We have seen how the relation $\psi(x) \sim x$ leads to $\pi(x) \sim x/\log x$. Here we apply the quantitative estimate of $\psi(x)$ of the preceding section to give a corresponding estimate of $\pi(x)$. The function $x \mapsto x/\log x$, however, is not the best choice for approximating $\pi(x)$; a better one is the so-called *logarithmic integral*. For $x > 1$, set

$$\operatorname{li} x := \lim_{\epsilon \to 0+} \left(\int_0^{1-\epsilon} + \int_{1+\epsilon}^x \right) \frac{du}{\log u}. \qquad (8.18)$$

For our application it is not important to start the integration at 0; we could have taken the integral from 2 or e, say.

The following table illustrates values of the functions we shall discuss. The entries have all been rounded off to the nearest integer.

x	$\pi(x)$	$\pi(x) - x/\log x$	$\operatorname{li} x - \pi(x)$	$\sqrt{x}$	$xe^{-\sqrt{\log x}}$
10^2	25	3	5	10	12
10^3	168	23	10	32	72
10^4	1,229	143	17	100	481
10^5	9,592	906	38	316	3,361
10^6	78,498	6,116	130	1,000	24,309
10^7	664,579	44,158	339	3,162	180,477
10^8	5,761,455	332,774	754	10,000	1,367,847
10^9	50,847,534	2,592,592	1701	31,623	10,543,124
10^{10}	455,052,511	20,754,029	3104	100,000	82,418,874

Table 8.1   P.N.T. DATA

The table suggests that li provides a good approximation to the prime counting function. We can indeed rephrase the P.N.T. as $\pi(x) \sim \operatorname{li} x$, for by l'Hospital's rule,

$$\lim_{x \to \infty} \operatorname{li} x / (x/\log x) = 1.$$

We give a simple approximation of the logarithmic integral. Write

$$\operatorname{li} x = \operatorname{li} e + \int_e^x \frac{du}{\log u},$$

where $\operatorname{li} e = 1.8951178\ldots$ (cf. (8.20) below). For $x > e$ and $N \in \mathbb{Z}^+$, we perform N integrations by parts to obtain

$$\operatorname{li} x = \sum_{n=1}^{N} \frac{(n-1)!\, x}{\log^n x} + \operatorname{li} e - e \sum_{n=0}^{N-1} n! + N! \int_e^x \frac{du}{\log^{N+1} u}.$$

By splitting the integral at $\sqrt{x}$, we see that the last term has the crude upper bound $N!\,\sqrt{x} + N!\, 2^{N+1}\, x\, (\log x)^{-N-1}$.

On the other hand, the last term is unbounded as $N \to \infty$. Thus we should stop the process after a finite number of steps, and obtain

$$\operatorname{li} x = \sum_{n=1}^{N} \frac{(n-1)!\, x}{\log^n x} + O_N\!\left(\frac{x}{\log^{N+1} x}\right).$$

This expression is an example of a nonconvergent series having a truncation error of the order of the first omitted term. A series of this type is called an *asymptotic expansion*.

We now obtain, by more careful integration by parts, another expansion of $\operatorname{li} x$ that is convergent and can be used for calculations.

Lemma 8.13 *For $x > 1$,*

$$\operatorname{li} x = \int_1^x \frac{1 - t^{-1}}{\log t}\, dt + \log \log x + \gamma,$$

where γ is Euler's constant.

Proof. In (8.18), the definition of $\operatorname{li} x$, replace u by $1/t$ in the first integral. We get

$$\operatorname{li} x = \lim_{\epsilon \to 0+} \left\{ -\int_{1/(1-\epsilon)}^{\infty} \frac{dt}{t^2 \log t} + \int_{1+\epsilon}^{x} \frac{du}{\log u} \right\}.$$

Now

$$0 < \int_{1+\epsilon}^{1/(1-\epsilon)} \frac{dt}{t^2 \log t} < \frac{1}{\log(1+\epsilon)} \left\{ \frac{1}{1+\epsilon} - (1-\epsilon) \right\} < \frac{\epsilon^2}{\log(1+\epsilon)},$$

which goes to 0 as $\epsilon \to 0+$, and so

$$
\begin{aligned}
\text{li}\, x &= \lim_{\epsilon \to 0+} \left\{ \int_{1+\epsilon}^{x} \frac{dt}{\log t} - \int_{1+\epsilon}^{\infty} \frac{dt}{t^2 \log t} \right\} \\
&= \int_{1}^{x} \frac{1 - t^{-2}}{\log t}\, dt - \int_{x}^{\infty} \frac{dt}{t^2 \log t} \\
&= \int_{1}^{x} \frac{1 - t^{-1}}{\log t}\, dt + \int_{1}^{x} \frac{1 - t^{-1}}{t \log t}\, dt - \int_{x}^{\infty} \frac{dt}{t^2 \log t}.
\end{aligned}
\tag{8.19}
$$

Combining (8.19) and (6.7), we obtain the claimed formula. $\qquad\qquad\square$

For small x, we can calculate $\text{li}\, x$ with a rapidly converging power series. Making a change of variable and a power series expansion, we get

$$
\int_{1}^{x} \frac{1 - t^{-1}}{\log t}\, dt = \int_{0}^{\log x} \frac{e^u - 1}{u}\, du = \sum_{n=1}^{\infty} \frac{(\log x)^n}{n \cdot n!}.
$$

Thus, for example,

$$
\text{li}\, e = \sum_{n=1}^{\infty} \frac{1}{n \cdot n!} + \gamma.
\tag{8.20}
$$

For large x, we can calculate the last integral by repeated integration by parts, as we now show. It is convenient to set $X = \log x$; then

$$
\int_{1}^{x} \frac{1 - t^{-1}}{\log t}\, dt = \int_{0}^{X} \frac{e^u - 1}{u}\, du.
$$

Lemma 8.14   *Let $X > 0$ and $N \in \mathbb{Z}^+$. Then*

$$
\int_{0}^{X} \frac{e^u - 1}{u}\, du = \sum_{n=1}^{N} (n-1)! \left\{ e^X - \sum_{j=0}^{n} X^j / j! \right\} / X^n + E_N,
$$

where

$$
E_N := E_N(X) := N! \int_{0}^{X} \left\{ e^u - \sum_{j=0}^{N} u^j / j! \right\} u^{-N-1}\, du.
$$

Proof.   Integrate by parts N times, each time integrating the expression containing e^u and inserting an integration constant of -1. At the first step we get

$$\int_0^X \frac{e^u - 1}{u} du = \frac{e^u - (u+1)}{u} \bigg|_0^X + \int_0^X \frac{e^u - (u+1)}{u^2} du.$$

Note that $(e^u - 1 - u)/u = O(|u|) \to 0$ as $u \to 0$. Continue inductively. $\square$

We conclude this discussion with a brief analysis of the error term E_N. Note first that $E_N(X) > 0$. If we expand the integrand of E_N as a Maclaurin series and integrate termwise, we find that

$$E_N(X) = \frac{X}{N+1} + \frac{X^2}{2(N+1)(N+2)} + \frac{X^3}{3(N+1)(N+2)(N+3)} + \cdots.$$

Thus we see that $E_N(X)$ is increasing in X and decreasing in N.

For $N + 1 > X > 0$, the preceding series yields

$$E_N(X) < \sum_{k=1}^\infty \frac{1}{k} \left(\frac{X}{N+1} \right)^k = \log \frac{N+1}{N+1-X} = o(1)$$

as $N \to \infty$. The preceding two lemmas yield the convergent expression

$$\mathrm{li}\, x = \sum_{n=1}^\infty (n-1)! \left\{ x - \sum_{j=0}^n (\log x)^j / j! \right\} / (\log x)^n + \log \log x + \gamma.$$

For $X \geq 2$ and $N \ll X/\log X$ we have a bound for $E_N(X)$. Write

$$E_N(X) = \left\{ \int_0^{X - \log X} + \int_{X - \log X}^X \right\} f_N(u)\, du,$$

where

$$f_N(u) := N! \left\{ e^u - \sum_{j=0}^N u^j / j! \right\} u^{-N-1} = N! \sum_{j=N+1}^\infty u^{j-N-1} / j!$$

$$= \frac{1}{N+1} + \frac{u}{(N+1)(N+2)} + \frac{u^2}{(N+1)(N+2)(N+3)} + \cdots,$$

an increasing function of u for $u > 0$. Thus

$$\int_0^{X - \log X} f_N(u)\, du \leq (X - \log X) f_N(X - \log X)$$

$$\leq X N! \, e^{X - \log X} / (X - \log X)^{N+1}.$$

Also, more simply,

$$\int_{X-\log X}^{X} f_N(u)\, du \leq N!\,(X - \log X)^{-N-1} \int_{X-\log X}^{X} e^u\, du$$

$$< N!\, e^X / (X - \log X)^{N+1}.$$

It follows that

$$E_N(X) \leq 2N!\, e^X X^{-N-1} \left\{ 1 - \frac{\log X}{X} \right\}^{-N-1},$$

and the last factor is bounded for $N \ll X/\log X$. In this case, we have $E_N(X) \ll N!\, e^X / X^{N+1}$.

PROBLEM 8.11   Find a value of N for which the sum in Lemma 8.14 approximates $\operatorname{li}(10^6)$ to within 1% of its true value.

We now establish de la Vallée Poussin's approximation of $\pi(x)$ and also an estimate under the assumption of the R.H. We shall improve the latter result in Theorem 8.26.

Theorem 8.15   *As $x \to \infty$,*

$$\pi(x) = \operatorname{li} x + O\{ x \exp(-c\sqrt{\log x}) \}.$$

Here c is the positive number occurring in Theorem 8.12. Further, if the Riemann hypothesis is true, then for any $\epsilon > 0$,

$$\pi(x) = \operatorname{li} x + O(x^{(1/2)+\epsilon}).$$

Proof.   Recall from (2.9) that

$$\sum_{n \leq x} \kappa(n) = \pi(x) + \frac{1}{2}\pi(x^{1/2}) + \frac{1}{3}\pi(x^{1/3}) + \cdots,$$

and hence

$$0 \leq \sum_{n \leq x} \kappa(n) - \pi(x) \leq \frac{1}{2}\pi(x^{1/2}) + \frac{1}{3}\pi(x^{1/3})(\log x / \log 2),$$

since $\pi(x^{1/n}) = 0$ if $n > \log x / \log 2$. Thus

$$\sum_{n \leq x} \kappa(n) = \pi(x) + O(\sqrt{x}/\log x),$$

and it suffices to estimate $\sum \kappa(n)$.

Now if we make some simple manipulations and appeal to Theorem 8.12, we obtain

$$\sum_{n \le x} \kappa(n) = \int_2^x \frac{du}{\log u} + \int_{2-}^x \frac{d\psi(u) - du}{\log u}$$

$$= \operatorname{li} x - \operatorname{li} 2 + \frac{\psi(x) - x}{\log x} + \frac{2}{\log 2} + \int_2^x \frac{\psi(t) - t}{t \log^2 t} \, dt$$

$$= \operatorname{li} x + O(x e^{-c\sqrt{\log x}}) + \int_2^x e^{-c\sqrt{\log t}} \log^{-2} t \, dt.$$

The integral can be treated by noting that

$$\int_2^x e^{-c\sqrt{\log t}} \, dt = \int_2^x e^{(1/2) \log t - c\sqrt{\log t}} t^{-1/2} \, dt$$

$$= O\left(e^{(1/2) \log x - c\sqrt{\log x}}\right) \int_2^x t^{-1/2} \, dt$$

$$= O(x \exp\{-c\sqrt{\log x}\}).$$

If the R.H. holds, then we apply the last part of Theorem 8.12 in the above argument (and omit the preceding integral estimate).   □

PROBLEM 8.12   Show that $\operatorname{li} x$ is indeed a better asymptotic approximation to $\pi(x)$ than $x/\log x$.

PROBLEM 8.13   Show that there exists a unique number B such that

$$\pi(x) - \frac{x}{\log x - B} = O\left(\frac{x}{\log^3 x}\right).$$

Find its value. Approximate $\pi(x)$ by this expression for some values of x listed in Table 8.1.

PROBLEM 8.14   Show that for $n \ge 2$,

$$p_n = n(\log n + \log \log n + O(1)).$$

8.8 Estimation of M

In §5.3 we deduced the estimate

$$M(x) = \sum_{n \leq x} \mu(n) = o(x)$$

from the relation $\psi(x) - x = o(x)$. In a similar way, one can deduce a quantitative estimate of M from the results for $\psi(x) - x$ of §8.6. However a much better estimate can be obtained by using the Perron inversion formula for M. We have already assembled most of the needed data in §§8.4–8.5. We give a further result in

Lemma 8.16 *Let K be the constant of Theorem 8.8 and let α be a fixed number in $(0,1)$. Then*

$$1/\zeta(s) = O\{\log(|t| + 2)\} \quad for \quad \sigma > 1 - \frac{\alpha K}{\log(|t| + 2)}.$$

Proof. The function $|\zeta(s)|$ has a positive lower bound on

$$\{s : \sigma > 1 - \alpha K / \log(|t| + 2), \ |t| \leq 2\},$$

and hence $1/\zeta$ is bounded there. We henceforth assume that $t > 2$.

Next, suppose that $\sigma \geq 1 + \log^{-1} t$. In this case we have

$$|\zeta^{-1}(s)| \leq \sum |\mu(n)| n^{-\sigma} \leq \zeta(\sigma) < 1 + \frac{1}{\sigma - 1} = O(\log t).$$

Finally, suppose that

$$1 - \frac{\alpha K}{\log(t + 2)} < \sigma < 1 + \frac{1}{\log t} := \sigma_0.$$

We have then

$$\log|\zeta(s)| = \log|\zeta(\sigma_0 + it)| + \int_{\sigma_0}^{\sigma} \Re \frac{\zeta'}{\zeta}(u + it) du$$

$$\geq -\log\log t + O(1).$$

The boundedness of the integral follows from Theorem 8.11 and the fact that the integration path is of length $O(\log^{-1} t)$. $\square$

If we now apply Lemma 7.14 in essentially the same way as we did in the proof of Theorem 8.12, we obtain

Theorem 8.17 *There exists a positive number c such that*

$$M(x) = O\{x \exp(-c\sqrt{\log x})\}.$$

On the assumption of the R.H. one can show that $1/\zeta(s) = O\{(|t|+2)^\epsilon\}$ for $\sigma \geq \sigma_0 > 1/2$ and $\epsilon > 0$ and hence that $M(x) = O(x^{1/2+\epsilon})$. However, the methods we have used do not yield the needed estimate of ζ^{-1}. The following problem shows what we can achieve with our available techniques under the assumption of the R.H.

PROBLEM 8.15   Assuming that the R.H. is true, show that

$$1/\zeta(s) = O\{(|t|+2)^A\}$$

holds uniformly on $\{s : \sigma > \frac{1}{2} + \epsilon\}$ for some positive A depending on an arbitrary positive ϵ. Also, deduce that there exists a constant $c < 1$ such that $M(x) = O(x^c)$. Hint. Introduce a suitable smoothing factor to allow application of the inversion formula. Then apply a differencing argument to recover $M(x)$ (cf. Problem 7.18 at the end of §7.8).

PROBLEM 8.16   Show that $\int_1^\infty |M(x)|x^{-2}dx < \infty$. Deduce from this that the Dirichlet series for $1/\zeta(s)$ converges on the line $\Re s = 1$.

PROBLEM 8.17   Show that there exists a constant $c > 0$ such that

$$Q(x) := \sum_{n \leq x} |\mu(n)| = \frac{6}{\pi^2}x + O\{\sqrt{x}\exp(-c\sqrt{\log x})\}.$$

PROBLEM 8.18   Show that

$$\sum_{n=1}^{\infty} \frac{\mu(n)\log n}{n} = -1.$$

PROBLEM 8.19   Assuming Theorem 8.17 deduce that

$$m(x) := \sum_{n \leq x} \frac{\mu(n)}{n} = O(\sqrt{\log x}\,\exp\{-c\sqrt{\log x}\,\}).$$

(Recall that $\sum_{n=1}^{\infty} \mu(u)/n = 0$.)

PROBLEM 8.20   (L. A. Rubel). Show that there exists a finite absolute bound B such that for all $x \geq 1$,

$$\sum_{n \leq x} \frac{1}{n}\left|m\left(\frac{x}{n}\right)\right| \leq B.$$

Hint. Express the sum as

$$\sum_{n\le\sqrt{x}}\frac{1}{n}\left|m\left(\frac{x}{n}\right)\right| + \sum_{1\le k<\sqrt{x}}|m(k)|\sum_{x/(k+1)<n\le x/k}\frac{1}{n}.$$

How weak an upper estimate for $|m(x)|$ can be used to solve the problem?

PROBLEM 8.21 (L. A. Rubel). Let $f \in \mathcal{A}$ and assume $f(n) \to L$ as $n \to \infty$. Prove that $\sum_{n\le x}(f * \mu)(n)/n \to L$. Hint. First consider $f = 1$. Next consider $f(n) \to 0$ as $n \to \infty$.

8.9   The density of zeros in the critical strip

For $T > 0$, let $N(T)$ denote the number of zeros of zeta in the rectangle $0 \le \sigma \le 1$, $0 \le t \le T$. A zero of multiplicity $k > 1$ (if any exists!) will be counted k times. We showed in §8.3 that $N(T)$ is unbounded, but $N(T)$ is finite for finite T because zeta is meromorphic on $\mathbb{C}$. We shall establish an asymptotic formula for $N(T)$ in

Theorem 8.18   *As $T \to \infty$,*

$$N(T) = \frac{T}{2\pi}\log\frac{T}{2\pi} - \frac{T}{2\pi} + O(\log T).$$

The theorem asserts that the average density of zeros in a rectangle $0 \le \sigma \le 1$, $T \le t \le T+1$ is asymptotic to $(2\pi)^{-1}\log T$ and gives an O-estimate for $N(T+1) - N(T)$. We shall establish the O-estimate directly as a step in proving the theorem.

Lemma 8.19

$$N(T+1) - N(T) = O(\log T).$$

Proof of the lemma.   We apply Lemma 8.10 as we did in obtaining formula (8.15), but this time with $s_0 = 6/5 + iT$, $T \ge 10$. We again take $r = 9/5$ and note that $|\zeta(s)/\zeta(s_0)| < \exp(A\log T)$ in the disc $|s - s_0| \le r$. If ρ runs through the zeros of ζ satisfying $|\rho - s_0| \le 9/10$, then we have

$$\Re\sum_{\rho}\frac{1}{s - \rho} < \frac{36A\log T}{9/5} + \Re\frac{\zeta'}{\zeta}(s_0).$$

Of course $(\zeta'/\zeta)(s_0)$ is bounded.

We claim that each of the terms in the last sum is positive and not too small. Indeed, we have (with $\rho = \beta + i\gamma$)

$$\Re \frac{1}{s_0 - \rho} = \frac{6/5 - \beta}{|s_0 - \rho|^2} > \frac{1/5}{(9/10)^2} > \frac{1}{5}.$$

It follows that there are $O(\log T)$ zeros of zeta in the disc $\{s : |s - s_0| \leq 9/10\}$. The rectangle $\{s : 1/2 \leq \sigma \leq 1, |t - T| \leq 1/2\}$ is contained in this disc and hence there are $O(\log T)$ zeros of zeta in the rectangle. Since the zeros of zeta are located symmetrically with respect to the critical line, $N(T + \frac{1}{2}) - N(T - \frac{1}{2}) = O(\log T)$, which is equivalent to our assertion. $\square$

PROBLEM 8.22 Show that Lemma 8.19 also follows from Jensen's formula [TiTF, §3.61] applied to the entire function ξ.

Proof of Theorem 8.18. It is convenient to use ξ in place of ζ at the beginning of the argument, for the same reasons as in the proof of Theorem 8.5. Let R denote the rectangle with vertices 2, $2 + iT$, $-1 + iT$, and -1. We may assume that T is not the ordinate of a zero, since $N(T)$ and the main term in the formula to be proved are each continuous from the right. Also, assume that $T \geq 3$.

The number of zeros of zeta within R is $N(T)$. Indeed, since

$$(1 - 2^{1-\sigma})\zeta(\sigma) = 1 - 2^{-\sigma} + 3^{-\sigma} - 4^{-\sigma} \pm \cdots > 0$$

for $\sigma > 0$, it follows that ξ has no zeros on this ray. By the functional equation, there are no zeros of ξ on the real axis; indeed ξ is positive there. Now by the argument principle,

$$2\pi N(T) = \Delta_R \arg \xi,$$

where $\Delta_R \arg \xi$ denotes the change in the argument of $\xi(s)$ as s traverses R in the positive sense.

There is no change in $\arg \xi$ along the base of R, since ξ is positive on the real axis. Let L denote the polygonal path consisting of line segments from 2 to $2 + iT$ to $1/2 + iT$. In view of the functional equation of ξ and the reflection principle we have $\pi N(T) = \Delta_L \arg \xi$.

Recalling the definition of ξ, we write

$$\Delta_L \arg \xi(s) = \Delta_L \arg \frac{1}{2}s(s - 1) + \Delta_L \arg \exp\left(-\frac{s}{2}\log \pi\right)$$

$$+ \Delta_L \arg \Gamma(s/2) + \Delta_L \arg \zeta(s).$$

The first two terms can be evaluated at once:

$$\Delta_L \arg \frac{1}{2}s(s-1) = \text{Arg}\left(iT + \frac{1}{2}\right) + \text{Arg}\left(iT - \frac{1}{2}\right) = \pi,$$

$$\Delta_L \arg \exp\left(-\frac{s}{2}\log \pi\right) = \Delta_L \left(-\frac{t}{2}\log \pi\right) = -\frac{T}{2}\log \pi.$$

Here and below Arg and Log denote the principal branches of arg and log respectively. We evaluate Γ using Stirling's formula (cf. Appendix):

$$\Delta_L \arg \Gamma\left(\frac{s}{2}\right) = \Delta_L \Im\left\{\text{Log}\,\Gamma\left(\frac{s}{2}\right)\right\} = \Im\left\{\text{Log}\,\Gamma\left(\frac{1}{4} + \frac{iT}{2}\right)\right\}$$

$$= \Im\left\{\left(-\frac{1}{4} + \frac{iT}{2}\right)\text{Log}\left(\frac{Ti}{2}\right) - \frac{Ti}{2} + \frac{1}{2}\log 2\pi\right\} + O(T^{-1})$$

$$= \frac{T}{2}\log\frac{T}{2} - \frac{\pi}{8} - \frac{T}{2} + O\left(\frac{1}{T}\right).$$

Combining the estimates we have

$$N(T) = \frac{T}{2\pi}\log\frac{T}{2\pi} - \frac{T}{2\pi} + \frac{7}{8} + S(T) + O(T^{-1}), \qquad (8.21)$$

where $S(T) = \pi^{-1}\Delta_L \text{Arg}\,\zeta(s)$. It suffices to show that $S(T) = O(\log T)$. We estimate the change in Arg ζ along $L_1 = L \cap \{s : \sigma = 2\}$ by noting that

$$\Re\zeta(2+it) \geq 1 - 2^{-2} - 3^{-2} - \cdots = 2 - \zeta(2) > 0.$$

It follows that $|\Delta_{L_1} \arg \zeta(s)| \leq \pi/2$.

We shall use the following representation of ζ'/ζ to estimate $\Delta \arg \zeta$ along $L_2 = L \cap \{s : t = T\}$:

$$\frac{\zeta'}{\zeta}(s) = \sum_{|\rho - s_0| \leq 9/8} \frac{1}{s - \rho} + O(\log T), \qquad (8.22)$$

where $s_0 = 5/4 + iT$ and $|s - s_0| \leq 3/4$. This follows from Lemma 8.10 and the fact that $\zeta(s) = O(t^A)$ for some A if $\Re s \geq -1$, $t \geq 2$. The last O-bound for zeta can be deduced from the estimates of §8.2 with the aid of the functional equation.

If we integrate (8.22) from $s = 2 + iT$ to $s = \frac{1}{2} + iT$ and take imaginary parts we obtain

$$\Delta_{L_2} \arg \zeta(s) = \sum_{|\rho - s_0| \leq 9/8} \Delta_{L_2} \arg(s - \rho) + O(\log T).$$

Now each term $\Delta_{L_2} \arg(s - \rho)$ is at most π in absolute value and there are $O(\log T)$ such terms by Lemma 8.19. It follows that $\Delta_{L_2} \arg \zeta(s)$ and hence $S(T)$ is $O(\log T)$. $\qquad \square$

PROBLEM 8.23   Let γ denote the ordinate of a nontrivial zero of zeta. Show that

$$\sum_{0 < \gamma \leq T} \gamma^{-1} = O(\log^2 T). \tag{8.23}$$

PROBLEM 8.24   Let γ_n denote the ordinate of the nth zero of zeta in the upper half plane, listed in increasing order. Show that

$$\gamma_n \sim 2\pi n / \log n \quad (n \to \infty).$$

8.10   An explicit formula for ψ_1

We have seen in Chapter 7 that summatory functions of polynomial growth can be represented by the Perron inversion formula. In most applications we then shifted the contour to the left until we met an obstacle such as a large number of singularities or considerable growth of the M.t. If the singularities and growth are "manageable," then we can push the contour arbitrarily far to the left. We would thereby achieve a representation of a summatory function in terms of the singularities in $\mathbb{C}$ of the integrand in the Perron formula. Such a representation is called an *explicit formula*. We have used this technique earlier to establish (7.22), the Fourier series of the sawtooth function.

Here we express

$$\psi_1(x) := \int_1^x \psi(t) dt$$

by its explicit formula. This representation is useful for making upper estimates of remainder terms and for oscillation theorems.

The first step in the contour deformation is to find certain horizontal line segments on which we can obtain an estimate of ζ'/ζ.

Lemma 8.20   *There exist sequences $T_2, T_3, \ldots, T_2', T_3', \ldots,$ and a positive number ν, such that $m < T_m,\ T_m' < m + 1$ and*

$$\frac{\zeta'}{\zeta}(\sigma + iT_m) = O(\log^2 T_m) \quad \text{and} \quad \frac{1}{\zeta(\sigma + iT_m')} = O(T_m'^{\nu})$$

hold uniformly for $-1 \le \sigma \le 2$ and $m = 2, 3, \ldots.$

Proof.   First we apply Lemma 8.10 in the same manner as was used in proving (8.22) to obtain the formula

$$\frac{\zeta'}{\zeta}(s) = \sum_{|\rho - s_0| \le 6} \frac{1}{s - \rho} + O(\log m),$$

where $s_0 = 2 + im$ and $|s - s_0| \le 4$. The sum, taken over zeros ρ of ζ for which $|\rho - s_0| \le 6$, includes all terms for which $m - 1 \le \Im\rho \le m + 2$. If we restrict s to have imaginary part between m and $m + 1$, the sum over the remaining terms can be absorbed into the error term by Lemma 8.19. Thus we have

$$\frac{\zeta'}{\zeta}(s) = \sum_{m-1 \le \Im\rho \le m+2} \frac{1}{s - \rho} + O(\log m)$$

for $-1 \le \Re s \le 2,\ m \le \Im s \le m + 1$. If there are $n - 1$ zeros ρ with $m - 1 \le \Im\rho \le m + 2$, then there is a number T_m in $(m, m + 1)$ such that ζ has no zero with imaginary part between $T_m - 1/(2n)$ and $T_m + 1/(2n)$. It follows from Lemma 8.19 that

$$\frac{\zeta'}{\zeta}(\sigma + iT_m) = O(n^2) + O(\log m) = O(\log^2 m) = O(\log^2 T_m).$$

The second estimate of the lemma will be useful in proving Lemma 11.17. We remark that if the right side of the estimate for ζ'/ζ were $\log T_m$ instead of $\log^2 T_m$, then integration along a horizontal segment would immediately give the desired lower estimate for ζ. However, we do not know whether the estimate for ζ'/ζ holds with the exponent 1. Instead we shall give an argument based on the observation that the values of a function cannot everywhere exceed the average value.

We take $s = \sigma + it$, where t is not the ordinate of any zeta zero $\rho = \beta + i\gamma$, and for $-1 \leq \sigma \leq 2$, $m \leq t \leq m+1$ we have

$$\log \zeta(s) = \log \zeta(2+it) + \int_{2+it}^{s} \frac{\zeta'}{\zeta}(u)\, du$$

$$= O(1) + \int_{2+it}^{s} \sum_{m-1 \leq \gamma \leq m+2} (u-\rho)^{-1} du + O(\log m)$$

$$= \sum_{m-1 \leq \gamma \leq m+2} \log(s-\rho) + O(\log m)$$

for suitable branches of the logarithm. Hence

$$\log |\zeta(s)| \geq \sum_{m-1 \leq \gamma \leq m+2} \log|t-\gamma| + O(\log m).$$

Note that the right side of the last estimate does not depend on σ, as long as σ lies in $[-1, 2]$. Hence

$$\min_{-1 \leq \sigma \leq 2} \log |\zeta(\sigma+it)| \geq \sum_{m-1 \leq \gamma \leq m+2} \log|t-\gamma| + O(\log m).$$

Now we have for $m-1 \leq \gamma \leq m+2$

$$\int_{m}^{m+1} \left| \log|t-\gamma| \right| dt \leq 2 \int_{0}^{2} |\log w|\, dw = 4\log 2,$$

and hence

$$\int_{m}^{m+1} \min_{-1 \leq \sigma \leq 2} \log |\zeta(\sigma+it)|\, dt \geq \sum_{m-1 \leq \gamma \leq m+2} \int_{m}^{m+1} \log|t-\gamma|\, dt + O(\log m)$$

$$\geq -\nu \log m$$

holds for a suitable positive constant ν.

It follows that there exists a number $T_m' \in (m, m+1)$ for which

$$|\zeta(\sigma+iT_m')| \geq T_m'^{-\nu}, \quad -1 \leq \sigma \leq 2. \qquad \square$$

Lemma 8.21 $\zeta'(s)/\zeta(s) = O\{\log(|s|+1)\}$ *on*

$$\{s : \sigma \leq -1, \quad |s-2n| \geq 1/2, \quad n = -1, -2, \ldots\}.$$

Proof. If we logarithmically differentiate the asymmetric functional equation for zeta, we obtain

$$\frac{\zeta'}{\zeta}(s) = \log 2\pi + \frac{\pi}{2}\cot\frac{\pi s}{2} - \frac{\Gamma'}{\Gamma}(1-s) - \frac{\zeta'}{\zeta}(1-s).$$

Now for s in the above region, $\Re(1-s) \geq 2$, and so

$$\left|\frac{\zeta'}{\zeta}(1-s)\right| \leq -\frac{\zeta'}{\zeta}(2), \quad \frac{\Gamma'}{\Gamma}(1-s) = O\{|\log(1-s)|\},$$

and $\cot \pi s/2$ is bounded.   □

Theorem 8.22   *For $x \geq 1$,*

$$\psi_1(x) = \frac{x^2}{2} - \sum_\rho \frac{x^{\rho+1}}{\rho(\rho+1)} - x\frac{\zeta'}{\zeta}(0) + \frac{\zeta'}{\zeta}(-1) - \sum_{r=1}^{\infty}\frac{x^{1-2r}}{2r(2r-1)}.$$

Here ρ extends over the nontrivial zeros of zeta (with appropriate multiplicity in case there are zeros of order exceeding one).

Remarks 8.23   For $x > 1$ and $\psi_0(x) := \frac{1}{2}\{\psi(x) + \psi(x-)\}$ there is an explicit formula

$$\psi_0(x) = x - \lim_{n\to\infty}\sum_{|\gamma|\leq T_n}\frac{x^\rho}{\rho} - \frac{\zeta'}{\zeta}(0) - \frac{1}{2}\log(1-x^{-2}).$$

This expression is the formal derivative of the formula for ψ_1. Because ψ_0 is not continuous, the convergence of $\sum x^\rho/\rho$ cannot be uniform near primes and prime powers. Thus, it is more convenient to work with ψ_1, whose corresponding series is uniformly convergent on any bounded interval.

Proof.   Consider the integral

$$I = I(m,x) = -\frac{1}{2\pi i}\int_C \frac{x^{s+1}}{s(s+1)}\frac{\zeta'}{\zeta}(s)ds,$$

where $C = C_m$ denotes the positively oriented rectangle having vertices $2 \pm iT_m$, $-2m - 1 \pm iT_m$. Here m is a positive integer and $\{T_m\}$ is the sequence from Lemma 8.20. Let $I_1(m)$ denote the integral over the line segment from $2 - iT_m$ to $2 + iT_m$ and let $I_2(m)$ denote the integral over the remainder of C.

By Perron's formula, $I_1(m) \to \psi_1(x)$ as $m \to \infty$. The estimates of ζ'/ζ of the preceding two lemmas imply that

$$I_2(m) = O\{x^3(\log m)^2 m^{-1}\} = o(1) \quad (m \to \infty).$$

On the other hand, by the residue theorem $I(m, x)$ equals the sum of the residues of $-x^{s+1}\zeta'(s)/\{\zeta(s)s(s+1)\}$ lying within C_m. Now let $m \to \infty$. The sum over the trivial zeros of $\zeta(s)$ gives the last series of the theorem; the sum over the nontrivial zeros of zeta converges absolutely by Lemma 8.19. Consequently we may drop the reference to the sequence T_m, and we have the desired representation for $\psi_1(x)$. $\qquad\square$

PROBLEM 8.25 Find the place in the proof where we used the condition that $x \geq 1$. What does the formula give for $x = 1$?

Corollary 8.24 *If the R.H. is true, then $\psi_1(x) = x^2/2 + O(x^{3/2})$.*

Proof. In this case we have

$$\left| \sum_\rho \frac{x^{3/2+i\gamma}}{\rho(\rho+1)} \right| \leq x^{3/2} \sum_\rho |\rho|^{-2} = O(x^{3/2}). \qquad\square$$

Corollary 8.25 *If the R.H. is true, then $\psi(x) = x + O(x^{1/2}\log^2 x)$.*

Proof. We difference the formula for ψ_1, obtaining on the one hand

$$\psi_1(x+1) - \psi_1(x) = \int_x^{x+1} \psi(u)du = \psi(x) + O(\log x).$$

On the other hand, we have

$$\psi_1(x+1) - \psi_1(x) = x + \frac{1}{2} - \sum_\rho \frac{(x+1)^{\rho+1} - x^{\rho+1}}{\rho(\rho+1)} - \frac{\zeta'}{\zeta}(0) + O\left(\frac{1}{x}\right).$$

If ω_ρ denotes the general term of the last series, then we have

$$\omega_\rho = \frac{1}{\rho} \int_x^{x+1} u^\rho du = O(x^{1/2}|\rho|^{-1}),$$

$$|\omega_\rho| \leq \frac{(x+1)^{3/2} + x^{3/2}}{|\rho(\rho+1)|} = O(x^{3/2}|\rho|^{-2}).$$

For $|\gamma| \leq x$, we use the first estimate and for $|\gamma| > x$ the second. Using Lemma 8.19 we find that

$$\left| \sum_\rho \omega_\rho \right| \leq A x^{1/2} \sum_{|\gamma| \leq x} |\rho|^{-1} + A x^{3/2} \sum_{|\gamma| > x} |\rho|^{-2}$$

$$\leq A' x^{1/2} \log^2 x$$

for some A and A'. This gives the desired estimate for ψ. $\qquad \square$

Theorem 8.26 *If the R.H. is true, then*

$$\frac{\pi(x) - \operatorname{li} x}{\sqrt{x} / \log x} = \frac{\psi(x) - x}{\sqrt{x}} + O(1) \tag{8.24}$$

and

$$\pi(x) = \operatorname{li} x + O(x^{1/2} \log x).$$

Proof. As in the proof of Theorem 8.15, we have

$$\Pi(x) := \sum_{n \leq x} \kappa(n) = \pi(x) + O(x^{1/2} / \log x)$$

and the identity

$$\Pi(x) = \operatorname{li} x + \int_{2-}^x \frac{d\psi - dt}{\log t} - \operatorname{li} 2.$$

If we combine these formulas and integrate by parts twice, we obtain

$$\pi(x) - \operatorname{li} x = \int_{2-}^x \frac{d\psi - dt}{\log t} + O(x^{1/2} / \log x)$$

$$= \frac{\psi(x) - x}{\log x} + \frac{\psi_1(x) - x^2/2}{x \log^2 x}$$

$$- \int_2^x \left\{ \psi_1(t) - \frac{1}{2} t^2 \right\} d\left(\frac{1}{t \log^2 t} \right) + O(x^{1/2} / \log x).$$

Now we apply the estimate $\psi_1(x) - x^2/2 = O(x^{3/2})$ to deduce (8.24). This formula and Corollary 8.25 yield $\pi(x) - \operatorname{li} x = O(x^{1/2} \log x)$. $\qquad \square$

PROBLEM 8.26 Assuming that $\zeta(1 + it) \neq 0$ for all real t, use the explicit formula to prove that $\psi_1(x) \sim x^2/2$ and then deduce the P.N.T.

PROBLEM 8.27   Let $\Theta = \sup\{\Re\rho : \zeta(\rho) = 0\}$. Assuming that $\frac{1}{2} < \Theta < 1$, give estimates for $\psi_1(x)$, $\psi(x)$, and $\pi(x)$.

8.11   Notes

8.1. The functional equation for zeta was first proved rigorously by Riemann in his famous paper on prime numbers cited in the Notes for §1.5. Over a hundred years before, Euler had arrived at this formula in the real domain by use of divergent series methods; cf. §§2.2–2.3 of [HarD]. For other proofs of the functional equation, see §§2.2–2.10 of [TiHB]. The proof we have presented seems to have been found first by M. Riesz; cf. H. Cramér, Thesis, Stockholm University, Almquist & Wiksell, Uppsala, 1917, pp. 9–10; also in Collected works, Springer-Verlag, Berlin, 1994, pp. 7–8.

8.3. Theorem 8.5 and most of the material in this section is due to Hadamard. Cf. J. Math. pures appl. (4), vol. 9 (1893), pp. 171–215.

8.4. Riemann's paper on prime numbers included six unproved assertions. Five of these were made without qualification, but the statement that all nonreal zeros of the zeta function lie on the line $\{s : \Re s = 1/2\}$ was preceded by the words "it is very likely that [es ist sehr wahrscheinlich dass] ...." This assertion, the so-called Riemann hypothesis ("R.H."), is the only one of Riemann's claims of that paper that remains unproved.

Theorem 8.8 was first obtained by de la Vallée Poussin, in Mémoires Couronnés de l'Acad. roy. des Sciences de Belgique, vol. 59 no. 1 (1899–1900). The proof given here (using Lemma 8.10) is due to Landau, Math. Zeit., vol. 20 (1924), pp. 98–104; also in [LanC], vol. 8, pp. 70–76. This important paper of Landau was reprinted in the 1953 edition of [LanH].

The rectangular portion of the largest known zerofree region was established computationally by S. Wedeniwski. The argument showed that the first 250 billion nontrivial zeros of the Riemann zeta function $\zeta(s)$ are simple and lie on the critical line $\sigma = 1/2$. This result was presented at the Conference in Number Theory in Honour of Professor H. C. Williams, Banff, Alberta, Canada, May 2003.

The largest currently known asymptotic zerofree region was established by K. Ford, in Number Theory for the Millennium, M. Bennett, et al. eds., vol. 2, AK Peters, Natick MA, 2002, pp. 25–56. Further discussion of zerofree estimates can be found in [TiHB], [Ivic], and [KarVo].

8.5. Theorem 8.11 is a slight modification of Theorem 20 of [Ing].

8.6. Cf. Theorems 21–23 of [Ing].

8.7. The first part of Theorem 8.15 was first obtained by de la Vallée Poussin and later more simply by Landau; cf. Notes for §8.4. The second part of this theorem is due to H. von Koch, Acta Math., vol. 24 (1901), pp. 159–182. The results on li x are given in Theorie des Integrallogarithmus, by N. E. Nielsen, B. G. Teubner, Leipzig, 1906.

More precise asymptotic estimates of p_n than that of Problem 8.14 have a rather complicated form; cf. M. Cipolla, Rend. Accad. Sci. Fis. Mat., Napoli (3), vol. 8 (1902), pp. 132–166. The following explicit one sided inequalities are known:

$$p_n > n \log n, \quad n \geq 1,$$

proved by J. B. Rosser in Proc. London Math. Soc. (2), vol. 45 (1938), pp. 21–44, and

$$p_n > n \left(\log n + \log \log n - 1 \right), \quad n \geq 2,$$

proved by P. Dusart, Math. Comp., vol. 68 (1999), pp. 411–415. The last estimate is the beginning of Cipolla's asymptotic series.

8.8. Theorem 8.17 is due to Landau, Rend. Circ. Mat. Palermo, vol. 26 (1908), pp. 169–302; also in [LanC], vol 3, pp. 411–544.

The assertion that the R.H. implies $M(x) = O(x^c)$ for every $c > 1/2$ was first proved by J. E. Littlewood, Comptes rendus Acad. Sci. Paris, vol. 154 (1912), pp. 263–266; also in Collected Papers of J. E. Littlewood, vol. 2, Oxford, 1982, pp. 793–796. See also [LanV], vol. 2, pp. 157–166.

8.9. Theorem 8.18 was conjectured by Riemann in his prime number paper, but with an erroneous error term. The result was first proved by H. von Mangoldt, Math. Annalen, vol. 60 (1905), pp. 1–19.

8.10. The idea of an explicit formula was introduced by Riemann, but without a rigorous proof. The explicit formula for ψ_0 was first established by H. von Mangoldt, J. reine angew. Math., vol. 114 (1895), pp. 255–305.

Chapter 9

Primes in Arithmetic Progressions

9.1 Residue characters

One of the early achievements of analytic number theory was Dirichlet's theorem on primes in arithmetic progressions, first proved in 1837. We formulate the theorem using the following notation. The letter k will be reserved for a positive integer denoting the common difference of successive terms in an arithmetic progression. If ℓ is also an integer, let $\mathcal{S}_{k\ell}$ denote the set of positive integers congruent to ℓ modulo k. Let $\mathcal{P}$ denote the set of primes.

If $(k, \ell) > 1$, $\mathcal{S}_{k\ell}$ can contain at most one prime. On the other hand, for some relatively prime pairs k, ℓ, straightforward elementary arguments show that $\mathcal{S}_{k\ell}$ contains infinitely many primes (see Notes). Dirichlet proved that, in fact, $\mathcal{S}_{k\ell}$ contains infinitely many primes whenever $(k, \ell) = 1$.

Theorem 9.1 *Let k and ℓ be relatively prime positive integers. Then $\mathcal{S}_{k\ell} \cap \mathcal{P}$ is infinite.*

For $k = 1$ or $k = 2$ this assertion is just Euclid's theorem that there are infinitely many primes. Proving the theorem for all $k > 2$ will be one of the main tasks of this chapter. For this purpose it is natural to consider the indicator function $e_{k\ell}$ of $\mathcal{S}_{k\ell}$. In other words, for positive integers k and ℓ, $e_{k\ell}$ is the arithmetic function satisfying

$$e_{k\ell}(n) = \begin{cases} 1 & \text{if} \quad n \equiv \ell \pmod{k}, \\ 0 & \text{if} \quad n \not\equiv \ell \pmod{k}. \end{cases}$$

PROBLEM 9.1   Suppose that $k \geq 3$ and $(k, \ell) = 1$. Show that $e_{k\ell}$ is not multiplicative. (The case $\ell \not\equiv 1 \pmod{k}$ is trivial.)

Dirichlet had the striking idea of expressing $e_{k\ell}$ as a linear combination of completely multiplicative arithmetic functions which, like $e_{k\ell}$, have period k and are zero on integers n for which $(n, k) > 1$. Such functions are called *residue characters modulo k*, or briefly, characters mod k.

In other words, a character modulo k is an arithmetic function χ with the following four properties:

$$\chi(mn) = \chi(m)\chi(n) \text{ for any positive integers } m \text{ and } n, \qquad (9.1)$$

$$\chi(n + k) = \chi(n) \text{ for any positive integer } n, \qquad (9.2)$$

$$\chi(n) = 0 \text{ for any positive integer } n \text{ with } (n, k) > 1, \qquad (9.3)$$

$$\chi(n_0) \neq 0 \text{ for some positive integer } n_0. \qquad (9.4)$$

PROBLEM 9.2   Let $\chi_3(n) := 2\sin(2\pi n/3)/\sqrt{3}$. Show that χ_3 is a character modulo 3. Show that χ_4 is a character modulo 4 if $\chi_4(n) := \sin(\pi n/2)$.

Let us determine all characters modulo 5. Since $2^4 \equiv 1 \pmod 5$, we must have $\chi(2)^4 = \chi(2^4) = \chi(1) = 1$ by requirements (9.1), (9.2), and (9.4). Thus $\chi(2)$ must be ± 1 or $\pm i$. Moreover, the value of $\chi(2)$ determines χ completely, since $\chi(1) = 1$, $\chi(3) = \chi(8) = \chi(2)^3$, $\chi(4) = \chi(2)^2$, $\chi(5) = 0$, and all other values are determined, by (9.2). On the other hand, each of these four choices for $\chi(2)$ yields a character modulo 5.

The Legendre symbol of elementary number theory provides a residue character modulo each odd prime p. Recall that if n is an integer, then

$$\left(\frac{n}{p}\right) = \begin{cases} 0 & \text{if } p \mid n, \\ 1 & \text{if } p \nmid n \text{ and } n \equiv x^2 \pmod p \text{ for some integer } x, \\ -1 & \text{if } p \nmid n \text{ and } n \not\equiv x^2 \pmod p \text{ for any integer } x. \end{cases}$$

We remark that (9.1) is a basic property of the Legendre symbol.

If $\chi_0(n) = 1$ when $(n, k) = 1$ and $\chi_0(n) = 0$ when $(n, k) > 1$, then χ_0 is a character modulo k; it is called the *principal character* mod k.

PROBLEM 9.3   If χ is a character modulo k, show that $\bar{\chi}$ is a character mod k, where $\bar{\chi}(n) = \overline{\chi(n)}$.

PROBLEM 9.4   Determine all characters modulo 8. Hint. If χ is a character modulo 8, show that $\chi(q^2) = 1$ for any odd integer q.

PROBLEM 9.5   Let k be an integer exceeding 1. Show that a periodic completely multiplicative function f is a character modulo its smallest positive period k. (That is, the requirement (9.3) is redundant in this case.) Hint. It suffices to prove that $f(p) = 0$ for all $p \mid k$.

Definition 9.2   A set of integers $n_1, n_2, \ldots, n_k$ is called a *complete residue system* modulo k if for every integer n there is precisely one n_i such that $n \equiv n_i \pmod{k}$. A set of integers $\{r_i\}$ is called a *reduced residue system* modulo k if (1) $(r_i, k) = 1$ for all i, (2) $r_i \not\equiv r_j \pmod{k}$ for all $i \neq j$, and (3) each integer n relatively prime to k is congruent to one of the members $\{r_i\}$ of the set.

For given k, there are $\varphi(k)$ integers in a reduced residue system modulo k. Thus there are $\varphi(k)$ distinct functions $e_{k\ell}$ with $(k, \ell) = 1$. These functions form a basis for the vector space $\mathcal{U}_k$ over $\mathbb{C}$ consisting of arithmetic functions which have period k and vanish on the integers n with $(n, k) > 1$. In order to express each $e_{k\ell}$ as a linear combination of characters modulo k, we shall show that the characters also form a basis for $\mathcal{U}_k$. We first show (Theorem 9.4) that any set of distinct characters modulo k is linearly independent. Later we shall show that there are sufficiently many characters modulo k to span $\mathcal{U}_k$. We begin by establishing an orthogonality relation between characters.

Theorem 9.3   *Let χ_1 and χ_2 be characters modulo k. Then*

$$\sum_{n \bmod k} \bar{\chi}_1(n)\chi_2(n) = \begin{cases} 0, & \chi_1 \neq \chi_2 \\ \varphi(k), & \chi_1 = \chi_2. \end{cases}$$

(The summation extends over a complete residue system modulo k.)

Proof.   We first show that $|\chi|$ can assume only the values 0 or 1. If $(n, k) = 1$, then the Euler–Fermat theorem gives $n^{\varphi(k)} \equiv 1 \pmod{k}$. It follows that

$$\chi(n)^{\varphi(k)} = \chi(n^{\varphi(k)}) = \chi(1) = 1,$$

and hence $\chi(n)$ is a $\varphi(k)$th root of unity. On the other hand, if $(n, k) > 1$, then by condition (9.3) we have $\chi(n) = 0$.

If $\chi_1 = \chi_2$, then

$$\sum_{n \bmod k} \bar{\chi}_1(n)\chi_2(n) = \sum_{n \bmod k} |\chi_1(n)|^2 = \sum_{\substack{n \bmod k \\ (n,k)=1}} 1 = \varphi(k).$$

Now suppose that $\chi_1 \neq \chi_2$. There must exist an integer n_0 for which $\chi_1(n_0) \neq \chi_2(n_0)$. Clearly, $(n_0, k) = 1$. Thus, when n runs through a complete residue system modulo k, nn_0 does likewise. It follows from the periodic and multiplicative properties of characters that

$$\sum_{n \bmod k} \bar{\chi}_1(n)\chi_2(n) = \sum_{n \bmod k} \bar{\chi}_1(n_0 n)\chi_2(n_0 n)$$

$$= \bar{\chi}_1(n_0)\chi_2(n_0) \sum_{n \bmod k} \bar{\chi}_1(n)\chi_2(n)$$

or

$$(1 - \bar{\chi}_1(n_0)\chi_2(n_0)) \sum_{n \bmod k} \bar{\chi}_1(n)\chi_2(n) = 0.$$

Now $1 - \bar{\chi}_1(n_0)\chi_2(n_0) \neq 0$, and thus the last sum must be zero. $\quad\square$

If $\chi = \chi_0$, the principal character modulo k, then

$$\sum_{n \bmod k} \chi_0(n) = \varphi(k),$$

while if χ is a nonprincipal character modulo k, Theorem 9.3 gives

$$\sum_{n \bmod k} \chi(n) = \sum_{n \bmod k} \chi_0(n)\chi(n) = 0. \qquad (9.5)$$

Now we establish the linear independence of characters.

Theorem 9.4 *If $\chi_1, \chi_2, \ldots, \chi_r$ are distinct characters modulo k and a_1, $a_2, \ldots, a_r$ are complex numbers such that $a_1\chi_1 + a_2\chi_2 + \cdots + a_r\chi_r$ is the zero function, then $a_1 = a_2 = \cdots = a_r = 0$.*

Proof. For each positive integer n we have

$$a_1\bar{\chi}_1(n)\chi_1(n) + a_2\bar{\chi}_1(n)\chi_2(n) + \cdots + a_r\bar{\chi}_1(n)\chi_r(n) = 0.$$

Summing over a complete residue system and applying Theorem 9.3, we get $a_1 = 0$. Similar reasoning shows that $a_2, \ldots, a_r$ are zero. $\quad\square$

In particular, since $\mathcal{U}_k$ has dimension $\varphi(k)$ and the collection of distinct characters modulo k is linearly independent by the preceding theorem, it follows that there are at most $\varphi(k)$ characters modulo k. After examining the multiplicative behavior of integers relatively prime to k, we shall show that there actually are $\varphi(k)$ characters modulo k.

9.2 Group structure of the coprime residue classes

If b and c are integers relatively prime to k, then bc is relatively prime to k. Moreover the residue class modulo k into which bc falls is determined by the residue classes to which b and c belong (rather than by the particular numbers b and c themselves). This defines a multiplication in the set of coprime residue classes, i.e., those residue classes consisting of integers relatively prime to k.

Since $b \cdot 1 \equiv b \pmod{k}$, the residue class containing 1 serves as a unity element for this multiplication. Given an integer b relatively prime to k, we know from elementary number theory that there is an integer b^* (also relatively prime to k) such that $bx \equiv 1 \pmod{k}$ if and only if $x \equiv b^*$ $\pmod{k}$. We usually refer to b^* as *the multiplicative inverse of b modulo* k, even though only the residue class containing b^* is determined.

Thus the residue classes relatively prime to k form an abelian (commutative) group under multiplication. This group is called the *coprime residue-class group modulo k* and will be denoted by $\mathcal{G}_k$.

Theorem 9.4 implies that the number of characters modulo k cannot exceed the dimension of $\mathcal{U}_k$, viz. $\varphi(k)$. We shall show that there are exactly $\varphi(k)$ characters modulo k with the aid of the following

Lemma 9.5 *There exist positive integers m, f_1, $f_2, \ldots, f_m$ and integers y_1, $y_2, \ldots, y_m$ satisfying $(y_i, k) = 1$ and $y_i^{f_i} \equiv 1 \pmod{k}$ for $i = 1$, $2, \ldots, m$ such that each coprime residue class of $\mathcal{G}_k$ contains exactly one element of the form $y_1^{j_1} y_2^{j_2} \cdots y_m^{j_m}$ with $0 \le j_i < f_i$ for $i = 1, 2, \ldots, m$.*

Proof. This result is a special case of the basis theorem for finite abelian groups, which is stated and proved in the Appendix. □

Another proof of the lemma can be given along the following lines. First suppose k is a prime or a power of a prime. If $k = 2$, 4, an odd prime, a power of an odd prime, or twice the power of an odd prime, then it is known that the coprime residue class group is cyclic and so the lemma holds with $m = 1$. In these cases y_1 is called a *primitive root* modulo k. On the other hand, if $k = 2^a$, where $a > 2$, then $\varphi(k) = 2^{a-1}$ but induction on a gives $x^{2^{a-2}} \equiv 1 \pmod{2^a}$ for any odd x; thus in this case no primitive root exists. However, one can show that when $k = 2^a$, where $a > 2$, the lemma holds with $m = 2$, $y_1 = -1$, $f_1 = 2$, $y_2 = 5$, and $f_2 = 2^{a-2}$.

If k is a number divisible by more than one prime, then we combine the preceding results with the Chinese remainder theorem. For example, if $k = 98000 = 2^4 \cdot 5^3 \cdot 7^2$, we can determine y_i and f_i, $1 \le i \le 4$, as follows. We choose y_1, y_2, y_3, y_4 satisfying

$$y_1 \equiv -1 \pmod{2^4}, \quad y_1 \equiv 1 \pmod{5^3}, \quad y_1 \equiv 1 \pmod{7^2},$$

$$y_2 \equiv 5 \pmod{2^4}, \quad y_2 \equiv 1 \pmod{5^3}, \quad y_2 \equiv 1 \pmod{7^2},$$

$$y_3 \equiv 1 \pmod{2^4}, \quad y_3 \equiv 2 \pmod{5^3}, \quad y_3 \equiv 1 \pmod{7^2},$$

$$y_4 \equiv 1 \pmod{2^4}, \quad y_4 \equiv 1 \pmod{5^3}, \quad y_4 \equiv 3 \pmod{7^2},$$

e.g. $y_1 = 36751$, $y_2 = 24501$, $y_3 = 89377$, and $y_4 = 54001$. We use the facts that 2 and 3 are primitive roots mod 5^3 and 7^2 respectively. Two congruences are required for 2^4 and one congruence for each of the moduli 5^3 and 7^2. Thus the assertion of the lemma holds with

$$m = 4, f_1 = 2, f_2 = 4, f_3 = 100 = \varphi(5^3), \text{ and } f_4 = 42 = \varphi(7^2).$$

9.3 Existence of enough characters

We show here that the set of characters modulo k forms a basis for the vector space $\mathcal{U}_k$ spanned by the functions $e_{k\ell}$, $(\ell, k) = 1$.

Theorem 9.6 *For any $k \in \mathbb{Z}$, there are $\varphi(k)$ characters modulo k.*

Proof. By Theorem 9.4 there can be at most $\varphi(k)$ distinct characters modulo k; we exhibit $\varphi(k)$ distinct characters modulo k. Our construction is motivated by the observation made in the proof of Theorem 9.3 that nonzero values of a residue character modulo k are $\varphi(k)$th roots of unity.

We use the result and notation of Lemma 9.5. Since the integers

$$y_1^{j_1} y_2^{j_2} \cdots y_m^{j_m} \quad (0 \le j_i < f_i; \quad i = 1, 2, \ldots, m)$$

represent each coprime residue class modulo k once and only once, it follows that $f_1 f_2 \cdots f_m = \varphi(k)$.

Let ρ_i denote an arbitrary f_ith root of unity for $i = 1, 2, \ldots, m$ and put

$$\chi(y_1^{j_1} \cdots y_m^{j_m}) = \rho_1^{j_1} \cdots \rho_m^{j_m}, \quad \chi(n) = 0 \text{ if } (n, k) > 1, \quad \chi(n + k) = \chi(n).$$

Then χ is a well-defined k-periodic completely multiplicative function supported on the positive integers relatively prime to k, and so is a character modulo k. Clearly, different m-tuples $\rho_1, \rho_2, \ldots, \rho_m$ give rise to different characters. Since there are f_i possible choices of ρ_i, for $1 \le i \le m$, the total number of characters modulo k is $f_1 f_2 \cdots f_m = \varphi(k)$. $\qquad\square$

If χ_1 and χ_2 are characters modulo k, we can define their product $\chi_1\chi_2$ by pointwise multiplication: $(\chi_1\chi_2)(n) = \chi_1(n) \cdot \chi_2(n)$. Under this binary operation the characters modulo k themselves form an abelian group C_k, which we have just proved to have order $\varphi(k)$. The identity element of C_k is the principal character modulo k and the inverse of a character χ is the character $\bar\chi$, whose values are the complex conjugates of those of χ.

Theorem 9.7 *Suppose $(\ell, k) = 1$ and set $c(\chi) := \bar\chi(\ell)/\varphi(k)$. Then*

$$e_{k\ell} = \sum_{\chi \in C_k} c(\chi)\chi.$$

Proof. Let $h = \varphi(k)$. Let $\chi_1, \chi_2, \ldots, \chi_h$ denote the distinct characters modulo k. Since these functions are linearly independent by Theorem 9.4, they form a basis for $\mathcal{U}_k$. In particular $e_{k\ell}$ can be expressed in the form

$$e_{k\ell} = c_1\chi_1 + c_2\chi_2 + \cdots + c_h\chi_h,$$

where the c's are suitable complex numbers.

To find c_j, first multiply the preceding equation by $\bar\chi_j$, obtaining

$$e_{k\ell}(n)\bar\chi_j(n) = c_1\chi_1(n)\bar\chi_j(n) + \cdots + c_h\chi_h(n)\bar\chi_j(n)$$

for any n. Then we sum this expression letting n run through a complete residue system modulo k. We find by Theorem 9.3

$$\bar\chi_j(\ell) = c_j \sum_{n \bmod k} |\chi_j(n)|^2 = c_j h. \qquad\square$$

PROBLEM 9.6 Suppose that $(\ell, k) = 1$ and $n \not\equiv \ell \pmod{k}$. Use the proof of Theorem 9.6 to show that there is a character χ^* such that $\chi^*(n) \ne \chi^*(\ell)$.

Using this fact and dualizing the proof of Theorem 9.3, show that

$$\sum_{\chi \in C_k} \bar{\chi}(\ell)\chi(n) = \varphi(k)e_{k\ell}(n)$$

9.4   L functions

Given a character χ modulo k, there is an associated D.s., called an L function, defined for $\sigma > 1$ by $L(s,\chi) := \sum \chi(n)n^{-s}$. The series is absolutely convergent for $\sigma > 1$, and since χ is completely multiplicative we can apply Corollary 6.5 to express L as a product

$$L(s,\chi) = \prod_{p}(1 - \chi(p)p^{-s})^{-1} = \prod_{p \nmid k}(1 - \chi(p)p^{-s})^{-1}$$

for $\sigma > 1$. Since there are $\varphi(k)$ distinct characters modulo k, there are $\varphi(k)$ distinct L functions modulo k.

Example 9.8   Let $k = 4$, let χ_0 be the principal character modulo 4, and let χ_1 be the character modulo 4 defined by $\chi_1(n) = -1$ for $n \equiv 3 \pmod 4$. Then for $\sigma > 1$ we have

$$L(s,\chi_0) = \sum_{k=0}^{\infty}(2k+1)^{-s} = (1 - 2^{-s})\zeta(s),$$

$$L(s,\chi_1) = \sum_{k=0}^{\infty}(-1)^k(2k+1)^{-s} = \prod_{p \equiv 1(4)}(1 - p^{-s})^{-1}\prod_{p \equiv 3(4)}(1 + p^{-s})^{-1}.$$

$$(9.6)$$

PROBLEM 9.7   Prove that the D.s. $\sum(-1)^{n-1}n^{-s}$ is not an L function.

For $\chi = \chi_0$, the principal character modulo k, we have for $\sigma > 1$

$$L(s,\chi_0) = \sum_{(n,k)=1} n^{-s} = \prod_{p \nmid k}(1 - p^{-s})^{-1} = \zeta(s)\prod_{p \mid k}(1 - p^{-s}). \qquad (9.7)$$

This function is called the *principal L function modulo k*. Knowledge of zeta provides extensive information about this function. In particular, $L(s,\chi_0)$ has an analytic continuation to $\mathbb{C}$ except for a simple pole at $s = 1$. The residue there equals $\varphi(k)/k$, where φ denotes Euler's function.

The remaining L functions have a larger half plane of convergence:

Lemma 9.9 *Let χ be a nonprincipal character. Then $\sigma_c\{L(\cdot,\chi)\} = 0$.*

Proof. Setting $S(x) = \sum_{n \le x} \chi(n)$, we have $L(s,\chi) = \widehat{S}(s)$. S is bounded since a sum of χ over a complete residue system modulo k yields zero. By Theorem 6.9 the D.s. for $L(s,\chi)$ converges for $\sigma > 0$. On the other hand, the series clearly diverges for $s = 0$.

Example 9.10 There exist infinitely many primes in each of the residue classes 1 (mod 4) and 3 (mod 4). We use the series in formula (9.6) for $L(s,\chi_1)$ and the alternating series inequalities

$$1 - 1/3 < 1 - 1/3^s < L(s,\chi_1) < 1 \quad (s > 1).$$

Suppose $\mathcal{P} \cap \{n : n \equiv 3 \ (\text{mod } 4)\}$ were finite. Then we would have $L(s,\chi_1) = \zeta(s)f(s)$, where f is a finite product of factors which do not vanish at $s = 1$. Letting $s \to 1+$, we obtain $|\zeta(s)f(s)| \to \infty$ while $|L(s,\chi_1)| \le 1$, which is impossible.

On the other hand, if $\mathcal{P} \cap \{n : n \equiv 1 \ (\text{mod } 4)\}$ were finite, then $L(s,\chi_1) = \{\zeta(s)/\zeta(2s)\}^{-1}g(s)$, where g is a finite product of factors each of which is bounded in a neighborhood of $s = 1$. Letting $s \to 1+$, we find that $\zeta(2s)g(s)/\zeta(s) \to 0$ while $L(s,\chi_1) \ge 2/3$, which is also impossible. $\square$

The essential elements in the preceding argument were the boundedness and nonvanishing of the nonprincipal L function at 1. These properties will be needed for the proof of Theorem 9.1. The first property follows from Lemma 9.9. We shall establish the second in

Lemma 9.11 *If χ_1 is a nonprincipal character, then $L(1,\chi_1) \ne 0$. If in addition χ_1 is real, then $L(1,\chi_1) > 0$.*

Proof. Let $P(s) = \prod L(s,\chi)$, the product extending over all characters χ mod k. If $L(1,\chi_1) = 0$, then the pole of the principal L function $L(s,\chi_0)$ at $s = 1$ would be cancelled and P would be analytic on $\{s : \Re s > 0\}$. We show by contradiction that P fails to be analytic at some points of this half plane.

There is a branch of $\log L(\cdot,\chi)$ satisfying

$$\log L(s,\chi) = \sum_p \sum_{\alpha=1}^{\infty} \frac{\chi(p^\alpha)}{\alpha p^{\alpha s}} = \sum_{n=1}^{\infty} \kappa(n)\chi(n)n^{-s} \qquad (9.8)$$

for $\sigma > 1$. This formula may be seen e.g. by the method of §6.6 in which we established an analogous formula for a logarithm of zeta. Alternatively, we can also establish (9.8) by taking logarithms in the Euler product for L and expanding each term of the resulting series. Note that the above series are absolutely convergent for $\sigma > 1$.

We set

$$Q(s) = \log P(s) = \sum_{\chi} \log L(s, \chi)$$

and, upon changing the summation order, obtain

$$Q(s) = \sum_{p} \sum_{\alpha} \sum_{\chi} \chi(p^{\alpha})/(\alpha p^{\alpha s})$$

$$= \varphi(k) \sum_{p^{\alpha} \equiv 1 \, (k)} 1/(\alpha p^{\alpha s}) = \sum_{n=1}^{\infty} f(n) n^{-s}, \text{ say.}$$

Here $f(1) := 0$ and for $n > 1$,

$$f(n) := \begin{cases} \varphi(k)/\alpha & \text{if } n = p^{\alpha} \equiv 1 \pmod{k}, \\ 0 & \text{otherwise.} \end{cases}$$

Let $g = \exp f$. Note that g assumes only nonnegative values. Then by Lemma 6.28 on operational calculus we have the D.s. representation

$$P(s) = \sum_{n=1}^{\infty} g(n) n^{-s}.$$

By Landau's oscillation theorem (Th. 6.31), this series must converge for all positive values of s, since P is assumed to be analytic on $\{s : \Re s > 0\}$.

On the other hand, $p^h \equiv 1 \pmod{k}$ for $p \nmid k$ and $h := \varphi(k)$ by Euler's theorem. Thus, $f(p^h) = 1$ for $p \nmid k$. Since $g = \exp f \geq f$, we have

$$\sum_{n=1}^{\infty} g(n) n^{-1/h} \geq \sum_{p} g(p^h) p^{-1} \geq \sum_{p \nmid k} f(p^h) p^{-1} = +\infty.$$

It follows that the abscissa of convergence of $\sum g(n) n^{-s}$ is strictly positive, and hence $L(1, \chi_1) \neq 0$.

Let χ_1 be real. By the Euler product representation, $L(\sigma, \chi_1) > 0$ for $\sigma > 1$. Also, $L(\sigma, \chi_1)$ is continuous on $[1, \infty)$ and $L(1, \chi_1) \neq 0$. It follows that $L(1, \chi_1) > 0$ in this case. $\qquad\square$

PROBLEM 9.8   Suppose χ_1 is a character which assumes some nonreal values. Modify the preceding argument to exploit the relation $L(\sigma, \bar{\chi}_1) = \overline{L(\sigma, \chi_1)}$ in place of Landau's theorem. Hint. If $L(1, \chi_1) = 0$, then $P(1) = 0$ and hence $Q(\sigma) \to -\infty$ as $\sigma \to 1+$.

PROBLEM 9.9   Let χ be the (real valued) character modulo 8 satisfying $\chi(3) = \chi(5) = -1$. By judicious grouping of terms, show that

$$L(1, \chi) = 1 - \frac{1}{3} - \frac{1}{5} + \frac{1}{7} + \frac{1}{9} - \frac{1}{11} - \frac{1}{13} + \frac{1}{15} + \cdots > \frac{3}{5}.$$

Finally, we note a few useful formulas for characters and L functions. Recall (Problem 2.25) that if f is a completely multiplicative arithmetic function and g and h are arbitrary arithmetic functions then

$$f \cdot (g * h) = (f \cdot g) * (f \cdot h). \tag{9.9}$$

In particular we have the formulas

$$(\chi \cdot \mu) * \chi = (\chi \cdot \mu) * (\chi \cdot 1) = \chi \cdot (\mu * 1) = \chi \cdot e_1 = e_1, \tag{9.10}$$

$$(\chi \cdot \Lambda) * \chi = (\chi \cdot \Lambda) * (\chi \cdot 1) = \chi \cdot (\Lambda * 1) = \chi \cdot L1 = L\chi. \tag{9.11}$$

(Here L is the logarithm operator!) These identities are equivalent to the following L function relations, valid for $\sigma > 1$:

$$\sum \chi(n)\mu(n)n^{-s} = 1/L(s, \chi), \tag{9.12}$$

$$\sum \chi(n)\Lambda(n)n^{-s} = -L'(s, \chi)/L(s, \chi). \tag{9.13}$$

9.5   Proof of Dirichlet's theorem

In this section we establish Theorem 9.1. Actually, we show that

$$\sum_{p \leq x} e_{k\ell}(p)(\log p)/p = \varphi(k)^{-1} \log x + O(1) \tag{9.14}$$

holds if $(\ell, k) = 1$. Recalling Mertens' relation $\sum_{p \leq x}(\log p)/p = \log x + O(1)$, we see that formula (9.14) expresses the deeper fact that the primes in each eligible residue class have a weighted density $1/\varphi(k)$. Further, by using the formula once at $x \geq 1$ and once at Mx for some suitably large $M > 1$ and differencing, we see that each eligible residue class contains a positive proportion of the primes.

We first multiply (9.11) by T^{-1} and apply (9.9), obtaining

$$(T^{-1}\chi\Lambda) * (T^{-1}\chi) = T^{-1}L\chi.$$

We apply iterated summation to this convolution and get

$$\sum_{m \le x} \frac{\chi(m)\Lambda(m)}{m} \sum_{\ell \le x/m} \frac{\chi(\ell)}{\ell} = \sum_{n \le x} \frac{\chi(n)\log n}{n}. \qquad (9.15)$$

For χ a nonprincipal character, $\sum_{n \le x} \chi(n) = O(1)$ by (9.5). By partial summation, the right side of (9.15) is bounded for such χ and, further,

$$L(1,\chi) - \sum_{n \le x/m} \chi(n)/n = \sum_{n > x/m} \chi(n)/n = O(m/x).$$

Putting these estimates into (9.15) we obtain

$$\sum_{m \le x} \frac{\chi(m)\Lambda(m)}{m} \left\{ L(1,\chi) + O\left(\frac{m}{x}\right) \right\} = O(1).$$

Now Chebyshev's upper bound for ψ, Lemma 4.5, implies that

$$\sum_{m \le x} \frac{\chi(m)\Lambda(m)}{m} O\left(\frac{m}{x}\right) = O(1).$$

Also, Lemma 9.11 asserts that $L(1,\chi) \ne 0$. Thus we have for nonprincipal characters χ the estimate

$$\sum_{m \le x} \frac{\chi(m)\Lambda(m)}{m} = O(1).$$

On the other hand, if $\chi = \chi_0$, Mertens' estimate (Lemma 4.8) gives

$$\sum_{m \le x} \frac{\chi_0(m)\Lambda(m)}{m} = \sum_{\substack{m \le x \\ (m,k)=1}} \frac{\Lambda(m)}{m} + O(1) = \log x + O(1).$$

We combine these estimates with Theorem 9.7, the representation of

$e_{k\ell}$, to obtain

$$\sum_{n \leq x} e_{k\ell}(n) \frac{\Lambda(n)}{n} = \sum_{n \leq x} \frac{1}{\varphi(k)} \sum_{\chi \in C_k} \bar{\chi}(\ell)\chi(n) \frac{\Lambda(n)}{n}$$

$$= \frac{1}{\varphi(k)} \sum_{\chi \in C_k} \bar{\chi}(\ell) \sum_{n \leq x} \frac{\chi(n)\Lambda(n)}{n}$$

$$= \frac{1}{\varphi(k)} \log x + O(1).$$

Finally, we have

$$\sum_{n \text{ not prime}} \Lambda(n)/n = \sum (\log p)(p^{-2} + p^{-3} + \cdots) = \sum \frac{\log p}{p(p-1)} < \infty,$$

and hence (9.14) holds. $\qquad\square$

PROBLEM 9.10 Show that $\sum \chi(p)/p$ and $\prod_p (1 - \chi(p)/p)$ converge for χ nonprincipal.

PROBLEM 9.11 Show that if $(k, \ell) = 1$, then

$$\sum_{p \leq x} e_{k\ell}(p)/p = \frac{\log\log x}{\varphi(k)} + C_{k,\ell} + O(\log^{-1} x)$$

for some constant $C_{k,\ell}$.

PROBLEM 9.12 Let $(k, \ell) = 1$. Show that the density of squarefree integers congruent to ℓ modulo k is $6/\{\pi^2 k \prod_{p|k}(1 - p^{-2})\}$. What can you say when $(k, \ell) > 1$?

9.6 P.N.T. for arithmetic progressions

Let

$$\pi(x; k, \ell) := \sum_{\substack{p \leq x \\ p \equiv \ell\,(k)}} 1 \quad \text{and} \quad \psi(x; k, \ell) := \sum_{\substack{n \leq x \\ n \equiv \ell\,(k)}} \Lambda(n).$$

For $(k, \ell) = 1$, we have seen (cf. (9.14)) that

$$\int_1^x t^{-1} d\psi(t; k, \ell) = \frac{1}{\varphi(k)} \log x + O(1).$$

This formula raises the question of whether, asymptotically, the primes are equally distributed among the $\varphi(k)$ eligible residue classes. This is indeed the case, as we now shall prove.

Theorem 9.12 *Let $(k, \ell) = 1$. Then*

$$\pi(x; k, \ell) \sim \frac{x}{\varphi(k) \log x} \qquad (x \to \infty).$$

We deduce this result by applying the Wiener-Ikehara theorem to a linear combination of L functions modulo k. The key step is to establish

Lemma 9.13 *Let χ be a nonprincipal character mod k. Then $L(s, \chi) \neq 0$ for $\Re s \geq 1$.*

Proof. For $\sigma > 1$, $L(s, \chi)$ is represented by a convergent infinite product of nonzero factors, and therefore has no zeros on this (open) half plane. By Lemma 9.11, $L(1, \chi) \neq 0$. We now show that there is no real $t \neq 0$ for which $L(1 + it, \chi) = 0$.

Letting χ_0 denote the principal character modulo k and recalling formula (9.8) for $\log L(s, \chi)$, we have for $\sigma > 1$

$$3 \log |L(\sigma, \chi_0)| + 4 \log |L(\sigma + it, \chi)| + \log |L(\sigma + 2it, \chi^2)|$$

$$= \Re \sum_{\substack{n=1 \\ (n,k)=1}}^{\infty} n^{-\sigma} \kappa(n) \{3 + 4n^{-it} \chi(n) + n^{-2it} \chi^2(n)\}.$$

Now $n^{-it} \chi(n) = e^{i\theta}$ for some real $\theta = \theta(n)$, and $3 + 4 \cos \theta + \cos 2\theta = 2(1 + \cos \theta)^2 \geq 0$, and also $\kappa(n) \geq 0$. It follows that

$$|L^3(\sigma, \chi_0) L^4(\sigma + it, \chi) L(\sigma + 2it, \chi^2)| \geq 1. \qquad (9.16)$$

Now let t_0 be a fixed nonzero real number. We have

$$L(\sigma, \chi_0) = \zeta(\sigma) \prod_{p|k} (1 - p^{-\sigma}) \sim \frac{1}{\sigma - 1} \prod_{p|k} (1 - p^{-1})$$

as $\sigma \to 1+$. We have $\chi^2 = \chi_0$ if χ is real, and $\chi^2 \neq \chi_0$ otherwise. In either case, $L(s, \chi^2)$ is analytic on $\{s = \sigma + it, \sigma > 0, t \neq 0\}$. It follows that $L(\sigma + 2it_0, \chi^2)$ is bounded for $1 \leq \sigma \leq 2$, say. Using (9.16), we obtain

$$|L(\sigma + it_0, \chi)| \geq c(\sigma - 1)^{3/4}$$

for some $c > 0$ and $1 < \sigma \leq 2$, i.e. a zero of L at $1 + it_0$ can have order at most $3/4$. Since L is analytic at $1 + it_0$, we have $L(1 + it_0, \chi) \neq 0$. $\qquad\square$

Proof of Theorem 9.12. For $\sigma > 1$ we have

$$\int x^{-s} d\psi(x; k, \ell) = \sum_{n \equiv \ell (k)} n^{-s} \Lambda(n)$$

$$= \frac{1}{\varphi(k)} \sum_{n=1}^{\infty} n^{-s} \sum_{\chi} \chi(n) \bar{\chi}(\ell) \Lambda(n)$$

$$= -\frac{\bar{\chi}_0(\ell)}{\varphi(k)} \frac{L'}{L}(s, \chi_0) - \sum_{\chi \neq \chi_0} \frac{\bar{\chi}(\ell)}{\varphi(k)} \frac{L'}{L}(s, \chi).$$

Now

$$\frac{-L'}{L}(s, \chi_0) = \sum_{(n,k)=1} \Lambda(n) n^{-s} = -\frac{\zeta'}{\zeta}(s) - \sum_{(n,k)>1} \Lambda(n) n^{-s},$$

and the last series converges for $\sigma > 0$. For each $\chi \neq \chi_0$, the function $(L'/L)(s, \chi)$ is analytic on the closed half plane $\{s : \sigma \geq 1\}$ in consequence of the analyticity and nonvanishing of $L(s, \chi)$ on the same set. Thus

$$\int x^{-s} d\psi(x; k, \ell) = \frac{1}{\varphi(k)} \cdot \frac{1}{s - 1} + F(s),$$

where F is analytic on $\{s : \sigma \geq 1\}$.

Further, $\psi(x; k, \ell)$ is increasing in x and hence the Wiener-Ikehara theorem is applicable. We conclude that

$$\psi(x; k, \ell) \sim x/\varphi(k)$$

for $(k, \ell) = 1$. The passage to the estimate

$$\pi(x; k, \ell) \sim x/\{\varphi(k) \log x\}$$

is made in essentially the same way as the passage from ψ to π in §4.2. The principal change consists in replacing (4.2) by

$$\psi(x; k, \ell) \leq \sum_{\substack{p \leq x \\ p \equiv \ell (k)}} \log p + \sum_{\substack{p^\alpha \leq x \\ \alpha \geq 2}} \log p \leq \pi(x; k, \ell) \log x + O(x^{1/2}). \qquad\square$$

PROBLEM 9.13   Where does the proof of Theorem 9.12 first fail if k and ℓ are not relatively prime?

PROBLEM 9.14   Let $\sum^*$ denote summation over the cubefree positive integers, i.e., those positive integers not divisible by the cube of an integer exceeding 1.

(a) Prove that if χ is any character, then for $\Re s > 1$,

$$\sum\nolimits^* \chi(n)n^{-s} = L(s,\chi)/L(3s,\chi^3). \qquad (9.17)$$

(b) Prove that if χ is any nonprincipal character and $\Re s > 1/3$, then the series $\sum^* \chi(n)n^{-s}$ converges and (9.17) holds. Hint. For $\Re s > 1/3$,

$$1/L(3s,\chi^3) = \sum \mu(n)\chi^3(n)n^{-3s}.$$

9.7   Notes

9.1.   The "straightforward elementary arguments" for special integer pairs k, ℓ are essentially extensions of Euclid's proof of the infinitude of the set of all primes. While Euclid's proof considered the prime factors of $P + 1$, where P is the product of some finite set of known primes, these proofs involve more complicated polynomials than $x+1$. For example, when $k = 8$ and $\ell = 7$, one can use the polynomial $8x^2 - 1$, since each prime factor of $8P^2 - 1$ is congruent to 1 or 7 modulo 8, and at least one prime factor must be congruent to 7 modulo 8. It was shown by R. Murty that a proof of Euclid's type can be given only if $\ell^2 \equiv 1 \pmod{k}$. Murty's paper appears in the Journal of Madras University, Section B (1988), pp. 161–169. If k is a divisor of 24, then $\ell^2 \equiv 1 \pmod{k}$ holds for every ℓ relatively prime to k; for a proof of Euclid's type when $k = 24$, see P. T. Bateman and M. E. Low, Amer. Math. Monthly, vol. 72 (1965), pp. 139–143.

Dirichlet's proof of Theorem 9.1 is given in his paper in Abhandlungen Preuss. Akad. Wiss. 1837, pp. 45–71; also in Werke, vol. 1, Berlin, 1889 (reprinted by Chelsea Publishing Co., 1969), pp. 313–342. Residue characters were introduced in this paper.

9.6.   Theorem 9.12 was obtained independently by Hadamard and de la Vallée Poussin in their P.N.T. papers cited in the Notes for §1.3.

Chapter 10

Applications of Characters

10.1 Integers generated by primes in residue classes

We have seen that the primes are equitably distributed among the eligible residue classes. Thus for example, about half the primes in any interval $[1, x]$ are congruent to 1 (mod 4) and the remaining odd primes are congruent to 3 (mod 4). This suggests that $\prod_{p \equiv 1 \ (\text{mod } 4)} (1 - p^{-\sigma})^{-1}$ is in some rough sense nearly equal to $\zeta(\sigma)^{1/2}$.

The above Euler product is associated with the arithmetic function $f = 1_5 * 1_{13} * 1_{17} * \cdots$, the indicator of the set of positive integers whose prime factors are all congruent to 1 (mod 4). On the other hand, the arithmetic function associated with $\zeta^{1/2}$ has a summatory function (of x) which is asymptotic to $x(\log x)^{-1/2}/\Gamma(1/2)$ (cf. Problem 7.11). Thus we might guess that the summatory function of f has a similar asymptotic behavior.

More generally, let k be a given integer, $k \geq 2$, and let $h = \varphi(k)$. Let a_1, $a_2, \ldots, a_h$ be a reduced residue system modulo k. Let $b_1, b_2, \ldots, b_h$ be real numbers in $[0, 1]$, not all 0. Let f be a multiplicative function having the properties (1) $f(p) := b_j$ if $p \equiv a_j$ (mod k) and (2) $0 \leq f \leq 1$. The values of $f(p)$ for $p \mid k$ and the values of $f(p^\alpha)$ for $\alpha \geq 2$ are arbitrary, subject to condition (2). Let $\beta := b_1 + b_2 + \cdots + b_h$. In the following theorem we shall estimate the summatory function of f.

Theorem 10.1 *Let f be as above. There exists a positive number $c = c(f)$ such that*

$$\sum_{n \leq x} f(n) \sim c \, x (\log x)^{(\beta/h) - 1}.$$

Proof.   We apply the generalized Wiener-Ikehara theorem (Th. 7.7) to $\widehat{F}(s) := \sum f(n) n^{-s}$. To do this, we show that

$$\widehat{F}(s) = (s - 1)^{-\beta/h} F^*(s),$$

where $F^*(s)$ is analytic on $\{s : \sigma \geq 1\}$ and $F^*(1) \neq 0$. Since $f \geq 0$, it follows that $F^*(1) > 0$.

By the multiplicativity of f,

$$\widehat{F}(s) = \prod_{\ell=1}^{h} \prod_{p \equiv a_\ell(k)} \left(1 + \frac{b_\ell}{p^s} + \frac{f(p^2)}{p^{2s}} + \cdots \right) K \qquad (10.1)$$

with

$$K := K(s) := \prod_{p \mid k} \left\{ 1 + \frac{f(p)}{p^s} + \frac{f(p^2)}{p^{2s}} + \cdots \right\}.$$

Each factor of K converges for $\Re s > 0$ and is positive at $s = 1$. Since k has only a finite number of prime divisors, K converges and defines an analytic function for $\Re s > 0$. Suppressing subscripts and reference to the modulus k, we express a typical product

$$\prod_{p \equiv a} \{ 1 + b\, p^{-s} + f(p^2) p^{-2s} + \ldots \} =: G(s) H(s),$$

where

$$H(s) := H_{a,b}(s) := \prod_{p \equiv a} \left\{ (1 - p^{-s})^b \left(1 + \frac{b}{p^s} + \frac{f(p^2)}{p^{2s}} + \cdots \right) \right\}$$

and

$$G(s) := G_{ab}(s) := \prod_{p \equiv a} (1 - p^{-s})^{-b}.$$

Given $p \equiv a$ with $(a, k) = 1$, consider a factor

$$(1 - p^{-s})^b (1 + b\, p^{-s} + f(p^2) p^{-2s} + \cdots)$$

of H. The binomial expansion of $(1 - p^{-s})^b$ has coefficients all of size at most 1, as are the coefficients b, $f(p^2), \ldots$ of the second factor. Formally multiplying, we find that each factor of H has the form

$$1 + 0 p^{-s} + 3\theta_2 p^{-2s} + 4\theta_3 p^{-3s} + \cdots,$$

with each $|\theta_\nu| \leq 1$. Thus $H(s) = \prod_{p \equiv a}\{1 + O(p^{-2\sigma})\}$ converges uniformly for $\Re s > 2/3$, say, and hence is analytic on this half plane. Also $H(1) > 0$, since each factor is positive.

Finally, we examine G. We have

$$\log G(s) = b \sum_{p \equiv a} \log(1 - p^{-s})^{-1} = b \sum_{p \equiv a} \sum_{\alpha=1}^{\infty} p^{-\alpha s}/\alpha.$$

As a comparison function, consider G^*, where

$$\log G^*(s) := \frac{b}{h} \sum_\chi \overline{\chi(a)} \log L(s, \chi)$$

$$= \frac{b}{h} \sum_\chi \sum_{n=1}^{\infty} \kappa(n) n^{-s} \chi(n) \overline{\chi(a)}$$

$$= b \sum_{n \equiv a} \kappa(n) n^{-s} = b \sum_{p^\alpha \equiv a} p^{-\alpha s}/\alpha.$$

The last sum extends over all primes p and positive integers α satisfying $p^\alpha \equiv a \pmod{k}$. Let

$$b\varphi_a(s) := \log G(s) - \log G^*(s);$$

this is a Dirichlet series with bounded coefficients extending over (some) higher prime powers and thus is analytic for $\Re s > 1/2$. It follows that

$$G(s) = \exp\{b\varphi_a(s)\} \prod_\chi L(s, \chi)^{b\bar{\chi}(a)/h}.$$

Now we combine the preceding relations, restoring the subscripts. For $\chi \neq \chi_0$, $L(s, \chi)$ is analytic and nonzero on the closed half plane $\{s : \sigma \geq 1\}$, and thus all factors of $G_{a_i b_i}$ except $L(s, \chi_0)^{b_i/h}$ have the same property.

For χ_0 we multiply over *all* the reduced residue classes and obtain

$$\prod_{i=1}^{h} L(s, \chi_0)^{b_i/h} = L(s, \chi_0)^{\beta/h} = \left\{\zeta(s) \prod_{p|k}(1 - p^{-s})\right\}^{\beta/h}$$

$$= (s - 1)^{-\beta/h}\left\{(s - 1)\zeta(s) \prod_{p|k}(1 - p^{-s})\right\}^{\beta/h}.$$

Determining the constant c from the preceding calculation is quite complicated. However, knowing the existence of c, we can determine it by an abelian estimate analogous to the proof of Theorem 1.8. We find that

$$c = \frac{1}{\Gamma(\beta/h)} \prod_{p|k} \left\{ 1 + \frac{f(p)}{p} + \frac{f(p^2)}{p^2} + \cdots \right\} \times$$

$$\lim_{s \to 1+} (s-1)^{\beta/h} \prod_{i=1}^{h} \prod_{p \equiv a_i} \left\{ 1 + \frac{b_i}{p^s} + \frac{f(p^2)}{p^{2s}} + \cdots \right\}.$$

Now set

$$J_{a,b}(s) = H_{a,b}(s) \exp\{b\varphi_a(s)\} \prod_{\chi \neq \chi_0} L(s,\chi)^{b\bar{\chi}(a)/h}.$$

Then we have $\widehat{F}(s) = F^*(s)(s-1)^{-\beta/h}$, with

$$F^*(s) = K(s) \left\{ (s-1)\zeta(s) \prod_{p|k} (1-p^{-s}) \right\}^{\beta/h} \prod_{i=1}^{h} J_{a_i,b_i}(s),$$

an analytic function with no zeros in $\{s : \sigma \geq 1\}$. Also, $0 < \beta/h \leq 1$. Thus the generalized Wiener-Ikehara theorem applies, and we obtain

$$\sum_{n \leq x} f(n) \sim cx(\log x)^{(\beta-h)/h}$$

with $c := c_f := F^*(1)/\Gamma(\beta/h)$.   □

Example 10.2   We return to the problem, mentioned at the beginning of this section, of estimating the number of integers in $[1, x]$ whose prime factors are all congruent to 1 modulo 4. Let $f = 1_5 * 1_{13} * 1_{17} * \cdots$. We want to estimate F, the summatory function of f. For $\Re s > 1$, let

$$\widehat{F}(s) := \sum f(n)n^{-s} := \prod_{p \equiv 1(4)} (1 - p^{-s})^{-1}.$$

In the notation in Theorem 10.1 we have $k = 4$, $h = 2$, and $\beta = 1$. Let

$$L(s, \chi_1) = 1 - 3^{-s} + 5^{-s} - 7^{-s} + 9^{-s} - \cdots$$

$$= \prod_{p \equiv 1(4)} (1 - p^{-s})^{-1} \prod_{p \equiv 3(4)} (1 + p^{-s})^{-1}.$$

Then

$$\widehat{F}(s)^2 = \zeta(s)(1 - 2^{-s})L(s, \chi_1) \prod_{p \equiv 3(4)} (1 - p^{-2s}).$$

We find from Theorem 10.1 that

$$F(x) \sim c\, x(\log x)^{-1/2},$$

where

$$c = \frac{1}{\Gamma(1/2)}\Big\{(1 - 1/2)L(1, \chi_1) \prod_{p \equiv 3(4)} (1 - p^{-2})\Big\}^{1/2}$$

is found via Theorem 7.7. Now the series for $L(1, \chi_1)$ shows it to be equal to arc tan $1 = \pi/4$. The reflection formula for the gamma function (see Appendix) gives $\Gamma(1/2) = \sqrt{\pi}$, and $\prod\{(1 - p^{-2}) : p \equiv 3(4)\}$ can be calculated as a finite product times a remainder; cf. §6.2. We obtain

$$c = 8^{-1/2}\Big\{ \prod_{p \equiv 3(4)} (1 - p^{-2})\Big\}^{1/2} = .327129\ldots.$$

PROBLEM 10.1   Let χ_1 denote the nonprincipal character mod 4. Chebyshev conjectured that

$$f(r) := \sum_p \chi_1(p)r^p \to -\infty, \quad r \to 1-, \tag{10.2}$$

i.e. that (in this sense) there are more primes p in the residue class 3 (mod 4) than in 1 (mod 4). Use Landau's oscillation theorem to prove that (10.2) implies that $L(s, \chi_1) \neq 0$ for $\Re s > 1/2$, i.e. the R.H. holds for $L(s, \chi_1)$. Hint. Use the formula

$$\Gamma(s)n^{-s} = \int_0^\infty x^{-s-1}e^{-n/x}\, dx$$

to show that, for $\Re s > 1$,

$$\Gamma(s) \operatorname{Log} L(s, \chi_1) = \int_1^\infty x^{-s-1}f(e^{-1/x})\, dx + G(s),$$

where $G(s)$ is regular for $\Re s > 1/2$; also note that $L(s, \chi_1) > 0$ for positive real values of s.

10.2   Sums of squares

A famous theorem of Lagrange asserts that any positive integer is representable as the sum of four squares. It is known that a positive integer n is representable as a sum of three squares unless $n = 4^a(8b + 7)$ for some nonnegative integers a and b. It follows that

$$\#\{n \leq x : n = a^2 + b^2 + c^2,\ a, b, c \in \mathbb{Z}\} = \frac{5}{6}x + O(\log x).$$

(Prove this!)

Here we study the distribution of integers representable as a sum of two squares. We first characterize such integers in terms of their prime factors.

Lemma 10.3   (Aubry–Thue). *Suppose p is a prime and a is an integer not divisible by p. Then there exist integers x and y such that*

$$x \equiv ay \pmod{p}, \quad 0 < \max(|x|, |y|) < p^{1/2}.$$

Proof.   Consider the set S of ordered pairs of integers (u, v) such that

$$0 \leq u < p^{1/2}, \quad 0 \leq v < p^{1/2}.$$

For each pair (u, v) in S we note the residue class modulo p into which $u - av$ falls. Since there are exactly p residue classes modulo p and the number of elements of S is

$$([p^{1/2}] + 1)^2 > (p^{1/2})^2 = p,$$

we must have two distinct ordered pairs (u_1, v_1) and (u_2, v_2) in S such that

$$u_1 - av_1 \equiv u_2 - av_2 \pmod{p},$$

or

$$u_1 - u_2 \equiv a(v_1 - v_2) \pmod{p}.$$

Since

$$0 < \max(|u_1 - u_2|,\ |v_1 - v_2|) < p^{1/2},$$

the lemma is established.   □

Lemma 10.4   *A positive integer n is representable as a sum of two squares if and only if each prime $q \equiv 3 \pmod 4$ in the unique factorization of n occurs with an even multiplicity.*

Proof. First, suppose that q is a prime congruent to 3 modulo 4 and $q^\alpha \| n$, where $n = x^2 + y^2 > 0$. We show that α is even. Let $d = (x, y)$ and put $x = dx_0$, $y = dy_0$. Then $(x_0, y_0) = 1$ and $n = d^2(x_0^2 + y_0^2)$. If α were odd, then $x_0^2 + y_0^2 \equiv 0 \pmod{q}$. It would follow that $q \nmid y_0$, for otherwise the last congruence implies that $q | x_0$, which is impossible. Thus there would exist z_0 such that $y_0 z_0 \equiv 1 \pmod{q}$, e.g. we could take $z_0 = y_0^{p-2}$. Hence

$$(x_0 z_0)^2 + 1 \equiv (x_0^2 + y_0^2)z_0^2 \equiv 0 \pmod{q},$$

which is impossible, since -1 is not a quadratic residue for $q \equiv 3 \pmod 4$.

Second, we show that if each prime $q \equiv 3 \pmod 4$ in the factorization of n appears to an even power, then n is expressible as a sum of two squares. The identity

$$(x^2 + y^2)(u^2 + v^2) = (xu - yv)^2 + (xv + yu)^2$$

shows that the product of two integers each expressible as a sum of two squares is itself expressible as a sum of two squares. If we observe that $2 = 1^2 + 1^2$ and $q^2 = q^2 + 0^2$ for any prime q, we see that it suffices to prove that any prime p congruent to 1 modulo 4 is a sum of two squares. To see this, we apply the lemma of Aubry–Thue with a chosen so that

$$a^2 + 1 \equiv 0 \pmod{p}.$$

The existence of such an integer a follows from the fact that -1 is a quadratic residue of any prime $p \equiv 1 \pmod 4$; for example, one solution is $a = \{(p-1)/2\}!$. By the lemma, there are integers x and y with

$$x \equiv ay \pmod{p}, \quad 0 < \max(|x|, |y|) < p^{1/2}.$$

Thus

$$x^2 + y^2 \equiv (a^2 + 1)y^2 \equiv 0 \pmod{p}$$

and $0 < x^2 + y^2 < 2p$. Hence $x^2 + y^2 = p$. $\square$

Let B denote the counting function of integers representable as a sum of two squares, i:e.

$$B(x) = \#\{n \le x : n = u^2 + v^2\}.$$

We estimate B in

Theorem 10.5   *There is a constant $\beta \doteq .764224$ such that*

$$B(x) \sim \beta x / \sqrt{\log x}.$$

Proof.   Let $f \in \mathcal{A}$ be the indicator function of $\{n \in \mathbb{Z}^+ : n = u^2 + v^2\}$. By the preceding lemma we see that f is multiplicative and

$$f(p^\alpha) = \begin{cases} 1 & \text{if } p = 2 \text{ or } p \equiv 1 \ (\mathrm{mod}\, 4), \\ 1 & \text{if } p \equiv 3 \ (\mathrm{mod}\, 4) \text{ and } \alpha \text{ is even}, \\ 0 & \text{if } p \equiv 3 \ (\mathrm{mod}\, 4) \text{ and } \alpha \text{ is odd}. \end{cases}$$

Letting $\widehat{B}$ denote the associated D.s., we have

$$\widehat{B}(s) = (1 - 2^{-s})^{-1} \prod_p (1 - p^{-s})^{-1} \prod_q (1 - q^{-2s})^{-1} \quad (\sigma > 1),$$

where p runs through the primes congruent to 1 (mod 4) and q the primes congruent to 3 (mod 4). Now f satisfies the hypotheses of Theorem 10.1. As in Example 10.2, we have $k = 4$, $h = 2$, and $\beta = 1$. It follows that $s \mapsto \sqrt{s-1}\, \widehat{B}(s)$ is analytic on the closed half plane $\{s : \sigma \geq 1\}$ and that B has an asymptotic formula of the stated form.

We can evaluate the constant by representing $\widehat{B}$ in terms of some other functions which are more tractable. We have

$$\widehat{B}(s)^2 = \left\{ (1 - 2^{-s})^{-1} \prod_p (1 - p^{-s})^{-1} \prod_q (1 - q^{-s})^{-1} \right\} \times$$

$$(1 - 2^{-s})^{-1} \left\{ \prod_p (1 - p^{-s})^{-1} \prod_q (1 + q^{-s})^{-1} \right\} \prod_q (1 - q^{-2s})^{-1}$$

$$= \zeta(s)(1 - 2^{-s})^{-1} L(s, \chi_1) \prod_q (1 - q^{-2s})^{-1}, \tag{10.3}$$

where χ_1 is the nonprincipal character modulo 4. The last product converges for $\sigma > 1/2$.

Since the factors of (10.3) other than ζ are analytic on the open half plane $\{s : \sigma > 1/2\}$, we conclude from Theorem 7.7 that

$$B(x) \sim \left\{ 2L(1,\chi_1) \prod_{q \equiv 3(4)} (1 - q^{-2})^{-1} \right\}^{1/2} x(\log x)^{-1/2}/\Gamma(1/2)$$

$$\sim \prod_{q \equiv 3(4)} (1 - q^{-2})^{-1/2} x/\sqrt{2 \log x}.$$

We evaluated the last product in Example 10.2. Thus the constant β in the statement of the theorem is $1/(4c) \doteq .764224$. $\square$

One can show by complex variable methods (cf. Notes) that $B(x)$ has an asymptotic expansion of the form

$$\frac{\beta x}{\sqrt{\log x}} \left\{ 1 + \frac{c_1}{\log x} + \frac{c_2}{\log^2 x} + \cdots \right\}. \qquad (10.4)$$

The value of c_1 has been determined to be about .581949. In the following table, which was calculated by D. Shanks, we list some values of $B(x)$, $\ell_0(x) = \beta x(\log x)^{-1/2}$, and $\ell_1(x) = \ell_0(x)(1 + c_1/\log x)$ (with entries rounded to the nearest integer).

x	$B(x)$	$\ell_0(x)$	$\ell_1(x)$	ℓ_0/B	ℓ_1/B
2^6	29	24	27	.8270	.9427
2^8	97	83	92	.8565	.9464
2^{10}	337	297	322	.8820	.9561
2^{12}	1197	1085	1161	.9067	.9702
2^{14}	4357	4019	4260	.9225	.9778
2^{16}	16096	15039	15828	.9343	.9834
2^{18}	60108	56717	59362	.9436	.9876
2^{20}	226419	215225	224260	.9506	.9905
2^{22}	858696	820836	852161	.9559	.9924
2^{24}	3273643	3143562	3253531	.9603	.9939
2^{26}	12534812	12080946	12471056	.9638	.9949

Table 10.1   SUMS OF TWO SQUARES DATA

The table shows that the rate at which $\ell_0(x)/B(x) \to 1$ is rather slow.

How far would the table have to be extended to guarantee that

$$\ell_0(x)/B(x) > .99? \tag{10.5}$$

If we assume that $B(x) > \ell_1(x)$ for all $x \geq 100$ (which is supported by the table), then we can give a lower estimate for the x's for which (10.5) holds. If (10.5) holds, then we have

$$(1 + c_1/\log x)^{-1} = \ell_0(x)/\ell_1(x) > .99$$

or $x > 10^{25} > 2^{83}$.

PROBLEM 10.2   Use the lemma of Aubry–Thue to prove that any prime $p \equiv 1 \pmod 3$ can be expressed in the form $p = x^2 + 3y^2$, where x and y are integers. Hints. If $p \equiv 1 \bmod 3$, then there exists an integer a such that $a^2 + 3 \equiv 0 \pmod p$ (by quadratic reciprocity). Also, note that $x^2 + 3y^2$ is either odd or divisible by 4.

PROBLEM 10.3   Prove that a positive integer n is expressible in the form $x^2 + 3y^2$, where x and y are integers, if and only if each prime $q \equiv 2 \pmod 3$ in the unique factorization of n occurs with even multiplicity. Hint. Use the algebraic identity

$$(x^2 + 3y^2)(u^2 + 3v^2) = (xu - 3yv)^2 + 3(xv + yu)^2.$$

PROBLEM 10.4   (i) Prove that there is a constant $c > 0$ such that

$$\#\{n \leq x : n = a^2 + 3b^2, \ a, b \in \mathbb{Z}\} \sim c\, x(\log x)^{-1/2}.$$

(ii) For the same constant c, prove that

$$\#\{n \leq x : n = a^2 + ab + b^2, \ a, b \in \mathbb{Z}\} \sim c\, x(\log x)^{-1/2}.$$

Hint. Use the identities

$$x^2 + xy + y^2 = (x + y/2)^2 + 3(y/2)^2 = \left(\frac{x-y}{2}\right)^2 + 3\left(\frac{x+y}{2}\right)^2,$$

$$u^2 + 3v^2 = (v - u)^2 + (v - u)(v + u) + (v + u)^2$$

to show that the sets in parts (i) and (ii) are the same.

PROBLEM 10.5   Obtain analogs of the results of the last three problems for the quadratic form $x^2 + 2y^2$. Hint. If p is a given odd prime, there exists an integer a such that $a^2 + 2 \equiv 0 \pmod p$ if $p \equiv 1$ or 3 (mod 8) but not if $p \equiv 5$ or 7 (mod 8).

10.3 A measure of nonprincipality

Given a nonprincipal character χ, there exist integers n such that $\chi(n) \neq 0$ and $\chi(n) \neq 1$. For χ a real and nonprincipal character modulo k, we shall estimate the smallest positive n for which $\chi(n) = -1$. Let N_χ denote this integer n. For example, if $\chi(n) = \left(\frac{n}{23}\right)$, the Legendre symbol (mod 23), then $N_\chi = 5$ (cf. §1.2).

In the preceding example, the ratio N_χ/k is well above zero; however, this occurrence might be atypical. Assuming a modest zerofree region for the associated D.s. $L(\cdot, \chi)$ (one far weaker than that predicted by the analogue of the R.H. for L functions, which asserts that $L(s, \chi) \neq 0$ for all s with real part exceeding $1/2$), we show that, asymptotically, N_χ has a much smaller order of magnitude than k.

Theorem 10.6 *Let $k > K$, K a certain absolute constant, and let χ be any real nonprincipal character mod k. Suppose there exists a number θ in $[1/2, 1 - 40/\log\log k)$ such that $L(s, \chi)$ has no zeros in the domain*

$$\{s : \Re s > \theta, |\Im s| < (\log k)^{1/2}\}.$$

Then

$$N_\chi \leq (\log k)^{1/(1-\theta)} (\log\log k)^{4/(1-\theta)}.$$

Proof. We shall assume that

$$N_\chi > (\log k)^{1/(1-\theta)} (\log\log k)^{4/(1-\theta)} =: x$$

and deduce a contradiction. The assumption implies that $\chi(n) = 1$ for $1 \leq n \leq x$ provided that $(n, k) = 1$ and hence that

$$S(x) := \sum_{n \leq x} \left(1 - \frac{n}{x}\right) \Lambda(n)\chi(n) \geq \sum_{n \leq x} \left(1 - \frac{n}{x}\right) \Lambda(n) - \sum_{\substack{p^\alpha \leq x \\ p|k}} \log p.$$

Now

$$\sum_{\substack{p^\alpha \leq x \\ p|k}} \log p = \sum_{p|k} \left[\frac{\log x}{\log p}\right] \log p \leq 2\log x \cdot \log k$$

$$\leq x^{1-\theta} \log x \leq x^{1/2} \log x.$$

The last estimates follow from the definition of x. This and subsequent estimates require k to be sufficiently large. Also, the P.N.T. implies that

$$\sum_{n \le x} \left(1 - \frac{n}{x}\right) \Lambda(n) = \frac{1}{2}x + o(x).$$

Thus we have

$$S(x) \ge \frac{1}{2}x + o(x) - x^{1/2} \log x \ge \frac{9}{20}x, \qquad (10.6)$$

provided that k is sufficiently large. On the other hand, we now estimate $S(x)$ from above by the Perron formula to obtain a contradiction. We use the fact that $L(s, \chi)$ has no pole at $s = 1$ and the hypothesis that $L(s, \chi) \ne 0$ for s in a certain rectangular region.

We begin with an estimate of L'/L for use in the Perron formula. Since

$$|L(s, \chi)| = \left|s \int_1^\infty u^{-s-1} \sum_{n \le u} \chi(u) du\right| \le |s| \int_1^\infty \frac{k}{2} u^{-3/2} du \le k(|t| + 2)$$

for $1/2 \le \Re s \le 2$ and since

$$|L(2 + it, \chi)| \ge \prod_{p \nmid k}(1 + p^{-2})^{-1} > \frac{\zeta(4)}{\zeta(2)} > \frac{6}{\pi^2} > \frac{1}{2},$$

Lemma 8.9 (Borel-Carathéodory), applied to $\log L(s, \chi)$ with $z_0 = 2 + it$, $R = 2 - \theta$, $r = 2 - \Re s$, gives

$$\left|\frac{L'}{L}(s, \chi)\right| \le \frac{3 \log\{2k(|t| + 4)\}}{(\Re s - \theta)^2},$$

for $\theta < \Re s \le 2$, $|\Im s| \le \frac{1}{2}\sqrt{\log k} < \sqrt{\log k} - R$. Take $h = 1/\log x$. By the definition of x and the condition that $\log k > e$ we have

$$e^{1/(1-\theta)} < (\log k)^{1/(1-\theta)} < x,$$

and hence $h + \theta < 1$. Now consider the following smoothed inversion formula (cf. Problem 7.17):

$$S(x) = \frac{1}{2\pi i} \int_{1+h-i\infty}^{1+h+i\infty} \frac{-L'}{L}(s, \chi) \frac{x^s ds}{s(s + 1)}$$

$$= I_1 + I_2 + I_3 + I_4 + I_5,$$

where (with the same integrand)

$$I_1 + I_5 := \frac{1}{2\pi i} \int_{1+h-i\infty}^{1+h-\frac{1}{2}i\sqrt{\log k}} + \frac{1}{2\pi i} \int_{1+h+\frac{1}{2}i\sqrt{\log k}}^{1+h+i\infty},$$

$$I_2 + I_4 := \frac{1}{2\pi i} \int_{1+h-\frac{1}{2}i\sqrt{\log k}}^{\theta+h-\frac{1}{2}i\sqrt{\log k}} + \frac{1}{2\pi i} \int_{\theta+h+\frac{1}{2}i\sqrt{\log k}}^{1+h+\frac{1}{2}i\sqrt{\log k}},$$

$$I_3 := \frac{1}{2\pi i} \int_{\theta+h-\frac{1}{2}i\sqrt{\log k}}^{\theta+h+\frac{1}{2}i\sqrt{\log k}}.$$

We have

$$|I_1| = |I_5| \le \frac{1}{2\pi} \left(-\frac{\zeta'}{\zeta}(1+h) \right) x^{1+h} \int_{\frac{1}{2}\sqrt{\log k}}^{\infty} t^{-2} dt$$

$$\le \frac{1}{2\pi} \left(\frac{2}{h} \right) (xe) \frac{2}{\sqrt{\log k}} < \frac{2x \log x}{\sqrt{\log k}}.$$

Also,

$$|I_2| = |I_4| \le \frac{3}{2\pi} \log\{k(\sqrt{\log k} + 8)\} \frac{4}{\log k} \int_{\theta+h}^{1+h} \frac{x^\sigma}{(\sigma - \theta)^2} d\sigma$$

$$\le 2x^\theta \int_h^{1-\theta+h} \frac{x^u}{u^2} du = 2x^\theta \log x \int_{h \log x}^{(1-\theta+h) \log x} \frac{e^v}{v^2} dv$$

$$= 2x^\theta \log x \int_1^{(1-\theta) \log x + 1} \frac{e^v dv}{v^2} \le \frac{6x}{(1-\theta)^2 \log x}$$

$\left(\int_1^Y e^v v^{-2} dv \sim e^Y / Y^2 \text{ by l'Hospital's Rule} \right)$. Finally, since $\Re s > 1/2$ in I_3,

$$|I_3| \le \frac{x^{\theta+h}}{2\pi} \int_{-\frac{1}{2}\sqrt{\log k}}^{\frac{1}{2}\sqrt{\log k}} \frac{3 \log\{2k(|t|+4)\}}{h^2} \frac{dt}{3/4 + t^2}$$

$$\le \frac{3ex^\theta}{2\pi} \log\{k(\sqrt{\log k} + 8)\} \log^2 x \int_{-\infty}^{\infty} \frac{dt}{3/4 + t^2}$$

$$\le 5x^\theta (\log k) \log^2 x.$$

Recalling the definition of x and the hypothesized bounds on θ we obtain

$$\frac{40 \log x}{\log \log k} < (1 - \theta) \log x = \log \log k + 4 \log \log \log k.$$

Thus, for sufficiently large k we have

$$|I_1| \le \frac{2x \log x}{\sqrt{\log k}} < \frac{2x \log \log k(\log \log k + 4 \log \log \log k)}{40 \sqrt{\log k}} < \frac{x}{20},$$

$$|I_2| \le \frac{6x}{(1 - \theta)^2 \log x} < \frac{6x}{(1 - \theta) \log \log k} < \frac{3x}{20},$$

$$|I_3| \le \frac{5x(\log k)(\log^2 x)}{x^{1-\theta}} = \frac{5x \log^2 x}{(\log \log k)^4}$$

$$< \frac{5x(\log \log k + 4 \log \log \log k)^2}{1600(\log \log k)^2} < \frac{x}{20}.$$

Together, we obtain

$$S(x) < \frac{x}{20} + \frac{x}{20} + \frac{3x}{20} + \frac{3x}{20} + \frac{x}{20} = \frac{9x}{20},$$

in contradiction to (10.6). Thus, if $L(\cdot, \chi)$ is zerofree in the hypothesized region, it is false that $N_\chi > (\log k)^{1/(1-\theta)} (\log \log k)^{4/(1-\theta)}$. $\qquad \square$

PROBLEM 10.6   Under the hypotheses of Theorem 10.6, prove that

$$\log N_\chi < (\log \log k)^2 / 39$$

for each sufficiently large modulus k.

10.4   Quadratic excess

Throughout this section p denotes an odd prime. We write that a is a q.r. when a is a quadratic residue modulo p. Similarly q.n.r. denotes a quadratic nonresidue modulo p. Let E_p denote the number of q.r.'s lying in the interval $(0, p/2)$ minus the number of q.r.'s in $(p/2, p)$. For $p \equiv 1$ (mod 4) we have $E_p = 0$, since -1 is a quadratic residue in this case. Thus, for each q.r. a, the number $p - a \equiv -a$ (mod p) is a q.r. as well, and so the q.r.'s in $(0, p)$ are symmetrically distributed about $p/2$.

There are exactly $(p - 1)/2$ q.r.'s in $(0, p)$; hence $E_p \ne 0$ if $p \equiv 3$ (mod 4), since $(p - 1)/2$ is odd in this case. In this and the next section we shall prove the assertion of Theorem 1.1, namely that $E_p > 0$ for $p \equiv 3$ (mod 4). The quantity E_p is sometimes called the *quadratic excess* modulo p.

PROBLEM 10.7 Show that E_p equals the number of q.r.'s lying in the interval $(0, p/2)$ minus the number of q.n.r.'s in this interval.

In this section we show that E_p has the same sign as the finite sum

$$\sum_{j=1}^{p} \sin(2\pi j^2/p) = \Im G_p,$$

where for positive integer values of n,

$$G_n = \sum_{j=1}^{n} \exp(2\pi i j^2/n).$$

Sums of this type were first considered by Gauss and today are called *Gaussian sums*. Here we are primarily interested in the case where n is an odd prime p, although in the next section it will be just as easy to consider the case of positive integer values of n.

Recall that the Legendre symbol $\left(\frac{m}{p}\right)$ is defined to be 1 if m is a q.r.; -1 if m is a q.n.r.; and 0 if $p \mid m$. Most of the results in this section depend on the following lemma.

Lemma 10.7 *Let $m \not\equiv 0 \pmod{p}$. As j runs from 1 to $p-1$, the least positive residue of mj^2 modulo p runs twice through the q.r.'s in $(0, p)$ if $\left(\frac{m}{p}\right) = 1$ and twice through the q.n.r.'s in $(0, p)$ if $\left(\frac{m}{p}\right) = -1$.*

Proof. By the multiplicativity of the Legendre symbol, we have

$$\left(\frac{mj^2}{p}\right) = \left(\frac{m}{p}\right), \qquad 1 \le j \le p - 1.$$

Also, the numbers $m1^2, m2^2, \ldots, m((p-1)/2)^2$ are mutually incongruent modulo p, since $mj^2 - mi^2 = m(j-i)(j+i)$ is not divisible by p for $1 \le i < j \le (p-1)/2$. But $m(p-j)^2 \equiv mj^2 \pmod{p}$ for $1 \le j \le (p-1)/2$ and so the least positive residues modulo p of $m(p-1)^2, m(p-2)^2, \ldots, m((p+1)/2)^2$ are the same respectively as the least positive residues modulo p of $m1^2, m2^2, \ldots, m((p-1)/2)^2$. $\qquad\square$

Lemma 10.8 *G_p is real if $p \equiv 1 \pmod{4}$ and purely imaginary if $p \equiv 3 \pmod{4}$.*

Proof. For $p \equiv 1 \pmod{4}$ the integer -1 is a q.r. By the preceding lemma, the least positive residue modulo p of j^2, and that of $-j^2$, each runs twice through the q.r.'s in $(0, p)$ as j runs through $1, 2, \ldots, p-1$. Thus

$$G_p = 1 + \sum_{j=1}^{p-1} e^{2\pi i j^2/p} = 1 + \sum_{j=1}^{p-1} e^{-2\pi i j^2/p} = \overline{G_p},$$

and so G_p is real.

For $p \equiv 3 \,(\mathrm{mod}\, 4)$ the integer -1 is a q.n.r. In this case the least positive residue modulo p of $-j^2$ runs twice through the q.n.r.'s in $(0, p)$ and so

$$G_p + \overline{G_p} = 1 + \sum_{j=1}^{p-1} e^{2\pi i j^2/p} + 1 + \sum_{j=1}^{p-1} e^{-2\pi i j^2/p} = 2 \sum_{k=0}^{p-1} e^{2\pi i k/p} = 0;$$

thus G_p is purely imaginary in this case. $\qquad\square$

In the following three lemmas we express E_p in terms of the imaginary part of G_p with the aid of a Fourier series. As usual, let

$$\mathrm{sgn}\, t = \begin{cases} 1 & \text{if } t > 0, \\ 0 & \text{if } t = 0, \\ -1 & \text{if } t < 0. \end{cases}$$

Lemma 10.9

$$2E_p = \sum_{j=1}^{p-1} \mathrm{sgn}\{\sin(2\pi j^2/p)\}. \qquad (10.7)$$

Proof. By Lemma 10.7, the least positive residue of j^2 modulo p runs twice through the q.r.'s in $(0, p)$ as j runs from 1 to $p-1$. The least positive residue of j^2 modulo p lies in $(0, p/2)$ or $(p/2, p)$ according as $\sin(2\pi j^2/p)$ is positive or negative. $\qquad\square$

Lemma 10.10 *For any positive integer m,*

$$\sum_{j=1}^{p-1} \sin(2\pi m j^2/p) = \left(\frac{m}{p}\right)\Im G_p, \qquad (10.8)$$

where $\left(\frac{\cdot}{p}\right)$ denotes the Legendre symbol modulo p.

Proof. If m is a multiple of p, each side of (10.8) is zero. Henceforth we assume that m is a fixed positive integer not divisible by p. If $p \equiv 1 \,(\mathrm{mod}\, 4)$, the right hand side of (10.8) is zero by Lemma 10.8. If in addition $\left(\frac{m}{p}\right) = 1$, the least positive residue of mj^2 modulo p runs twice through the q.r.'s in $(0, p)$ when j runs from 1 to $p - 1$, by Lemma 10.7; since the q.r.'s in $(0, p)$ are located symmetrically about $p/2$, the left hand side of (10.8) is zero in this case. Similarly if $\left(\frac{m}{p}\right) = -1$ the least positive residue of mj^2 runs twice through the q.n.r.'s in $(0, p)$; since the q.n.r.'s in $(0, p)$ are also located symmetrically about $p/2$, the left hand side of (10.8) is also zero in this case. So much for the case $p \equiv 1 \pmod{4}$.

Now we assume $p \equiv 3 \pmod{4}$. By Lemma 10.7 the left hand side of (10.8) is the same for all m with $\left(\frac{m}{p}\right) = 1$, and in particular, equals the value for $m = 1$, namely $\Im G_p$. This establishes (10.8) for $\left(\frac{m}{p}\right) = 1$. Similarly, the left hand side of (10.8) is the same for all m with $\left(\frac{m}{p}\right) = -1$. Since $\left(\frac{-1}{p}\right) = -1$ for $p \equiv 3 \pmod{4}$, for $\left(\frac{m}{p}\right) = -1$ we have

$$\sum_{j=1}^{p-1} \sin(2\pi m j^2/p) = \sum_{j=1}^{p-1} \sin(-2\pi j^2/p) = -\Im G_p. \qquad \square$$

Lemma 10.11 *For real y,*

$$\text{sgn}(\sin 2\pi y) = \frac{4}{\pi} \sum_{m=1}^{\infty} \frac{\sin\{(2m-1)2\pi y\}}{2m-1}. \tag{10.9}$$

Proof. If y is an integer multiple of $1/2$, each side of (10.9) is zero. Otherwise

$$[2y] = 2[y] \quad \text{if} \quad 0 < y - [y] < \frac{1}{2},$$

$$[2y] = 2[y] + 1 \quad \text{if} \quad \frac{1}{2} < y - [y] < 1,$$

and so

$$2([y] - y + \frac{1}{2}) - ([2y] - 2y + \frac{1}{2}) = \frac{1}{2} \quad \text{if} \quad 0 < y - [y] < \frac{1}{2},$$

$$2([y] - y + \frac{1}{2}) - ([2y] - 2y + \frac{1}{2}) = -\frac{1}{2} \quad \text{if} \quad \frac{1}{2} < y - [y] < 1.$$

In view of Lemma 7.15, the preceding formulas give

$$\frac{1}{2} \text{sgn}(\sin 2\pi y) = 2S(y) - S(2y) = \frac{2}{\pi} \sum_{m=1}^{\infty} \frac{\sin\{(2m-1)2\pi y\}}{2m-1}.$$

Thus (10.9) holds for all real values of y. $\qquad\qquad\qquad\qquad$ □

Theorem 10.12 *If p is an odd prime, then*

$$E_p = 2\pi^{-1} L(1, \chi) \Im G_p,$$

where χ is the real nonprincipal character modulo $2p$ given in terms of the Legendre symbol as follows:

$$\chi(k) = \begin{cases} \left(\frac{k}{p}\right) & \text{if } k \text{ is odd,} \\ 0 & \text{if } k \text{ is even.} \end{cases}$$

Proof. Using Lemmas 10.9 and 10.11 we have

$$E_p = \sum_{j=1}^{p-1} \frac{2}{\pi} \sum_{m=1}^{\infty} \frac{\sin\{(2m-1)2\pi j^2/p\}}{2m-1}$$

$$= \frac{2}{\pi} \sum_{m=1}^{\infty} \frac{1}{2m-1} \sum_{j=1}^{p-1} \sin\{(2m-1)2\pi j^2/p\}.$$

By Lemma 10.10 this may be written as

$$E_p = \frac{2}{\pi} \sum_{m=1}^{\infty} \frac{1}{2m-1} \left(\frac{2m-1}{p}\right) \Im G_p. \qquad\qquad □$$

By Lemma 9.11 we know that $L(1, \chi) > 0$. We shall complete the proof of Theorem 1.1 in the next section by proving that $\Im G_p > 0$ for all $p \equiv 3 \pmod 4$.

10.5   Evaluation of Gaussian sums

In this section we evaluate the Gaussian sum G_n for every positive integer n. The absolute value of G_n is easily determined, as follows.

Lemma 10.13 $|G_n| = \sqrt{n}$ when n is odd, $G_n = 0$ when $n \equiv 2 \pmod 4$, and $|G_n| = \sqrt{2n}$ when $n \equiv 0 \pmod 4$.

Proof. We have

$$|G_n|^2 = \overline{G_n} G_n = \sum_{j=1}^{n} e^{-2\pi i j^2/n} \left\{ \sum_{k=1}^{n} e^{2\pi i (k+j)^2/n} \right\}$$

$$= \sum_{j=1}^{n} \sum_{k=1}^{n} e^{2\pi i (2jk+k^2)/n}$$

$$= \sum_{k=1}^{n} e^{2\pi i k^2/n} \sum_{j=1}^{n} e^{4\pi i jk/n}.$$

If n is odd, the inner sum is zero when $1 \le k \le n-1$ and is equal to n when $k = n$; thus $|G_n|^2 = n$ in this case. If n is even, the inner sum is zero when $1 \le k \le n/2 - 1$ or $n/2 + 1 \le k \le n - 1$, but is equal to n when $k = n/2$ or $k = n$; the assertions for even n follow easily. $\square$

PROBLEM 10.8 When $n \equiv 2 \,(\mathrm{mod}\, 4)$ show that $G_n = 0$ by comparing the j-th term and the $(j + n/2)$-th term for $j = 1, 2, \ldots, n/2$.

By Lemmas 10.8 and 10.13, $G_p = \pm\sqrt{p}$ if $p \equiv 1 \,(\mathrm{mod}\, 4)$ and $G_p = \pm i\sqrt{p}$ when $p \equiv 3 \,(\mathrm{mod}\, 4)$. It will turn out that the plus sign holds in both cases. This will follow from the following elegant reciprocity theorem.

Theorem 10.14 *If m and n are positive integers and mn is even, then*

$$\sum_{j=0}^{n-1} e^{\pi i m j^2/n} = \frac{1+i}{\sqrt{2}} \sqrt{\frac{n}{m}} \sum_{k=0}^{m-1} e^{-\pi i n k^2/m}.$$

Proof. We consider a contour integral involving the function

$$f(z) = \frac{e^{\pi i m z^2/n}}{e^{2\pi i z} - 1}.$$

Let R be a large positive number and $P_R := Re^{\pi i/4}$. The contour $\mathcal{C} = \mathcal{C}_R$ which we choose is the positively oriented parallelogram with vertices

$$P_R, \; -P_R, \; -P_R + n, \; \text{and} \; P_R + n,$$

except that we make semicircular detours to the left with radius less than 1 around the points 0 and n. (See Fig. 10.1.) Eventually, we shall let $R \to \infty$.

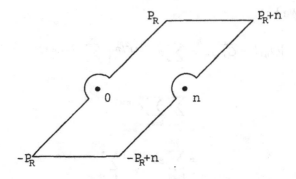

Fig. 10.1 CONTOUR C_R

On this contour $\text{Arg } z^2 \approx \pi/2$ for $|z|$ large, and so for such z, $|e^{\pi imz^2/n}|$ is small. Within C the function f has poles with residue $e^{\pi imj^2/n}/(2\pi i)$ at the points $z = j$ for $j = 0, 1, \ldots, n-1$. By the residue theorem,

$$\int_C f(z)dz = \sum_{j=0}^{n-1} e^{\pi imj^2/n}.$$

On the two horizontal parts of the contour C we have $z = \pm Re^{\pi i/4} + x$, where $0 \leq x \leq n$, and there

$$|e^{\pi imz^2/n}| = e^{-\pi m(R^2 \pm \sqrt{2}Rx)/n} \leq e^{-\pi m(R^2 - \sqrt{2}Rn)/n};$$

also it is immediate that $e^{2\pi iz} - 1$ is bounded away from zero on the horizontal parts of C. Thus as $R \to \infty$, each of the integrals of $f(z)$ along the horizontal parts of the contour is $o(1)$.

The sum of the integrals along the right and left sides of the contour is

$$\int_{-Re^{\pi i/4}+n}^{Re^{\pi i/4}+n} f(z)dz - \int_{-Re^{\pi i/4}}^{Re^{\pi i/4}} f(z)dz = \int_{-Re^{\pi i/4}}^{Re^{\pi i/4}} \{f(z+n) - f(z)\}dz,$$

where the integrals are taken on straight-line paths except for the aforementioned semicircular detours at the crossings with the real axis. But

$$f(z+n) - f(z) = \frac{e^{\pi im(z+n)^2/n} - e^{\pi imz^2/n}}{e^{2\pi iz} - 1}$$

$$= \frac{e^{\pi imz^2/n}(e^{2\pi imz+\pi imn} - 1)}{e^{2\pi iz} - 1}.$$

Since mn is even, this may be written

$$f(z+n) - f(z) = e^{\pi imz^2/n} \sum_{k=0}^{m-1} e^{2\pi ikz}$$

$$= \sum_{k=0}^{m-1} e^{-\pi ink^2/m} e^{\pi im(z+nk/m)^2/n}.$$

Thus $f(z+n) - f(z)$ is an entire function of z; accordingly the semicircular detours are unnecessary, and from now on all integrals will be taken along straight-line paths. The sum of the two integrals along the nonhorizontal parts of the contour C therefore equals

$$\sum_{k=0}^{m-1} e^{-\pi ink^2/m} \int_{-Re^{\pi i/4}}^{Re^{\pi i/4}} e^{\pi im(z+nk/m)^2/n} dz$$

$$= \sum_{k=0}^{m-1} e^{-\pi ink^2/m} \int_{-Re^{\pi i/4}+nk/m}^{Re^{\pi i/4}+nk/m} e^{\pi imz^2/n} dz.$$

Now for $0 < k < m$,

$$\int_{-Re^{\pi i/4}+nk/m}^{Re^{\pi i/4}+nk/m} e^{\pi imz^2/n} dz$$

$$= \left\{ \int_{-Re^{\pi i/4}}^{Re^{\pi i/4}} + \int_{Re^{\pi i/4}}^{Re^{\pi i/4}+nk/m} - \int_{-Re^{\pi i/4}}^{-Re^{\pi i/4}+nk/m} \right\} e^{\pi imz^2/n} dz,$$

and each of the last two integrals is bounded by

$$(nk/m)e^{-\pi m(R^2-\sqrt{2}Rnk/m)/n} < ne^{-\pi m(R^2-\sqrt{2}Rn)/n} = o(1)$$

as $R \to +\infty$. Thus

$$\sum_{j=0}^{n-1} e^{\pi ij^2 m/n} = \int_C f(z)dz$$

$$= \sum_{k=0}^{n-1} e^{-\pi ink^2/m} \int_{-Re^{\pi i/4}}^{Re^{\pi i/4}} e^{\pi imz^2/n} dz + o(1).$$

In the last integral we set $z = e^{\pi i/4} u \sqrt{n/m}$ with u real and get

$$\int_{-Re^{\pi i/4}}^{Re^{\pi i/4}} e^{\pi i m z^2/n} dz = e^{\pi i/4} \sqrt{\frac{n}{m}} \int_{-R\sqrt{n/m}}^{R\sqrt{n/m}} e^{-\pi u^2} du.$$

Letting R go to infinity we obtain

$$\sum_{j=0}^{n-1} e^{\pi i j^2 m/n} = K \frac{1+i}{\sqrt{2}} \sqrt{\frac{n}{m}} \sum_{k=0}^{m-1} e^{-\pi i n k^2/m},$$

where $K = \int_{-\infty}^{\infty} e^{-\pi u^2} du$. Taking $m = 1$ and $n = 2$ we obtain the evaluation $K = 1$. $\qquad\qquad\square$

Corollary 10.15 *For n a positive integer,*

$$G_n = \sum_{j=0}^{n-1} e^{2\pi i j^2/n} = \begin{cases} (1+i)\sqrt{n} & \text{if } n \equiv 0 \ (\mathrm{mod}\, 4), \\ \sqrt{n} & \text{if } n \equiv 1 \ (\mathrm{mod}\, 4), \\ 0 & \text{if } n \equiv 2 \ (\mathrm{mod}\, 4), \\ i\sqrt{n} & \text{if } n \equiv 3 \ (\mathrm{mod}\, 4). \end{cases}$$

Proof. By the preceding theorem with $m = 2$ we have

$$\sum_{j=0}^{n-1} e^{2\pi i j^2/n} = \frac{1+i}{2} \sqrt{n} \left(1 + e^{-\pi i n/2}\right). \qquad\qquad\square$$

We see, in particular, that $\Im G_p > 0$ for primes $p \equiv 3 \ (\mathrm{mod}\, 4)$. Thus $E_p > 0$, i.e. there are more quadratic residues $(\mathrm{mod}\, p)$ lying in the interval $(0, p/2)$ than in $(p/2, p)$. This completes the proof of Theorem 1.1. $\qquad\square$

10.6 Notes

10.1. A result similar to Theorem 10.1 but with more restrictive hypotheses was obtained by Landau, Amer. J. Math., vol. 31 (1909), pp. 86–102; also in [LanC], vol. 4, pp. 131–147. Cf. [LanH], §§177–183.

An efficient scheme for computing $\prod_{p \equiv 3(4)} (1 - p^{-2})$ is described in D. Shanks, Math. Comp., vol. 18 (1964), pp. 75–86.

10.2. A particularly elegant proof of Lagrange's four-square theorem is that by A. Brauer and R. L. Reynolds, Canad. J. Math., vol. 3 (1951), pp. 367–374; this proof is given in W. J. LeVeque's Topics in Number Theory, vol. 1, Addison–Wesley, 1956, pp. 133–136 and in LeVeque's Fundamentals of Number Theory, Addison–Wesley, 1977, pp. 179–180, 187–189.

The Aubry–Thue theorem was proved independently by L. Aubry, Mathesis (4), vol. 3 (1913), pp. 33–35, and A. Thue, Archiv for Mathematik og Naturvidenskab, vol. 34 (1917), no. 15; also in Selected Mathematical Papers, Universitetsforlaget, Oslo, 1977, pp. 555–559. The history of the theorem is given in some detail in the above-mentioned paper of Brauer and Reynolds.

The three-square theorem was first proved by Legendre, but the best-known proof is that of Dirichlet; this appeared in J. reine angew. Math., vol. 40 (1850), pp. 228–232 (in German) and in J. Math. pures appl. (2), vol. 4 (1859), pp. 233–240 (in French); also in Dirichlet's Werke, vol. 2 Berlin, 1897, pp. 89–96. This proof has been reproduced in several other places:

- [LanH], pp. 550–555,
- [LanV], Satz 187,
- L. E. Dickson's History of the Theory of Numbers, vol. 2, Carnegie Institution of Washington, 1920 (reprinted by Chelsea Publishing Co, 1952), pp. 263–264
- Dickson's Modern Elementary Theory of Numbers, Chicago, 1939, pp. 88–96.

Theorem 10.5 was first proved by Landau, using contour integration, in Archiv Math. Phys. (3), vol. 13 (1908), pp. 305–312; also in [LanC], vol. 4, pp. 59–66. Although nineteenth century mathematicians were probably aware that most positive integers are not representable as sums of two squares, there appears to have been no explicit discussion of the order of magnitude of the counting function $B(x)$ before Landau's 1908 paper; cf. pp. 225–254 of vol. 2 of Dickson's above cited History.

Landau's proof of Theorem 10.5 is sketched in [HarR], pp. 60–63. As Hardy remarks, Landau's argument leads to the asymptotic expansion (10.4). In Math. Comp., vol. 18 (1964), pp. 75–86, D. Shanks evaluated the coefficient c_1; the earlier calculation of this constant by G. K. Stanley in J. London Math. Soc., vol. 3 (1928), pp. 232–237 and vol. 4 (1929), p. 32, is incorrect.

10.3. Theorem 10.6 is essentially due to G. L. Miller, J. Comput. System Sci., vol. 13 (1976), pp. 300–317.

10.4. Theorem 10.12 is due to Dirichlet; cf. Notes for §1.2.

10.5. Our proof of Theorem 10.14 stems from L. J. Mordell, Messenger of Math., vol. 48 (1918), pp. 54–56; it was given again in Acta Math., vol. 61 (1933), pp. 323–360. This argument appears also in [Apos], pp. 195–200, and in [ChanI], pp. 35–39.

Chapter 11

Oscillation Theorems

11.1 Introduction

The main theme of the last few chapters has been the asymptotic approximation of summatory functions. Given a function $F \in \mathcal{V}$, we have sought a smooth approximating function φ such that $E := F - \varphi$ is small compared to $|F|$. We call E an *error term*. In several cases we have given explicit upper estimates of $|E|$. In this chapter we inquire how near these estimates are to the true size of E. This is an area in which analytic methods generally yield much more information than elementary ones.

We have encountered significant upper bound and oscillation estimates already in studying the weighted prime counting function ψ: On one hand, assuming the R.H., we have the O-estimate $\{\psi(x) - x\}/\sqrt{x} = O(\log^2 x)$ (Corollary 8.24), and on the other hand, in Theorem 8.7 we showed that $\{\psi(x) - x\}/\sqrt{x} \not\to 0$ as $x \to \infty$.

Let F and G be real valued functions on some half line (a, ∞) and assume that G is positive. We say that $F = \Omega_+(G)$ (or $F = \Omega_-(G)$) if there exists a positive constant c such that the relation $F(x)/G(x) > c$ (or $< -c$) holds for some arbitrarily large values of x. If both relations hold, we say that $F = \Omega_\pm(G)$. If at least one of the relations holds, i.e. $|F(x)|/G(x) > c$ for some arbitrarily large x, then we say that $F = \Omega(G)$. In other words, "Ω" is the negation of the o relation, and G provides a measure of what we call the oscillation. As simple examples, we have $x - N(x) = \Omega_+(1)$ and $x \sin x = \Omega_\pm(x)$. The result of Theorem 8.7 can be restated as $\psi(x) = x + \Omega(\sqrt{x})$.

Note that an Ω relation requires only that some inequality holds for a

261

sequence of x's tending to infinity, and not for all (sufficiently large) values of x, as an O relation would require. Consequently, it is in some sense easier to establish Ω estimates than O estimates. We would like to obtain for a given problem an Ω estimate which equals the O estimate, as is the case in the first two examples above. In many instances where there is a gap between the two estimates, it is plausible that the Ω relation is closer to the truth.

One way to obtain omega results is to form the contrapositive of a theorem containing an o-estimate (or an asymptotic relation) in the hypothesis. For example the relation $\psi(x) = x + \Omega(\sqrt{x})$ was established in this way.

PROBLEM 11.1   Establish an analogue of Lemma 7.1 for a function $F \in \mathcal{V}$ satisfying $F(x) = c\,x \log x + o(x \log x)$. Use this result to show that $\sum_{n \leq x} n(1_2 * 1_2)(n)$ is not asymptotic to an expression of the form $cx \log x$. (Recall that 1_2 is the indicator function of powers of 2. Cf. Problem 7.19.)

11.2 Approximate periodicity

Some oscillation theorems are based on the nearly periodic behavior of trigonometric polynomials. We shall establish such results using

Lemma 11.1   (Dirichlet approximation theorem). *Let M and $N \in \mathbb{Z}^+$ and $T > 0$. Let $a_1, \ldots, a_N$ be real numbers. There exists a real number $h \in [T, TM^N]$ such that*

$$\|a_i h\| \leq 1/M \quad (1 \leq i \leq N).$$

Here $\|\cdot\|$ denotes the distance to the nearest integer.

Proof. Let I^N denote the unit cube in N dimensional Euclidean space $\mathbb{R}^N$, and divide I^N into M^N cubelets of side $1/M$. Let $\underline{a} = (a_1, \ldots, a_N)$ denote a point in $\mathbb{R}^N$, and consider the sequence of points in I^N

$$nT\underline{a}\,(\mathrm{mod}\,1), \quad n = 1, 2, \ldots, M^N + 1.$$

For example, the second point is

$$(2Ta_1 - [2Ta_1], \ldots, 2Ta_N - [2Ta_N]).$$

There are $M^N + 1$ points and M^N cubelets, so by Dirichlet's box principle, at least one cubelet contains two points. Suppose that $n' < n$ and

$nT\underline{a} \pmod 1$ and $n'T\underline{a} \pmod 1$ lie in the same cubelet. For $1 \le i \le N$, the ith coordinates of $nT\underline{a} \pmod 1$ and $n'T\underline{a} \pmod 1$ lie in an interval $[j_i/M, (j_i+1)/M]$. Take $h = T(n - n')$. Then each coordinate of the point $h\underline{a} \pmod 1$ lies in $[-1/M, 1/M]$. $\qquad \square$

The following specialization of the lemma is familiar and often useful.

Corollary 11.2   *Let ξ be irrational and $M \in \mathbb{Z}^+$. There exists a positive integer $n \le M$ such that $\|\xi n\| < 1/M$.*

PROBLEM 11.2   Let ξ be irrational and $\alpha > 0$. Show that

$$\sum_{n=1}^{\infty} (n\|n\xi\|)^{-\alpha} = \infty.$$

Let f denote a real or complex valued function on $\mathbb{R}$ and let $\epsilon > 0$. A real number τ is called an ϵ *almost period* of f if

$$|f(x + \tau) - f(x)| < \epsilon \quad (x \in \mathbb{R}).$$

For example, $4\pi + \epsilon$ is an ϵ almost period of the sine function because

$$|\sin(4\pi + \epsilon + x) - \sin x| = |\sin(x + \epsilon) - \sin x|$$

$$= \left| \int_x^{x+\epsilon} \cos u \, du \right| < \epsilon.$$

More interesting examples are provided by trigonometric polynomials, i.e. functions of the form

$$P(x) = a_1 e^{ib_1 x} + \cdots + a_N e^{ib_N x}$$

for $a_1, \ldots, a_N \in \mathbb{C}$ and $b_1, \ldots, b_N \in \mathbb{R}$.

Lemma 11.3   *For each $\epsilon > 0$, all trigonometric polynomials have arbitrarily large ϵ almost periods.*

Proof.   Let $P(x)$ be a trigonometric polynomial. Let $T > 0$, $\epsilon > 0$, and $M \in \mathbb{Z}^+$, $M > 2\pi \sum |a_n|/\epsilon$. By Lemma 11.1, there exists a number $h \in [T, TM^N]$ such that

$$\|(b_i h/(2\pi))\| < \epsilon \Big/ \Big(2\pi \sum_{n=1}^{N} |a_n|\Big) \quad (1 \le i \le N).$$

Then, with the triangle inequality and the estimate $|\sin \pi u| \leq \pi \|u\|$, we have for all real x

$$|P(x+h) - P(x)| = \left| \sum_{n=1}^{N} a_n e^{ib_n(x+h/2)} (e^{ib_n h/2} - e^{-ib_n h/2}) \right|$$

$$\leq \sum_{n=1}^{N} |2a_n \sin(b_n h/2)|$$

$$\leq 2\pi \sum_{n=1}^{N} |a_n| \max_{1 \leq i \leq N} \|b_i h/(2\pi)\| < \epsilon.$$

Thus h is an ϵ almost period of P. $\qquad\square$

Corollary 11.4 *Let $\widehat{F}(s) := \sum a_n n^{-s}$ be a D.s. and let $\sigma_0 > \sigma_a(\widehat{F})$. Then for any $\epsilon > 0$ the function $t \mapsto \widehat{F}(\sigma_0 + it)$ has arbitrarily large ϵ almost periods.*

Proof. We can approximate $\widehat{F}$ uniformly on the line $\{s = \sigma_0 + it : -\infty < t < \infty\}$ to within $\epsilon/3$ by a trigonometric polynomial

$$P(t) = \sum_{n=1}^{N} a_n n^{-s} = \sum_{n=1}^{N} (a_n n^{-\sigma_0}) e^{-it \log n}.$$

By the preceding lemma there exist arbitrarily large numbers h such that

$$|P(t+h) - P(t)| < \epsilon/3 \quad (-\infty < t < \infty).$$

The desired conclusion follows by writing

$$\widehat{F}(\sigma_0 + it + ih) - \widehat{F}(\sigma_0 + it) = P(t+h) - P(t) + \sum_{n=N+1}^{\infty} a_n n^{-\sigma_0 - it - ih}$$

$$- \sum_{n=N+1}^{\infty} a_n n^{-\sigma_0 - it}$$

and applying the triangle inequality. $\qquad\square$

In §11.7 we shall require a form of Kronecker's theorem on simultaneous diophantine approximations. We state and prove this result here. We say that numbers $b_1, \ldots, b_N$ are *linearly independent over* $\mathbb{Q}$ if

$$b_1 c_1 + \cdots + b_N c_N = 0$$

for rational numbers $c_1, \ldots, c_N$ implies that $c_1 = \cdots = c_N = 0$.

Lemma 11.5 *Let $a_1, \ldots, a_N \in \mathbb{C}$, $b_1, \ldots, b_N$, $\theta \in \mathbb{R}$. Suppose that $b_1, \ldots, b_N$ are linearly independent over $\mathbb{Q}$. Then there exists a sequence of real numbers $x_j \to \infty$ such that*

$$\sum_{n=1}^{N} a_n e^{ib_n x_j} \to e^{i\theta} \sum_{n=1}^{N} |a_n|.$$

Proof. We can assume that $\theta = 0$, for otherwise we first replace each a_n by $a'_n = a_n e^{-i\theta}$. Let

$$f(x) := 1 + \sum_{n=1}^{N} a_n e^{ib_n x}.$$

We shall show that

$$\limsup_{x \to \infty} |f(x)| = 1 + \sum_{n=1}^{N} |a_n|. \tag{11.1}$$

We have $|f(x)| \leq 1 + \sum |a_n|$, and when equality (nearly) holds in the triangle inequality, then all nonzero terms in f must have (nearly) the same argument, namely 0. This will show that

$$\sum_{n=1}^{N} a_n e^{ib_n x_j} \to \sum_{n=1}^{N} |a_n|$$

for some sequence $x_j \to \infty$.

The proof of (11.1) is based on the principle that an integral of a high power of a function can be used to approximate the maximum of the function. For q a positive integer we have by the multinomial theorem

$$f(x)^q = \sum (c_{j_1, \ldots, j_N} a_1^{j_1} \ldots a_N^{j_N}) e^{ix(b_1 j_1 + \cdots + b_N j_N)}.$$

Here the sum extends over all N-tuples of nonnegative integers $(j_1, \ldots, j_N)$ with $j_1 + \cdots + j_N \leq q$. The c's denote multinomial coefficients. The only property of the c's that we use is their positivity. By the hypothesis of linear independence, all the exponents $b_1 j_1 + \cdots + b_N j_N$ are distinct. With an obvious change of notation we write

$$f(x)^q =: \sum \alpha_k e^{ix\beta_k}, \tag{11.2}$$

where the exponents β_k are distinct real numbers.

We now collect information about the α_k's, starting with $\sum |\alpha_k|$. In order to treat the absolute values, it is convenient to introduce an auxiliary function

$$g(t_1, \ldots, t_N)^q := (1 + |a_1|t_1 + \cdots + |a_N|t_N)^q$$

$$= \sum c_{j_1, \ldots, j_n} |a_1^{j_1} \cdots a_N^{j_N}| t_1^{j_1} \cdots t_N^{j_N}.$$

The same N-tuples $(j_1, \ldots, j_N)$ and multinomial coefficients occur here as in the representation of f^q. It follows that

$$\sum |\alpha_k| = \sum c_{j_1, \ldots, j_N} |a_1^{j_1} \cdots a_N^{j_N}|$$

$$= g(1, \ldots, 1)^q = (1 + |a_1| + \cdots + |a_N|)^q.$$

The number of coefficients $\alpha_k = c_{j_1, \ldots, j_N} a_1^{j_1} \cdots a_N^{j_N}$ in f^q is at most $(q+1)^N$ since $0 \le j_i \le q$ for $i = 1, \ldots, N$.

Finally, we estimate the size of the individual coefficients α_k with the aid of the elementary formula

$$\lim_{X \to \infty} \frac{1}{X} \int_0^X e^{in x} dx = \begin{cases} 1, & \eta = 0, \\ 0, & \eta \in \mathbb{R} \smallsetminus \{0\}. \end{cases}$$

If β_ℓ is an exponent occurring in (11.2) we have

$$\lim_{X \to \infty} \frac{1}{X} \int_0^X f(x)^q e^{-i\beta_\ell x} dx = \alpha_\ell.$$

Thus each coefficient α_ℓ in (11.2) satisfies

$$|\alpha_\ell| \le \limsup_{x \to \infty} |f(x)|^q.$$

We now combine the preceding estimates to obtain

$$(1 + |a_1| + \cdots + |a_N|)^q = \sum |\alpha_k| \le (q+1)^N \limsup_{x \to \infty} |f(x)|^q,$$

or upon taking qth roots

$$\limsup_{x \to \infty} |f(x)| \ge (1 + |a_1| + \cdots + |a_N|)(q+1)^{-N/q}.$$

If we let $q \to \infty$ and recall that $|f(x)| \le 1 + \sum |a_n|$, we obtain (11.1). $\square$

PROBLEM 11.3   Deduce the following alternative form of Kronecker's theorem. Let $d_1, \ldots, d_N, b_1, \ldots, b_N$ be real and suppose that $b_1, \ldots, b_N$ are linearly independent over $\mathbb{Q}$. Prove that there exists a sequence $x_j \to \infty$ such that for $1 \leq n \leq N$,

$$\|b_n x_j - d_n\| \to 0.$$

11.3   The use of Landau's oscillation theorem

The relation $F = \Omega_+(G)$ is equivalent to the negation of the assertion

$$\limsup_{x \to \infty} F(x)/G(x) \leq 0.$$

The latter formulation is suitable for application of Theorem 6.31.

Example 11.6   Let G be the summatory function of Example 6.34:

$$G(x) = \sum_{n \leq x} g(n) = \sum_{n \leq x} (2e_1 - 1)^{*-1}(n).$$

We showed in §7.2.1 that $G(x) \sim x^\rho/\{-\rho\zeta'(\rho)\}$ where $\rho \doteq 1.728647\ldots$ denotes the positive root of the equation $\zeta(s) = 2$.

Here we shall prove that

$$G(x) = x^\rho/\{-\rho\zeta'(\rho)\} + \Omega_\pm(x^{\rho-\epsilon}) \tag{11.3}$$

for any fixed $\epsilon > 0$. The key step in the demonstration of (11.3) is

Lemma 11.7   *The equation* $\zeta(s) = 2$ *has roots* ρ_j $(j = 1, 2, \ldots)$ *with* $\Re\rho_j \to \rho$ *and* $\Im\rho_j \to \infty$. *All nonreal roots satisfy* $\Re\rho_j < \rho$.

Proof.   We apply Rouché's theorem. The zero of $\zeta(s) - 2$ at $s = \rho$ is simple, since

$$\zeta'(\rho) = -\sum (\log n) n^{-\rho} \neq 0.$$

Let δ be a positive number less than $\rho - 1$ and such that there are no other zeros of $\zeta(s) - 2$ inside or on the circle $|s - \rho| = \delta$. Take

$$\epsilon = \min_{|s-\rho|=\delta} |\zeta(s) - 2|.$$

Next, we show that there exist arbitrarily large positive numbers $T = T(\delta, \epsilon)$ such that

$$|\zeta(s) - \zeta(s + iT)| < \epsilon \quad \text{on} \quad \{s : |s - \rho| \le \delta\}. \qquad (11.4)$$

The proof parallels that of Corollary 11.4. Given any positive integer N,

$$|\zeta(s) - \zeta(s + iT)| \le \left| \sum_{n=1}^{N} n^{-s}(1 - e^{-iT \log n}) \right| + 2 \sum_{N+1}^{\infty} |n^{-s}|$$

$$=: I + II, \quad \text{say.}$$

Since $\rho - \delta > 1$, we can first choose N so large that

$$II \le 2 \sum_{n \ge N+1} n^{-\rho+\delta} < \epsilon/2.$$

Then, by Lemma 11.1 with $a_n = (\log n)/(2\pi)$, there exist arbitrarily large numbers T such that

$$I \le \sum_{n=1}^{N} n^{-\rho+\delta} |1 - e^{-iT \log n}| < \epsilon/2.$$

These estimates are valid uniformly for $|s - \rho| \le \delta$; thus (11.4) holds there.

With such choice of T, we have

$$\zeta(s + iT) - 2 = \{\zeta(s) - 2\} + \{\zeta(s + iT) - \zeta(s)\}.$$

The first term on the right side has a simple zero at $s = \rho$ and dominates the second term on the circle $\{s : |s - \rho| = \delta\}$. It follows by Rouché's theorem that $\zeta(z + iT) = 2$ for some z satisfying $|z - \rho| < \delta$.

It is clear that $|\zeta(s)| < 2$ for $\Re s > \rho$. Also, as we showed in (7.13), $|\zeta(\rho + it)| < 2$ for t real and nonzero. Thus any nonreal zero ρ_j must satisfy $\Re \rho_j < \rho$. $\qquad \square$

Proof of (11.3).   Let $c = \{-\rho \zeta'(\rho)\}^{-1}$ and let A be real and ϵ positive. We have

$$H(s) := \frac{1}{s(2 - \zeta(s))} - \frac{c}{s - \rho} - \frac{A}{s - \rho + \epsilon}$$

$$= \int x^{-s-1}\{G(x) - cx^{\rho} - Ax^{\rho-\epsilon}\}dx.$$

H is analytic on the ray $\{s : s \text{ real}, s > \rho - \epsilon\}$. On the other hand, the preceding lemma shows that H has singularities arbitrarily close to the line $\{s : \sigma = \rho\}$, and hence the abscissa of convergence of the integral is $\geq \rho$.

By Landau's theorem,

$$x \longmapsto \int_1^x \{G(t) - ct^\rho - At^{\rho-\epsilon}\}t^{-1}dt$$

is not ultimately monotone, and hence $G(x) - cx^\rho - Ax^{\rho-\epsilon}$ changes sign infinitely often as $x \to \infty$. It follows upon taking $A = +1$ that

$$G(x) = cx^\rho + \Omega_+(x^{\rho-\epsilon}).$$

The choice $A = -1$ gives Ω_-. $\qquad\qquad\square$

11.4 A quantitative estimate

Results like that of §11.3 are of a qualitative nature. Here we obtain a numerical oscillation estimate from knowledge of a singularity of an M.t. at a nonreal point.

Theorem 11.8 *Let F be a real valued function in $\mathcal{V}$. Suppose that $\widehat{F}$ exists and has a continuation as a meromorphic function having a pole of order $m \geq 1$ at some point $\beta + i\gamma$ with $\beta > 0$ and $\gamma > 0$ and that the principal part there is*

$$\sum_{j=-1}^{-m} c_j(s - \beta - i\gamma)^j.$$

Moreover, suppose that there exists no singularity of $\widehat{F}$ on the real segment $[\beta, \infty)$. Then

$$\limsup_{x\to\infty} \frac{F(x)}{x^\beta(\log x)^{m-1}} \geq \frac{|c_{-m}|}{|\beta + i\gamma|(m-1)!} \qquad (11.5)$$

and

$$\liminf_{x\to\infty} \frac{F(x)}{x^\beta(\log x)^{m-1}} \leq \frac{-|c_{-m}|}{|\beta + i\gamma|(m-1)!}. \qquad (11.6)$$

Remark 11.9 Landau's oscillation theorem tells only that F changes sign infinitely often.

Proof. It suffices to establish (11.6) since (11.5) follows upon replacing F by $-F$. Suppose that c is a positive number such that

$$F(x) + cx^\beta (\log x)^{m-1} / \Gamma(m)$$

is positive for all sufficiently large values of x. If no such c exists then the left side of (11.6) equals $-\infty$ and there is nothing further to show.

Let $\sigma > \beta$. By Landau's theorem we have the representation

$$g(s) := \frac{\widehat{F}(s)}{s} + \frac{c}{(s - \beta)^m}$$

$$= \int_1^\infty x^{-s} \left\{ F(x) + cx^\beta \frac{(\log x)^{m-1}}{\Gamma(m)} \right\} x^{-1} dx =: \int x^{-s} f(x) dx.$$

Since f is positive for all $x \geq X$, with some $X \geq 1$,

$$|g(s)| \leq \int_1^X x^{-\sigma} |f(x)| dx + \int_X^\infty x^{-\sigma} f(x) dx$$

$$\leq \int_1^X x^{-\sigma} \{|f(x)| - f(x)\} dx + g(\sigma) = g(\sigma) + O(1).$$

Now take $s = \sigma + i\gamma$ and let $\sigma \to \beta+$. We have

$$\frac{|c_{-m}|}{|\sigma + i\gamma|(\sigma - \beta)^m} \{1 + o(1)\} = |g(\sigma + i\gamma)|$$

$$\leq g(\sigma) + O(1) = \frac{c}{(\sigma - \beta)^m} + O(1).$$

By the last inequality, $c \geq |c_{-m}| / |\beta + i\gamma|$. If we choose $c' < |c_{-m}| / |\beta + i\gamma|$, then we find that

$$F(x) + c'x^\beta (\log x)^{m-1} / \Gamma(m) \leq 0$$

for some arbitrarily large values of x. This establishes (11.6). $\qquad \square$

We showed in Theorem 8.7 that $\{\psi(x) - x\} / \sqrt{x} \neq o(1)$ as $x \to \infty$. Now we can give explicit oscillation estimates for this expression.

Corollary 11.10 *Let $\frac{1}{2} + i\gamma_1$ denote the zero of zeta having smallest positive imaginary part. Then*

$$\varlimsup_{x \to \infty} \frac{\psi(x) - x}{x^{1/2}} \geq \frac{1}{\left|\frac{1}{2} + i\gamma_1\right|} > \frac{1}{15}, \quad \varliminf_{x \to \infty} \frac{\psi(x) - x}{x^{1/2}} \leq \frac{-1}{\left|\frac{1}{2} + i\gamma_1\right|}.$$

Proof. Let $F(x) = \psi(x) - (x - 1)$. Then

$$\widehat{F}(s) = -\frac{\zeta'}{\zeta}(s) - \frac{1}{s-1},$$

which has no (nonremovable) singularities on the positive real axis. On the other hand, it is known that the zero of the Riemann zeta function having smallest positive imaginary part is a simple zero located at $\rho_1 = \frac{1}{2} + i\gamma_1$, where $\gamma_1 \doteq 14.134725$. Thus $\widehat{F}$ has a simple pole with residue 1 at ρ_1, and the claimed inequality follows from the last theorem. $\qquad\square$

Remark 11.11 We have seen that

$$Q(x) := \sum_{n \le x} |\mu(n)| = \frac{6}{\pi^2}x + O(x^{1/2}).$$

From the representation $\widehat{Q}(s) = \zeta(s)/\zeta(2s)$ and the preceding data we have

$$\varlimsup_{x \to \infty} x^{-1/4}\left\{Q(x) - \frac{6}{\pi^2}x\right\} \ge \left|\frac{\zeta(\rho_1/2)}{\rho_1\zeta'(\rho_1)}\right| \ge .100403$$

and that the limit inferior is at most $-|\zeta(\rho_1/2)/\{\rho_1\zeta'(\rho_1)\}|$.

PROBLEM 11.4 Assume the hypotheses of Theorem 11.8, except that $\widehat{F}$ has a pole at the real point β as well as at $\beta + i\gamma \ne \beta$. State and prove an analogous theorem for this case.

PROBLEM 11.5 In §7.9 we showed that

$$B(x) := \sum_{\substack{n \le x \\ n = 2^\alpha 3^\beta}} n \sim \frac{x \log x}{\log 2 \log 3}.$$

Use the preceding theorem to give an estimate of the oscillation of

$$B(x) - x\{\log(\sqrt{6}x) - 1\}/\{\log 2 \log 3\}.$$

What size oscillation can you show in an elementary way?

PROBLEM 11.6 Let $f = T1_2$, i.e. $f(2^k) = 2^k$, $k = 0, 1, 2, \ldots$, and $f(n) = 0$ if n is not a power of 2. Let

$$g_1(x) := \frac{1}{x}\sum_{n \le x} f(n).$$

Determine $\limsup_{x \to \infty} g_1(x)$ and $\liminf_{x \to \infty} g_1(x)$. What are the best estimates for $g_1(x)$ that you can deduce from Theorem 11.8?

PROBLEM 11.7   We have proved in Corollary 8.24 that under the R.H.
$\psi_1(x) := \int_1^x \psi(t)dt = x^2/2 + O(x^{3/2})$. Assuming the R.H., show that $x^{3/2}$
is the true order of the error term.

11.5   The use of many singularities

Suppose that an M.t. has many poles on its line of convergence but no
other singularities there. We can apply Theorem 11.8 using any one of the
nonreal singularities—but only one. Can one achieve better estimates by
using information about many singularities at the same time? The following
important result shows that in many cases the answer is affirmative.

Theorem 11.12   (A. E. Ingham).   *Let F be a real valued function in
$\mathcal{V}$ and satisfy $F(x) < \log^\beta x$ or $F(x) > -\log^\beta x$ for some $\beta < 1$ and all
sufficiently large x. Let*

$$G(s) := \int_1^\infty x^{-s-1} F(x) dx \quad (= \widehat{F}(s)/s)$$

*converge for $\sigma > 0$. Let $T > 0$ and let F^* and F_T^* be real trigonometric
polynomials*

$$F^*(u) = \sum_{-N}^N \alpha_n e^{i\gamma_n u}, \quad F_T^*(u) = \sum_{-N}^N \alpha_n \left(1 - \frac{|\gamma_n|}{T}\right) e^{i\gamma_n u}$$

with γ_n real, $|\gamma_n| < T$, $\gamma_{-n} = -\gamma_n$, and $\alpha_{-n} = \overline{\alpha_n}$. For $\sigma > 0$ let

$$G^*(s) := \int_0^\infty e^{-su} F^*(u) du = \sum_{-N}^N \alpha_n (s - i\gamma_n)^{-1}.$$

*Suppose that $G - G^*$ has a continuation as a continuous function on the
closed strip $\{s : \sigma \geq 0, |t| \leq T\}$. Then*

$$\int_1^\infty F(u) K_{T/2}(y - \log u) u^{-1} du = \pi F_T^*(y) + o_T(1) \quad (y \to +\infty) \quad (11.7)$$

and

$$\varliminf_{u \to \infty} F(u) \leq \inf_u F_T^*(u) \leq \sup_u F_T^*(u) \leq \varlimsup_{u \to \infty} F(u). \quad (11.8)$$

Here $K_{T/2}$ denotes the Fejér kernel (cf. §7.2).

Remarks 11.13 The proof is related to that of the Wiener-Ikehara theorem; here too some limiting argument would be simplified if Lebesgue theory were used. If neither of the one sided bounds hypothesized for F holds, then (11.8) is trivial. The theorem could be proved using a weaker one sided bound for F by replacing the Fejér kernel by another function which vanishes more swiftly at infinity. The trigonometric polynomial F^* is chosen to make G^* cancel the poles of G at points $i\gamma$ for $-T < \gamma < T$.

Proof. For $\sigma > 0$ we have

$$\int_{-T}^{T} \int_{1}^{\infty} u^{-s-1} F(u) du \left(1 - \frac{|t|}{T}\right) e^{ity} dt$$

$$= \int_{-T}^{T} \left(1 - \frac{|t|}{T}\right) e^{ity} (G - G^*)(s) dt$$

$$+ \int_{-T}^{T} \left(1 - \frac{|t|}{T}\right) e^{ity} \sum_{-N}^{N} \frac{\alpha_n}{s - i\gamma_n} dt. \qquad (11.9)$$

We are going to change the integration order on the left side of (11.9). We first express this integral as

$$\int_{t=-T}^{T} \int_{u=1}^{U} + \int_{t=-T}^{T} \int_{u=U}^{\infty},$$

and show that (for each $\sigma > 0$) the last integral is uniformly small for large U and $-T \le t \le T$.

As usual, let $F^{\pm}(u) = \max(\pm F(u), 0)$, so that $F = F^+ - F^-$ and $|F| = F^+ + F^-$. By hypothesis

$$I(F) := \int_{1}^{\infty} u^{-\sigma-1} F(u) du$$

is finite and by the one sided condition on F, one of $I(F^+)$, $I(F^-)$ has this property as well. Since $I(F) = I(F^+) - I(F^-)$, the other integral also is finite and so

$$I(|F|) = I(F^+) + I(F^-) < \infty.$$

It follows that

$$\left| \int_{U}^{\infty} u^{-\sigma-1-it} F(u) du \right| \le \int_{U}^{\infty} u^{-\sigma-1} |F(u)| du$$

is uniformly small for sufficiently large U.

Thus the exchange of integration order is justified on the left side of (11.9) and we have

$$\int_{-T}^{T}\int_{1}^{\infty} u^{-\sigma-1}F(u)e^{it(y-\log u)}du\left(1-\frac{|t|}{T}\right)dt$$

$$=\int_{1}^{\infty} u^{-\sigma-1}F(u)\left\{\int_{-T}^{T}\left(1-\frac{|t|}{T}\right)e^{it(y-\log u)}dt\right\}du$$

$$=2\int_{1}^{\infty} u^{-\sigma-1}F(u)K_{T/2}(y-\log u)du.$$

We look next at the right side of (11.9). For any real numbers γ, σ, and y satisfying $|\gamma| < T$, $\sigma > 0$, and $y > 0$ we have

$$\int_{-T}^{T}\frac{1}{s-i\gamma}\left(1-\frac{|t|}{T}\right)e^{ity}dt=\left(1-\frac{|\gamma|}{T}\right)\int_{-T}^{T}\frac{e^{ity}}{s-i\gamma}dt$$

$$+\int_{-T}^{T}\frac{|\gamma|-|t|}{\sigma+i(t-\gamma)}\frac{e^{ity}}{T}dt.$$

Now

$$\int_{-T}^{T}\frac{e^{ity}}{s-i\gamma}dt=\frac{e^{-\sigma y}}{i}\int_{\sigma-iT}^{\sigma+iT}\frac{e^{sy}ds}{s-i\gamma}=\frac{e^{y(i\gamma-\sigma)}}{i}\int_{y(\sigma-iT-i\gamma)}^{y(\sigma+iT-i\gamma)}\frac{e^{z}dz}{z}$$

$$=2\pi e^{-\sigma y+i\gamma y}+O\left\{\frac{1}{y(T-|\gamma|)}\right\}.$$

The last equation follows from deforming the integration contour as was done in the first case of Lemma 7.11. The constant implied by the O symbol is absolute.

Also, we have

$$\frac{|\gamma|-|t|}{\sigma+i(t-\gamma)}\to\frac{|\gamma|-|t|}{i(t-\gamma)}\quad(\sigma\to 0+)$$

uniformly in any fixed set $\{t\in[-T,T]:|t-\gamma|\geq\epsilon>0\}$, and we have the uniform bound

$$\left|\frac{|\gamma|-|t|}{\sigma+i(t-\gamma)}\right|\leq 1\quad(-T\leq t\leq T).$$

It follows that as $\sigma \to 0+$,

$$\int_{-T}^{T} \frac{e^{ity}}{T} \frac{|\gamma| - |t|}{\sigma + i(t - \gamma)} dt \to \int_{-T}^{T} \frac{e^{ity}}{T} \frac{|\gamma| - |t|}{i(t - \gamma)} dt$$

uniformly with respect to y, and we have for each of a finite number of γ's

$$\lim_{\sigma \to 0+} \int_{-T}^{T} \frac{1}{s - i\gamma} \left(1 - \frac{|t|}{T}\right) e^{ity} dt$$

$$= \left(1 - \frac{|\gamma|}{T}\right) 2\pi e^{i\gamma y} + \int_{-T}^{T} \frac{e^{ity}}{T} \frac{|\gamma| - |t|}{i(t - \gamma)} dt + O\left(\frac{1}{yT}\right).$$

Also, as $\sigma \to 0+$,

$$\int_{-T}^{T} \left(1 - \frac{|\gamma|}{T}\right) (G - G^*)(s) e^{ity} dt \to \int_{-T}^{T} \left(1 - \frac{|\gamma|}{T}\right) (G - G^*)(it) e^{ity} dt,$$

since the continuation of $G - G^*$ is continuous on $\{s : \sigma \geq 0, |t| \leq T\}$.

By (11.9) and the preceding estimates we obtain

$$\lim_{\sigma \to 0+} 2 \int_{1}^{\infty} u^{-\sigma-1} F(u) K_{T/2}(y - \log u) du$$

$$= \sum_{|\gamma_n| < T} \left(1 - \frac{|\gamma_n|}{T}\right) 2\pi \alpha_n e^{i\gamma_n y} + 2H(y) + O\left(\frac{1}{yT}\right), \quad (11.10)$$

where

$$2H(y) = \int_{-T}^{T} \left(1 - \frac{|t|}{T}\right) e^{ity} (G - G^*)(it) dt + \sum_{|\gamma_n| < T} \alpha_n \int_{-T}^{T} \frac{e^{ity}}{T} \frac{|\gamma_n| - |t|}{i(t - \gamma_n)} dt.$$

We wish to take the limit inside the integral in (11.10). To this end, we set $F = F^+ - F^-$ and assume, without loss of generality, that $F(x) < \log^\beta x$ holds for all $x \geq x_0$. This estimate and inequality (7.3) for $K_{T/2}$ imply that

$$\int_{u=1}^{\infty} u^{-1} F^+(u) K_{T/2}(y - \log u) du < \infty.$$

Given $\epsilon > 0$, choose U so large that (for fixed y and any $\sigma > 0$)

$$0 < \int_{U}^{\infty} u^{-\sigma} F^+(u) K_{T/2}(y - \log u) \frac{du}{u} < \int_{U}^{\infty} F^+(u) K_{T/2}(y - \log u) \frac{du}{u} < \epsilon.$$

For fixed $y > 0$, we have, as $\sigma \to 0+$,

$$u^{-\sigma-1}F^+(u)K_{T/2}(y - \log u) \to u^{-1}F^+(u)K_{T/2}(y - \log u)$$

uniformly for $1 \le u \le U$. Thus

$$\lim_{\sigma \to 0+} \int_1^\infty u^{-\sigma}F^+(u)K_{T/2}(y - \log u)\frac{du}{u} = \int_1^\infty F^+(u)K_{T/2}(y - \log u)\frac{du}{u}.$$

It follows from (11.10) and the preceding limit that

$$\lim_{\sigma \to 0+} \int_1^\infty u^{-1-\sigma}F^-(u)K_{T/2}(y - \log u)du$$

exists. Arguing as we did to take the limit inside the integral in equation (7.10), we obtain

$$\lim_{\sigma \to 0+} \int_1^\infty u^{-\sigma}F^-(u)K_{T/2}(y - \log u)\frac{du}{u} = \int_1^\infty F^-(u)K_{T/2}(y - \log u)\frac{du}{u}.$$

Putting the results together, we obtain

$$\int_1^\infty u^{-1}F(u)K_{T/2}(y - \log u)du = \pi F_T^*(y) + H(y) + O\left(\frac{1}{yT}\right).$$

Now $H(y) \to 0$ as $y \to \infty$ by the Riemann-Lebesgue lemma (Lemma 7.5), and thus (11.7) is established.

In particular, taking $F = F^* = 1$, we get $G(s) = G^*(s) = 1/s$. Then,

$$\int_1^\infty u^{-1}K_{T/2}(y - \log u)du = \pi + o(1) \qquad (y \to \infty)$$

or

$$\int_{-\infty}^\infty K_{T/2}(v)dv = \pi. \tag{11.11}$$

The preceding calculation and (11.7) show that F_T^* is asymptotically an average of F. Thus we have

$$\varliminf_{u \to \infty} F(u) \le \varliminf_{u \to \infty} F_T^*(u), \qquad \varlimsup_{u \to \infty} F(u) \ge \varlimsup_{u \to \infty} F_T^*(u).$$

Finally, Lemma 11.3 implies that

$$\varliminf_{u \to \infty} F_T^*(u) = \inf_u F_T^*(u), \qquad \varlimsup_{u \to \infty} F_T^*(u) = \sup_u F_T^*(u).$$

Thus (11.8) holds. $\qquad\qquad\qquad\qquad\qquad\qquad\qquad\qquad\qquad\qquad\qquad\qquad$ $\square$

11.5.1 Applications

Recall that Liouville's function λ is completely multiplicative and equal to -1 on all primes. It was conjectured by G. Pólya that

$$L(x) := \sum_{n \leq x} \lambda(n) \leq 0 \qquad (x \geq 2).$$

This conjecture was supported by calculations for all numbers in the range $2 \leq x \leq 10^6$. However, it was disproved by C. B. Haselgrove using (essentially) Theorem 11.12 and a computer search for large values of the trigonometric polynomial F_T^*.

We have the formulas

$$s \int_1^\infty x^{-s-1} L(x) \, dx = \int_{1-}^\infty x^{-s} dL(x) = \zeta(2s)/\zeta(s) \qquad (\sigma > 1),$$

$$\int_1^\infty x^{-s-3/2} L(x) \, dx = \zeta(2s+1) \Big/ \Big\{ \Big(s + \frac{1}{2}\Big) \zeta\Big(s + \frac{1}{2}\Big) \Big\} \qquad (\sigma > 1/2).$$

If Pólya's conjecture were true, then by Landau's theorem, the last representation would hold for $\sigma > 0$ and the R.H. would be true.

Haselgrove found suitable values of the parameter T and the variable u so as to make the associated function $L_T^*(u) > 0$. An explicit value of x for which $L(x) > 0$ was later found by R. S. Lehman, who showed that $L(906, 180, 359) = +1$.

PROBLEM 11.8   Suppose that the hypotheses of Theorem 11.12 are satisfied for any (fixed) $T > 0$. Show that

$$\sum |\alpha_n|^2 \leq \limsup_{u \to \infty} |F(u)|^2.$$

Hint. Consider

$$\lim_{X \to \infty} \frac{1}{2X} \int_{-X}^X |F_T^*(u)|^2 du.$$

PROBLEM 11.9   Use the preceding estimate to show that

$$\sum_{\substack{n \leq x \\ n=2^\alpha 3^\beta}} n - \frac{x \log x}{\log 2 \log 3} \neq O(x).$$

Cf. §7.9 (for the associated generating function) and Problem 11.5.

11.6   Sign changes of $\pi(x) - \operatorname{li} x$

We have seen both theoretically and from Table 8.1 that $\operatorname{li} x$ is a good approximation to $\pi(x)$. To study the oscillation of their difference, we begin by examining the difference of the related functions

$$\Pi(x) := \sum_{n \le x} \kappa(n) = \pi(x) + \frac{1}{2}\pi(x^{1/2}) + \frac{1}{3}\pi(x^{1/3}) + \cdots$$

and (recalling Lemma 8.13)

$$L_1(x) := \int_1^x (1 - t^{-1})\, dt / \log t = \operatorname{li} x - \log \log x - \gamma.$$

Lemma 11.14   *Let $\beta + i\gamma$ denote a zero of ζ with $0 < \beta < 1$ and let $\theta < \beta$. Then $\Pi(x) - L_1(x) = \Omega_\pm(x^\theta)$.*

Proof.   Let $\sigma > 1$ and let $C = 1$ or $C = -1$. By (6.8),

$$\log \zeta(s)\frac{s-1}{s} - \frac{Cs}{s-\theta} = s \int_1^\infty x^{-s-1}\{\Pi(x) - L_1(x) - Cx^\theta\}dx. \quad (11.12)$$

Since $\log \zeta$ has a singularity at $\beta + i\gamma$, the abscissa of convergence of the M.t. in (11.12) is at least β. On the other hand, the left side of (11.12) is analytic at all points of the real line to the right of θ, and $\theta < \beta$. Thus, by Landau's oscillation theorem, $\Pi(x) - L_1(x) - Cx^\theta$ changes sign infinitely often. Taking $C = +1$ ($C = -1$) yields the Ω_+ (Ω_-) result. $\square$

It was observed that $\operatorname{li} x > \pi(x)$ for all $x \ge 2$ in the range of existing tables of primes. Moreover, there were theoretical grounds for believing that this inequality held generally, since $\pi(x) < \Pi(x)$ for $x \ge 4$ and $\Pi(x)$ is the naturally occurring function that is approximated by $\operatorname{li} x$. The preceding lemma implies that $\Pi(x) - \operatorname{li} x$ changes sign infinitely often. Does $\pi(x) - \operatorname{li} x$ have sign changes (aside from the trivial one at $\operatorname{li}^{-1}(0) = 1.451369\ldots$)? The proof of an infinitude of sign changes by J. E. Littlewood in 1914 was a major accomplishment. We shall establish this result using Theorem 11.12 and approximate periodicity.

Theorem 11.15   *Let $F_1(x) := (\pi(x) - \operatorname{li} x)/(\sqrt{x}/\log x)$. Then we have $\limsup_{x \to \infty} F_1(x) = +\infty$ and $\liminf_{x \to \infty} F_1(x) = -\infty$.*

Proof.   First, we may assume the truth of the R.H. Otherwise, the preceding lemma would give $\pi(x) - \operatorname{li} x = \Omega_\pm(x^\theta)$, for some $\theta > \frac{1}{2}$ and there would be nothing further to show.

Let us define $F(x) := \{\psi(x) - x\}/\sqrt{x}$ and recall Theorem 8.26: under the R.H. we have $F_1(x) = F(x) + O(1)$. We showed in Theorem 8.7 that $F \neq o(1)$, and in Corollary 11.10 that

$$\limsup_{x \to \infty} F(x) > 1/15, \quad \liminf_{x \to \infty} F(x) < -1/15.$$

Here we show that F is in fact unbounded from above and below. We shall give a proof by contradiction, assuming that F satisfies a one sided bound. This enables us to apply Theorem 11.12.

We have

$$-\frac{1}{s + \frac{1}{2}} \frac{\zeta'}{\zeta}\left(s + \frac{1}{2}\right) - \frac{1}{s - \frac{1}{2}} = \int_1^\infty x^{-s} F(x) \frac{dx}{x} =: G(s),$$

and the singularities are simple poles at the zeros of $\zeta(s + 1/2)$. In particular, under the R.H., the only nonreal singularities are the poles at points $s = \pm i\gamma_n$, $n = 1, 2, 3, \ldots$, with appropriate repetitions made for any multiple zeros of zeta which may exist. (None has been found to date.)

The integral defining G converges for $\sigma > 0$. This can be seen by applying Landau's oscillation theorem to the expression

$$G(s) + \frac{K}{s} = \int x^{-s} \{F(x) + K\} x^{-1} dx,$$

with an appropriate value of K chosen to make $F + K$ be of one sign.

Let $T \geq 30$ be given, not the ordinate of any zeta zero. We set

$$F^*(u) = -\sum_{-N}^{N} \frac{e^{i\gamma_n u}}{\frac{1}{2} + i\gamma_n}, \quad F_T^*(u) = -\sum_{-N}^{N} \left(1 - \frac{|\gamma_n|}{T}\right) \frac{e^{i\gamma_n u}}{\frac{1}{2} + i\gamma_n},$$

where $N = N(T)$ denotes the number of zeros $\rho = \frac{1}{2} + i\gamma$ of zeta satisfying $0 < \gamma \leq T$. Also, we set

$$G^*(s) = -\sum_{-N}^{N} \frac{(s - i\gamma_n)^{-1}}{\frac{1}{2} + i\gamma_n}.$$

Now $G - G^*$ is continuable to the closed region $\{s : \sigma \geq 0, -T \leq t \leq T\}$, and thus Theorem 11.12 applies.

We have

$$\left| \sum_{|\gamma| \leq T} \left(1 - \frac{|\gamma|}{T}\right) e^{i\gamma u} \left\{ \frac{1}{\frac{1}{2} + i\gamma} - \frac{1}{i\gamma} \right\} \right| \leq \sum_\gamma \gamma^{-2} = B < \infty$$

by Lemma 8.19. Note that B is independent of T. Thus

$$F_T^*(u) = -\sum_{|\gamma|<T} \left(1 - \frac{|\gamma|}{T}\right) \frac{e^{i\gamma u}}{i\gamma} + \theta B$$

$$= -2 \sum_{0<\gamma<T} \left(1 - \frac{\gamma}{T}\right) \frac{\sin \gamma u}{\gamma} + \theta B,$$

where $\theta = \theta(y, T)$ and $|\theta| \le 1$. If we choose $u = 1/T$, then we have

$$\frac{2}{T} \sum_{0<\gamma<T} \left(1 - \frac{\gamma}{T}\right) \frac{\sin \gamma/T}{\gamma/T} \ge \frac{1}{T} \sum_{0<\gamma\le T/2} \frac{2}{\pi} > c \log T.$$

We have used the estimates $x^{-1} \sin x \ge 2/\pi$ for $0 < x \le \pi/2$ and $N(T/2) > (c\pi/2)T \log T$ for some $c > 0$ and all $T \ge 30$. (As was mentioned in the proof of Corollary 11.10, $\zeta(1/2 + i\gamma_1) = 0$, where $\gamma_1 \doteq 14.134725$.)

It follows that if T is large, then $F_T^*(1/T)$ assumes a large negative value. Also, since

$$-2 \sum_{0<\gamma<T} \left(1 - \frac{\gamma}{T}\right) \frac{\sin \gamma u}{\gamma}$$

is an odd function of u, a large positive value is assumed at $u = -1/T$. By Lemma 11.3, the trigonometric polynomial $F_T^*(u)$ assumes large positive and large negative values for arbitrarily large positive values of u.

By Theorem 11.12 we have

$$\varlimsup_{x\to\infty} F(x) \ge \sup_x F_T^*(x) \ge c \log T - B$$

and

$$\varliminf_{x\to\infty} F(x) \le \inf_x F_T^*(x) \le -c \log T + B.$$

As T is arbitrary, the limit superior is $+\infty$ and the limit inferior $-\infty$. $\square$

11.7 The size of $M(x)/\sqrt{x}$

We remarked (after Theorem 8.17) that if the R.H. is true, then

$$M(x) := \sum_{n \le x} \mu(n) = O(x^c)$$

for any $c > 1/2$. Conversely, if this bound holds, then

$$1/\zeta(s) = s \int_1^\infty x^{-s-1} M(x)\, dx$$

provides an analytic continuation of $1/\zeta$ to the half plane $\{s : \sigma > 1/2\}$ and hence the R.H. is true (cf. Theorem 6.9).

Stieltjes thought that he had proved that $M(x) = O(x^{1/2})$; indeed, he asserted that it was probable that

$$|M(x)| \le \sqrt{x} \quad (0 < x < \infty).$$

This hypothesis was subsequently studied by Mertens and has been named for him. It is supported by calculations for all $x \le 10^8$, but was disproved by Odlyzko and te Riele (see Notes).

Let

$$M^+ = \varlimsup_{x \to \infty} \frac{M(x)}{\sqrt{x}}, \qquad M^- = \varliminf_{x \to \infty} \frac{M(x)}{\sqrt{x}}.$$

The question of whether M^+ or M^- is finite is still undecided. We shall show several conditions which would have to be satisfied for at least one of M^+, M^- to be finite.

Lemma 11.16 *If $M^+ < \infty$ or $M^- > -\infty$, then the Riemann hypothesis is true and all the nontrivial zeros of zeta are simple.*

Proof. If the R.H. is false and if $\zeta(\beta + i\gamma) = 0$ for some $\beta > 1/2$, then by Theorem 11.8,

$$\varlimsup_{x \to \infty} M(x)/x^\beta > 0 \quad \text{and} \quad \varliminf_{x \to \infty} M(x)/x^{-\beta} < 0.$$

Similarly, if the R.H. is true but $1/2 + i\gamma$ is a zero of multiplicity $m > 1$, then

$$\varlimsup_{x \to \infty} M(x)/(x^{1/2} \log^{m-1} x) > 0 \quad \text{and} \quad \varliminf_{x \to \infty} M(x)/(x^{1/2} \log^{m-1} x) < 0.$$

In either case, we have $M^+ = +\infty$ and $M^- = -\infty$. $\qquad \square$

Following Ingham, we shall show that if M^+ or M^- is finite, then there are many nontrivial linear relations among the imaginary parts of the zeros of the zeta function. The argument we give depends on the following

Lemma 11.17   *Let ρ range through the nonreal zeros of zeta, and assume that all such zeros are simple. Then*

$$\sum |\rho \zeta'(\rho)|^{-1} = +\infty.$$

Proof.   Let $T > 3$ be a number for which Lemma 8.20 applies: there exists an absolute constant ν such that

$$|\zeta(\sigma + iT)| > T^{-\nu} \quad (-1 \le \sigma \le 2).$$

Let $C = C(T)$ be the rectangular contour with vertices

$$2 + i, \ 2 + iT, \ -1 + iT, \ \text{and} \ -1 + i,$$

traversed in that order. Label the sides of C as C_1, C_2, C_3, and C_4, starting with the right side.

The residue theorem gives (with $\gamma = \Im\rho$)

$$\frac{1}{2\pi i} \int_C \frac{ds}{s\zeta(s)} = \sum_{0 < \gamma < T} \frac{1}{\rho\zeta'(\rho)},$$

and we would like to show that the integral is large for large T. It is convenient, however, to introduce a convergence factor. Let n be a positive integer larger than $\nu - 1$, and let

$$I(T) := \frac{1}{2\pi i} \int_C \left(1 - \frac{s}{iT}\right)^n \frac{ds}{s\zeta(s)} = \sum_{0 < \gamma < T} \left(1 - \frac{\rho}{iT}\right)^n \frac{1}{\rho\zeta'(\rho)}.$$

We estimate $I(T)$ over each side separately. On C_1 we have $\Re s = 2$, and the D.s. for $1/\zeta$ converges uniformly there, justifying termwise integration. Expanding $(1 - s/iT)^n$ by the binomial theorem and integrating we obtain

$$\int_{C_1} = \sum_{j=0}^{n} \sum_{m=1}^{\infty} \binom{n}{j} \mu(m) \int_{2+i}^{2+iT} \left(\frac{-s}{iT}\right)^j m^{-s} \frac{ds}{s}.$$

For $1 \le j \le n$ we have

$$\left| \int_{2+i}^{2+iT} s^{j-1} m^{-s} ds \right| \le m^{-2} \int_1^T |2 + it|^{j-1} dt$$

$$\le m^{-2} 3^{j-1} \int_1^T t^{j-1} dt \le m^{-2} 3^{j-1} T^j / j.$$

For $j = 0$ and $m \geq 2$ we make a contour integral with vertices at

$$2 + i, \ T + i, \ T + iT, \ \text{and} \ 2 + iT$$

and apply Cauchy's theorem; we find that

$$\left| \int_{2+i}^{2+iT} m^{-s} ds/s \right| < 2 \int_2^T m^{-\sigma} d\sigma + m^{-T} = O\left(\frac{m^{-2}}{\log m} \right),$$

with an absolute O constant. Thus

$$\int_{\mathcal{C}_1} = \int_{2+i}^{2+iT} \frac{ds}{s} + \sum_{j=1}^n \sum_{m=1}^\infty O(m^{-2} 3^{j-1} T^j / j) + \sum_{m=2}^\infty O(m^{-2}/\log m)$$

$$= \log T + O_n(1).$$

On $\mathcal{C}_2$, $\Im s = T$ and $1 - s/iT = -\sigma/iT = O(1/T)$. Thus

$$\int_{\mathcal{C}_2} = O(T^{-n-1}) \max_{-1 \leq \sigma \leq 2} \frac{1}{|\zeta(\sigma + iT)|} = O_n(1).$$

For $\int_{\mathcal{C}_3}$ take $s = 2 - it$ in the functional equation given in Theorem 8.1:

$$1/\zeta(1-s) = 2^s \pi^{s-1} \sin(\pi s/2) \Gamma(1-s)/\zeta(s).$$

The weak form of Stirling's formula given in Lemma 8.3 then yields

$$|1/\zeta(-1+it)| \leq K t^{-3/2} \qquad (t \geq 3).$$

For $|t| \leq 3$, $1/|\zeta(-1+it)| = O(1)$, since ζ is continuous and nonzero here. Also, the convergence factor is bounded on $\mathcal{C}_3$. Thus $\int_{\mathcal{C}_3} = O_n(1)$.

It is known that $1/|\zeta|$ is bounded on $\mathcal{C}_4$. (We could avoid using this fact by noting that zeros of zeta are isolated, by analyticity, and hence $\mathcal{C}_4$ could be chosen with imaginary part *near* 1 to avoid any zeros.) The convergence factor is also bounded on $\mathcal{C}_4$ and thus $\int_{\mathcal{C}_4} = O_n(1)$. It follows that

$$I(T) = \frac{\log T}{2\pi i} + O_n(1).$$

We are restricting our attention to zeros of zeta $\rho = 1/2 + i\gamma$ with imaginary part satisfying $1 < \gamma < T$. Thus we have

$$\left| 1 - \frac{\rho}{iT} \right| = \left| 1 - \frac{\gamma}{T} + \frac{i}{2T} \right| < 1,$$

and hence

$$\sum_{0<\gamma<T} \frac{1}{|\rho\zeta'(\rho)|} \geq \frac{\log T}{2\pi} + O_n(1).$$

Since T can be taken arbitrarily large,

$$\sum |\rho\zeta'(\rho)|^{-1} = +\infty. \qquad \qquad \square$$

Theorem 11.18 *Let $\gamma_1 < \gamma_2 < \cdots$ denote the imaginary parts of the zeros of zeta in the upper half plane. If there are no linear relations*

$$\sum_{n=1}^{N} c_n\gamma_n = 0 \qquad (c_n \in \mathbb{Z}, \ c_n \ \text{not all} \ 0), \qquad (11.13)$$

or only a finite number of such relations, then $M^+ = +\infty$ and $M^- = -\infty$.

Remark 11.19   A strong reason for doubting that $M(x)/\sqrt{x}$ is bounded is that not even one linear relation among the γ_n's has been found to date.

Proof.   We shall assume that $M^+ < +\infty$ or $M^- > -\infty$ and deduce a contradiction. In view of Lemma 11.16 we may assume the R.H. and the simplicity of the zeros of zeta. We have

$$(1/2+s)^{-1}\zeta(1/2+s)^{-1} = \int x^{-s}\frac{M(x)}{\sqrt{x}}\frac{dx}{x} =: G(s);$$

the conditions of Theorem 11.12 hold if we take

$$F(x) := M(x)/\sqrt{x} \quad \text{and} \quad F^*(x) := \sum_{|\gamma|<T} \frac{1}{\rho\zeta'(\rho)}e^{i\gamma x}$$

for some $T > 0$. We have then

$$M^+ \geq \sup F_T^*(x), \quad M^- \leq \inf F_T^*(x),$$

where

$$F_T^*(x) := \sum_{|\gamma|<T} \left(1 - \frac{|\gamma|}{T}\right)\frac{1}{\rho\zeta'(\rho)}e^{i\gamma x}. \qquad (11.14)$$

If there are no linear relations among the γ_n's, Lemma 11.5 gives

$$\sup F_T^*(x) = 2 \sum_{0<\gamma_n<T} \left(1 - \frac{\gamma_n}{T}\right) |\rho\zeta'(\rho)|^{-1},$$

$$\inf F_T^*(x) = -2 \sum_{0<\gamma_n<T} \left(1 - \frac{\gamma_n}{T}\right) |\rho\zeta'(\rho)|^{-1}.$$

The argument is essentially the same in case there are a finite number of relations of the type (11.13). Then there is a last γ, say γ_M, which is so involved, and we apply Kronecker's theorem in the range $\gamma_M < \gamma < T$.

Finally, we have from the foregoing lemma that $\sum |\rho\zeta'(\rho)|^{-1} = +\infty$. It follows upon letting $T \to \infty$ that $M^+ = +\infty$ and $M^- = -\infty$. $\qquad\square$

PROBLEM 11.10 Put $g(x) = \sum_{n\leq x} \mu(n)/n$. Under the hypotheses of Theorem 11.18, prove that

$$\liminf_{x\to\infty} x^{1/2}g(x) = -\infty \quad \text{and} \quad \limsup_{x\to\infty} x^{1/2}g(x) = +\infty.$$

11.7.1 Numerical calculations

We conclude with a description of some estimates of $M^\pm$, the limit superior and limit inferior of $M(x)/\sqrt{x}$.

If we apply Theorem 11.8, using the zero of zeta with smallest positive imaginary part, we find that

$$M^+, -M^- \geq 1/|\rho\zeta'(\rho)| = .089145\ldots.$$

Estimates made with other individual zeros of zeta appear to give smaller lower bounds for $|M^\pm|$.

W. B. Jurkat observed that, with little calculation, one can obtain $M^- < -.5$, $M^+ > .49$. His argument is based on the explicit formula

$$\{M(x) + 2M^*(1/x)\}x^{-1/2} = \sum_\rho x^{i\gamma}/(\rho\zeta'(\rho)) =: H(x), \qquad (11.15)$$

where the sum defining $H(x)$ extends over the nontrivial zeros $\rho = \frac{1}{2} + i\gamma$ of zeta, grouped into suitable blocks, and

$$M^*(x) := 1 + \sum_{n=1}^\infty \frac{(-1)^n(2\pi x)^{2n}}{2n(2n)!\zeta(2n+1)},$$

an entire function of x. If we assume the R.H. and the simplicity of all zeros of zeta, then formula (11.15) holds for all positive, noninteger x (cf. [TiHB], Th. 14.27).

The right side of (11.15) is not a trigonometric polynomial, but is almost periodic in a distributional sense (cf. Jurkat's article cited in the Notes). Thus, a large value of $H(x)$ will be nearly repeated for arbitrarily large values of x. Also, $2M^*(1/x)x^{-1/2} \to 0$ as $x \to \infty$. Thus we have

$$M^+ = \sup H(x), \quad M^- = \inf H(x).$$

By evaluating the power series for $M^*(1)$, we find that

$$M^- \le H(1-) = 0 + 2M^*(1) \doteq -.505491,$$

$$M^+ \ge H(1+) = 1 + 2M^*(1) \doteq +.494509.$$

The best oscillation estimates for $M(x)/\sqrt{x}$ to date have been achieved with the inequalities

$$M^+ \ge \varlimsup_{x \to \infty} F_T^{**}(x), \quad M^- \le \varliminf_{x \to \infty} F_T^{**}(x),$$

where F_T^{**} is a trigonometric polynomial analogous to the function given in Theorem 11.12. The trigonometric polynomial is evaluated at judiciously chosen values of T and x. In this way Odlyzko and te Riele showed that

$$M^+ \ge 1.06 \quad \text{and} \quad M^- \le -1.009.$$

A key step in their argument is the application of the small vectors algorithm of Lenstra, Lenstra, and Lovasz to make the judicious choices.

None of the methods described provide specific values of x for which $M(x)/\sqrt{x}$ is large positive or large negative. Odlyzko and te Riele estimate, however, that the smallest value of x with $M(x) > \sqrt{x}$ exceeds 10^{20}.

11.8 The error term in the divisor problem

We conclude this chapter with a brief discussion of the quality of error estimates in the Dirichlet divisor problem. Recall that the goal here is to estimate

$$N_2(x) := \sum_{n \le x} (1 * 1)(n)$$

and in Corollary 3.32 we showed that

$$\Delta(x) := N_2(x) - \{x \log x + (2\gamma - 1)x\} = O(x^\alpha) \tag{11.16}$$

with $\alpha = 1/2$. This is far from the truth; G. F. Voronoï proved that (11.16) holds with $\alpha = 1/3 + \epsilon$ for any $\epsilon > 0$, and exponential sum methods have led to estimates with $\alpha \approx 0.324$; the best bound to date is $\alpha = 7/22 + \epsilon$ achieved by H. Iwaniec and C. J. Mozzochi via a hybrid method. In the other direction, G. H. Hardy showed that $\Delta(x) = \Omega(x^{1/4})$. Proofs of this result using an argument based on Plancherel's formula (the Fourier transform analogue of the Parseval identity for Fourier series) are given in [TiHB], §12.6, and [Ivic], §13.6.

11.9 Notes

11.2. Our proof of Lemma 11.5 (Kronecker's theorem) is based on one of several by Harald Bohr, in Proc. London Math. Soc. (2), vol. 21 (1922), pp. 315–316; also in Bohr's Collected Math. Works, vol. 3, Item D1, Dansk Mat. Forening, København, 1952.

11.4. The idea of Theorem 11.8 goes back to E. Schmidt, Math. Annalen, vol. 57 (1903), pp. 195–204. Remark 11.11 is essentially due to H. M. Stark, Proc. A.M.S., vol. 17 (1966), pp. 1211–1214.

11.5. The breakthrough result of Ingham presented here appeared in Amer. J. Math., vol. 64 (1942), pp. 313–319.

Haselgrove's disproof of the Pólya conjecture that $\sum_{n \leq x} \lambda(n) \leq 0$ for all $x \geq 2$ appeared in Mathematika, vol. 5 (1958), pp. 141–145. Haselgrove's calculations were verified by R. S. Lehman, Math. Comp., vol. 14 (1960), pp. 311–320; also, Lehman found numerous values of x less than 10^9 for which the preceding sum is positive.

P. Turán established in Danske Vid. Selsk. Mat.–Fys. Medd., vol. 24 no. 17 (1948) a number of sufficient conditions for the truth of the R.H. One theorem asserted that *if no partial sum of the D.s. for zeta vanishes in the half plane* $\{s : \sigma > 1\}$, *then the R.H. is true.* Turán deduced from his hypothesis that the partial sums of the series $\sum \lambda(n)/n$ were all nonnegative, and then the R.H. follows by an application of Landau's theorem.

H. L. Montgomery showed in Studies in Pure Math., Birkhäuser, Basel, 1983, pp. 497–506, that if $0 < c < 4/\pi - 1$ and $N > N_0(c)$, then $\sum_{n \leq N} n^{-s}$

in fact has zeros in the half plane $\sigma > 1 + c \log\log N / \log N$. From this result we see that the hypothesis of Turán's theorem is not satisfied.

11.6. Littlewood's proof of Theorem 11.15 first appeared in Comptes rendus Acad. Sci. Paris, vol. 158 (1914), pp. 1869–1872; also in Littlewood's Collected Papers, vol. 2, Oxford, 1982, pp. 829–832. Many proofs have been given of this celebrated result; a small sampling includes:

- Hardy and Littlewood, Acta Math., vol. 41 (1918), pp. 119–196, particularly §5 (also in Hardy's Collected Papers, vol. 2, Oxford, 1967, pp. 20–99)
- [LanV], vol. 2, pp. 123–150
- [Ing], Theorem 35
- Diamond, l'Enseignement Math. (2), vol. 21 (1975), 1–14.

It was shown by H. J. J. te Riele that a nontrivial sign change of $\pi(x) -$ li x occurs for some $x < 10^{361}$, in Math. Comp., vol. 48 (1987), pp. 323–328. (There is a trivial change of sign of $\pi(x) - $ li x at $x \doteq 1.451369$.)

11.7. In a letter to Ch. Hermite dated 11 July 1885, T. J. Stieltjes claimed to have proved that $M(x) = O(\sqrt{x})$ (No. 79 in Corresp. d'Hermite et de Stieltjes, B. Baillaud and H. Bourget, eds., Gauthier–Villars, Paris, 1905, pp. 160–164). In his research announcement in Comptes rendus Acad. Sci. Paris, vol. 101 (13 July 1885), pp. 153–154, Stieltjes asserted "merely" that he had established $M(x) = O(x^c)$ for any $c > 1/2$.

Theorem 11.18 was established by Ingham in the paper cited in the notes for §11.5.

Jurkat's article appeared in Proc. Symposia Pure Math vol. XXIV, H. G. Diamond, ed., American Math Soc., Providence, R.I., 1973, pp. 147–158.

Odlyzko and te Riele's disproof of Mertens' conjecture that $|M(x)| \le \sqrt{x}$ appeared in J. reine angew. Math., vol. 357 (1985), pp. 138–160. Detailed information on estimates of the ratio $M(x)/\sqrt{x}$ is given in this paper and in an essay by te Riele in T. J. Stieltjes, Collected Papers, vol. 1, Springer–Verlag, Berlin, 1993, pp. 69–79.

11.8. Voronoï's paper was published in J. reine angew. Math., vol. 126 (1903), pp. 241–282. The paper of Iwaniec and Mozzochi is contained in J. Number Theory, vol. 29 (1988), pp. 60–93. Hardy's result appeared in Proc. London Math. Soc. (2), vol. 13 (1916), pp. 1–25 and 192–213; also in Collected Papers, vol. 2, Oxford, 1967, pp. 268–315.

Chapter 12

Sieves

12.1 Introduction

Sieve problems in number theory are typically of the following type. We are given a finite sequence $\mathcal{A}$ of integers and a finite set $\mathcal{P}$ of pairwise coprime natural numbers, e.g. the primes (usually!) or fixed powers of the primes. We seek to estimate the number of elements of $\mathcal{A}$ that remain when we delete those falling into certain residue classes modulo members of $\mathcal{P}$.

In this chapter, we develop two methods for making upper bound sieve estimates which arise naturally in many arithmetical investigations. The first, the Brun-Hooley method, is combinatorial. It is effective in problems in which $\mathcal{P}$ consists of primes and a small number of residue classes are being removed for each prime in $\mathcal{P}$. This method also gives nontrivial lower bounds in many interesting cases, but we shall merely sketch it. The second method, a form of the so-called large sieve, employs Fourier analysis; it restricts attention to cases in which $\mathcal{A}$ is the set of integers in an interval, but has a particularly elegant formulation and yields upper bound estimates even when many residue classes are removed per prime in $\mathcal{P}$. In the next chapter, we describe several applications of these methods.

An example of a sieve occurs in §1.4, where we estimated the number of squarefree integers in $[1, X]$. We deleted ("sieved out") those integers in this interval that are divisible by the square of a prime. (This is our one example in which the sieving set, which we have nevertheless called $\mathcal{P}$, does not consist of primes.) The sieve argument works without too much effort because $\prod_p (1 - p^{-2})$ converges, which implies that a positive proportion of the original set survived. In this chapter, we shall consider problems in which the proportion of elements remaining is very small. The large number of elements being deleted makes it difficult to take an accurate census of the surviving population.

A simple example of a sieve in which most numbers are removed occurs in the proof of Lemma 4.3 that $\pi(x) = o(x)$. In this case, we take $\mathcal{A}$ to be the set of all integers in an interval $[1, x]$ and $\mathcal{P} = \mathcal{P}_r = \{2, 3, \ldots, p_r\}$, the first r primes. We remove an element a from $\mathcal{A}$ if $a \equiv 0 \pmod{p}$ for some $p \in \mathcal{P}$. If we let $S(\mathcal{A}, \mathcal{P})$ denote the number of elements of $\mathcal{A}$ that remain, then $\pi(x) \le r + S(\mathcal{A}, \mathcal{P})$, and the rest of the proof consists in showing that $\limsup_{x \to \infty} S(\mathcal{A}, \mathcal{P})/x$ is small for r large.

Another example, which we shall discuss in §§13.1, 13.4, and 13.6 is the so-called twin prime problem. A pair of *twin primes* is a couple of positive integers $(n, n + 2)$ such that each of n and $n + 2$ is prime. For example, $(11, 13)$ is a pair of twin primes. Note that 5 occurs in both the pairs $(3, 5)$ and $(5, 7)$, and that no other prime occurs in two pairs.

The main open problem for twin primes is to give an accurate asymptotic estimate of $TP(x)$, the number of pairs of twin primes $(p, p + 2)$ satisfying $p \le x$. At present we do not know even whether there are infinitely many twin primes. Calculation and heuristic arguments (cf. §13.6) suggest that $TP(x) \sim Kx/\log^2 x$ for some constant K.

To see why we consider this to be a sieve problem, let us restate it. For $N > 2$, let $\mathcal{A} = \{n(n + 2) : 1 \le n \le N\}$. We delete from $\mathcal{A}$ all those elements which are divisible by some prime $p \le \sqrt{N + 2}$. The collection of remaining elements, which we call $\mathcal{A}'$, consists of all products of twin primes $(p, p + 2)$ satisfying $\sqrt{N + 2} < p \le N$. Indeed, if either of n, $n + 2$ is composite or if $n \le \sqrt{N + 2}$, then there is some $p \le \sqrt{N + 2}$ such that $p \mid n(n + 2)$.

It follows that

$$|\mathcal{A}'| = TP(N) - TP(\sqrt{N + 2}),$$

whence

$$|\mathcal{A}'| \le TP(N) \le |\mathcal{A}'| + \sqrt{N + 2}.$$

Thus upper and lower bounds for $|\mathcal{A}'|$ provide upper and lower bounds for $TP(N)$.

12.2 The sieve of Eratosthenes and Legendre

Here we attempt to estimate the prime counting function $\pi(x)$ by a method based on the famous sieve of Eratosthenes. The argument is instructive for subsequent work even though, as we shall see, our success will be limited. This time, as earlier, we take $\mathcal{A}$ to be the set of all integers in an interval $[1, x]$, $\mathcal{P} = \mathcal{P}_r = \{2, 3, \ldots, p_r\}$, P to be the product of the primes in $\mathcal{P}$, and $S(\mathcal{A}, \mathcal{P})$ to be the number of elements of $\mathcal{A}$ that remain after deletion of those divisible by a prime of $\mathcal{P}$.

We compute $S(\mathcal{A}, \mathcal{P})$ by means of the formula

$$(1 * \mu)(n) = \sum_{d|n} \mu(d) = e(n) = \begin{cases} 1, & \text{if } n = 1, \\ 0, & \text{if } n > 1, \end{cases} \tag{12.1}$$

the characteristic property of the Möbius function. An element $a \in \mathcal{A}$ remains after sieving by elements of $\mathcal{P}$ if and only if $(a, P) = 1$. Thus we can express the indicator function of the survivors by

$$\sum_{d|(a,P)} \mu(d), \quad a \in \mathcal{A},$$

and their number by

$$S(\mathcal{A}, \mathcal{P}) = \sum_{a \in \mathcal{A}} \sum_{d|(a,P)} \mu(d).$$

Now $d|(a, P)$ precisely when both $d|a$ and $d|P$, so we can reorder the sum as

$$S(\mathcal{A}, \mathcal{P}) = \sum_{d|P} \mu(d) \sum_{\substack{a \in \mathcal{A} \\ d|a}} 1 = \sum_{d|P} \mu(d) \left[\frac{x}{d}\right]. \tag{12.2}$$

It follows that

$$S(\mathcal{A}, \mathcal{P}) = x \sum_{d|P} \frac{\mu(d)}{d} + \sum_{d|P} \mu(d)\left(\left[\frac{x}{d}\right] - \frac{x}{d}\right)$$

$$= x \prod_{p \in \mathcal{P}} \left(1 - \frac{1}{p}\right) + \sum_{d|P} \mu(d)\left(\left[\frac{x}{d}\right] - \frac{x}{d}\right). \tag{12.3}$$

The last result is often called the formula of Eratosthenes–Legendre, and formula (12.2) is an instance of the inclusion–exclusion principle:

$$S(\mathcal{A}, \mathcal{P}) = [x] - \sum_{i \leq r} \left[\frac{x}{p_i}\right] + \sum_{i < j \leq r} \left[\frac{x}{p_i p_j}\right] \mp \cdots .$$

If r, the number of primes in $\mathcal{P}$, is suitably small, then the last sum in (12.3) can be regarded as an error term, for

$$\left| \sum_{d|P} \mu(d) \left(\frac{x}{d} - \left[\frac{x}{d}\right]\right) \right| \leq \sum_{d|P} 1 = 2^r. \qquad (12.4)$$

For example, if $r \leq 1.4 \log x$, then the last sum is less than $x^{0.98}$, and in this case the main term in (12.3) provides an asymptotic estimate of $S(\mathcal{A}, \mathcal{P})$. For slightly larger values of r, the term 2^r is greater than x and *a fortiori* exceeds what we might have expected to be the main term of (12.3).

Could one perhaps obtain an asymptotic formula

$$S(\mathcal{A}, \mathcal{P}) \sim x \prod_{p \in \mathcal{P}} \left(1 - \frac{1}{p}\right) \qquad (12.5)$$

for larger sets $\mathcal{P}$ by using the cancellation μ induces in the last sum in (12.3)? For a sufficiently large set $\mathcal{P}$, the answer is No, as we now show.

Example. Take $\mathcal{P}$ to be the set of all primes not exceeding $x^{1/2}$ and $\mathcal{A}$ still to be $\mathbb{Z} \cap [1, x]$. Then

$$S(\mathcal{A}, \mathcal{P}) = \pi(x) - \pi(x^{1/2}) + 1, \qquad (12.6)$$

for by Eratosthenes' sieve, any element counted in $S(\mathcal{A}, \mathcal{P})$ is either 1 or a prime in $(x^{1/2}, x]$. By the P.N.T., the right side of (12.6) is asymptotic to $x/\log x$. On the other hand, by the Mertens product formula (Cor. 6.19),

$$x \prod_{p \leq \sqrt{x}} \left(1 - \frac{1}{p}\right) \sim e^{-\gamma} \frac{x}{\log \sqrt{x}} = 2e^{-\gamma} \frac{x}{\log x}.$$

Since $2e^{-\gamma} = 1.1229 \ldots \neq 1$, (12.5) cannot hold in this case.

Let us consider, from a probabilistic point of view, why (12.5) succeeds for small sets $\mathcal{P}$ and fails for $|\mathcal{P}|$ large. We continue to take $\mathcal{A} = \mathbb{Z} \cap [1, x]$. For b a positive integer, the number of multiples of b in $\mathcal{A}$ is $[x/b]$. If we let $\Pr(\mathcal{A}, b)$ denote the probability that a randomly chosen element of $\mathcal{A}$ is divisible by b, then

$$\Pr(\mathcal{A}, b) = \frac{[x/b]}{x} = \frac{1}{b} + O\left(\frac{1}{x}\right).$$

For p, p' distinct primes in $\mathcal{P}$, we have

$$\Pr(\mathcal{A}, pp') = \frac{1}{pp'} + O\left(\frac{1}{x}\right),$$

$$\Pr(\mathcal{A}, p) \cdot \Pr(\mathcal{A}, p') = \frac{1}{pp'} + O\left(\frac{1}{xp} + \frac{1}{xp'} + \frac{1}{x^2}\right),$$

and thus

$$\frac{\Pr(\mathcal{A}, pp')}{\Pr(\mathcal{A}, p) \cdot \Pr(\mathcal{A}, p')} = 1 + O\left(\frac{pp'}{x}\right) \approx 1,$$

if pp' is small compared with x. In this case we say, in probabilistic language, that the "events" of an integer in $\mathcal{A}$ being divisible respectively by primes p and p' are nearly independent. On the other hand, this relation fails if $pp' \gg x$.

The probability of an integer in $\mathcal{A}$ not being divisible by a (small) prime p is about $1 - 1/p$. If the primes of $\mathcal{P}$ are suitably small compared with x, then the independence relation applies and we expect (12.5) to hold. On the other hand, if $\mathcal{P}$ contains many large primes, then the independence relation is poor, and it is not surprising that (12.5) should fail.

Each of the preceding examples contains a valuable hint. For $\mathcal{P}$ the set of primes up to $\sqrt{x}$, the quantities on each side of (12.5) are of the same order of magnitude, even though they are not asymptotic; this suggests that we might seek to establish O-estimates when asymptotic relations appear hopeless. For $\mathcal{P}$ a suitably small set of primes, the "error term" in (12.3) does not exceed the "main term"; the theory we shall develop replaces μ by another function that yields a "main term" of comparable size, but whose smaller support serves to control the size of the "error term."

12.3 Sieve setup

To use notation that is standard in sieve theory, we must rename an arithmetic function. *Let $\nu(n)$ denote the number of distinct prime divisors of n.* The name previously used, $\omega(n)$, is being reassigned here.

Let $\mathcal{A}$ be a finite *sequence* (to allow possible repetitions) of integers, $\mathcal{P}$ a finite set of primes, $P = \Pi_{p \in \mathcal{P}} \, p$, and

$$S(\mathcal{A}, \mathcal{P}) = \#\{a \in \mathcal{A} : (a, P) = 1\}.$$

We assume that there is a number X and a multiplicative function ω with the following properties: for $d | P$, $\omega(d)/d$ is the proportion of elements of $\mathcal{A}$ that are divisible by d in the sense that the quantities

$$\#\{a \in \mathcal{A} : a \equiv 0 \pmod{d}\} - \frac{\omega(d)}{d} X =: r_{\mathcal{A}}(d), \quad d | P, \tag{12.7}$$

are small—at least on average—for some range of numbers d; for $p \notin \mathcal{P}$, set $\omega(p) = 0$; for higher prime powers, $\omega(p^\alpha)$ can be defined arbitrarily, since these numbers will play no role in what follows. Taking $d = 1$ in (12.7), we see that X is an approximation of $|\mathcal{A}|$.

The probability that a residue class in $\mathcal{A}$ is not sieved out by a prime $p \in \mathcal{P}$ is $1 - \omega(p)/p$. Our goal will be to compare $S(\mathcal{A}, \mathcal{P})/X$ with

$$V(\mathcal{P}) := \prod_{p \in \mathcal{P}} \left(1 - \frac{\omega(p)}{p}\right).$$

In the twin prime example, take $\mathcal{A} = \{n(n+2) : 1 \le n \le N\}$ and $\mathcal{P}$ to be all primes not exceeding some number $N_1 \le \sqrt{N+2}$. Then $X = N$ and the congruence $n(n+2) \equiv 0 \pmod{p}$ has solutions $n \equiv 0, -2 \pmod{p}$. Thus we take $\omega(2) = 1$ and $\omega(p) = 2$ for $3 \le p \le N_1$. If $d = p_{\nu_1} \cdots p_{\nu_r}$, a product of r distinct primes from $\mathcal{P}$, then by the Chinese remainder theorem, $n(n+2) \equiv 0 \pmod{d}$ is equivalent to

$$n(n+2) \equiv 0 \pmod{p_{\nu_i}}, \quad 1 \le i \le r.$$

Thus $\omega(d) = \prod_{p | d} \omega(p)$. In this case $r_{\mathcal{A}}(1) = 0$, and by easy calculations

$$|r_{\mathcal{A}}(2)| = |[N/2] - N/2| \le 1,$$

$$|r_{\mathcal{A}}(p)| = |[N/p] + [(N+2)/p] - 2N/p| \le 2, \quad 3 \le p \le N_1,$$

and in general,

$$|r_{\mathcal{A}}(p_{\nu_1} \cdots p_{\nu_r})| \le 2^r. \tag{12.8}$$

On repeating the calculation of §12.2, we get the following extension of the Eratosthenes-Legendre formula:

$$S(\mathcal{A}, \mathcal{P}) = \sum_{d|P} \mu(d) \sum_{\substack{a \in \mathcal{A} \\ d|a}} 1 = X \sum_{d|P} \mu(d) \frac{\omega(d)}{d} + \sum_{d|P} \mu(d) r_{\mathcal{A}}(d). \qquad (12.9)$$

The leading term on the right side is

$$X \prod_{p \in \mathcal{P}} \left(1 - \frac{\omega(p)}{p}\right) =: XV(\mathcal{P}),$$

but the remainder again threatens to overwhelm the leading term if $|\mathcal{P}|$ is large. Indeed, even when $|r_{\mathcal{A}}(d)| \leq 1$ for all $d \mid P$, we cannot make a better estimate than

$$|\sum_{d|P} \mu(d) r_{\mathcal{A}}(d)| \leq \sum_{d|P} 1 = 2^{|\mathcal{P}|},$$

which exceeds X for $|\mathcal{P}| > \log X / \log 2$. To make effective estimates in interesting situations, we want to develop a method of "tricking" the Möbius function, i.e. to replace (12.1) by some approximation that keeps the leading term comparable in size with $XV(\mathcal{P})$ but converts the remainder term in (12.9) into a "genuine" error term.

The following estimate of V. Brun, which we shall establish presently, provides a simple two sided approximation to (12.1) and restricts the number of terms that occur in the remainder term in (12.9).

Proposition 12.1 *Let $0 \leq k < \nu(n)$. Then*

$$m_k(n) := \sum_{\substack{d|n \\ \nu(d) \leq k}} \mu(d) \begin{cases} > 0, & \text{if } k \text{ is even,} \\ < 0, & \text{if } k \text{ is odd,} \end{cases}$$

and, for each positive integer n,

$$m_\ell(n) \leq \sum_{d|n} \mu(d) \leq m_k(n) \qquad (12.10)$$

holds for any positive odd integer ℓ and any nonnegative even integer k.

Remark 12.2 If $k \geq \nu(n)$ in the preceding proposition, then $m_k(n) = e(n)$, i.e. $m_k(n) = 0$ for $n > 1$ and $m_k(1) = 1$.

The proof of Proposition 12.1 follows easily from

Lemma 12.3 *Let n be a squarefree positive integer and h any multiplicative function. Also, write $p(d)$ for the least prime divisor of d for $d > 1$. Then, for any nonnegative integer k,*

$$\sum_{d|n} h(d) = \sum_{\substack{d|n \\ \nu(d) \leq k}} h(d) + \sum_{\substack{d|n \\ \nu(d) = k+1}} h(d) \prod_{\substack{p|n \\ p < p(d)}} (1 + h(p)). \qquad (12.11)$$

Proof. If $k \geq \nu(n)$, there is nothing to prove. Otherwise, we have

$$\sum_{d|n} h(d) - \sum_{\substack{d|n \\ \nu(d) \leq k}} h(d) = \sum_{\substack{d|n \\ \nu(d) \geq k+1}} h(d),$$

and each argument d on the right can be written uniquely in the form $d = tm$ with t composed of the $k+1$ largest prime factors of d. (If $\nu(d) = k+1$, then $m = 1$.) Now the last sum equals

$$\sum_{\substack{t|n \\ \nu(t)=k+1}} h(t) \sum_{\substack{m|n \\ p|m \Rightarrow p < p(t)}} h(m) = \sum_{\substack{t|n \\ \nu(t)=k+1}} h(t) \prod_{\substack{p|n \\ p < p(t)}} (1 + h(p)). \qquad \square$$

Corollary 12.4 *Let ω be the multiplicative function from equation (12.7), n a squarefree positive integer, and k a nonnegative integer. Then*

$$\sum_{d|n} \mu(d)\frac{\omega(d)}{d} = \sum_{\substack{d|n \\ \nu(d) \leq k}} \mu(d)\frac{\omega(d)}{d} - (-1)^k \sum_{\substack{d|n \\ \nu(d)=k+1}} \frac{\omega(d)}{d} \prod_{\substack{p|n \\ p < p(d)}} \left(1 - \frac{\omega(p)}{p}\right).$$

Corollary 12.5 *Let n be a squarefree positive integer and k a nonnegative integer. Then*

$$\sum_{d|n} \mu(d) = \sum_{\substack{d|n \\ \nu(d) \leq k}} \mu(d) - (-1)^k \sum_{\substack{d|n, \, \nu(d)=k+1 \\ p(d)=p(n)}} 1.$$

Proof. Take $h = \mu$ in Lemma 12.3. The product in (12.11) is 0 unless it is empty, i.e. unless $p(d) = p(n)$. $\square$

Proposition 12.1 follows immediately from the last corollary, since $m_k(n) = m_k(\text{sqf}(n))$, where $\text{sqf}(n) := \prod_{p|n} p$, the squarefree part of n.

12.4 The Brun-Hooley sieve

Partition $\mathcal{P}$ into a collection of disjoint subsets $\mathcal{P}_j$, $1 \le j \le r$, for some r, and set $P_j = \prod_{p \in \mathcal{P}_j} p$. Then Hooley's extension of Brun's sieve starts with the inequality

$$\sum_{d|(a,P)} \mu(d) = \prod_{j=1}^{r} \sum_{d_j|(a,P_j)} \mu(d_j) \le \prod_{j=1}^{r} \sum_{\substack{d_j|(a,P_j) \\ \nu(d_j) \le k_j}} \mu(d_j), \qquad (12.12)$$

valid for any choice of even nonnegative integers $k_1, \ldots, k_r$. It follows from (12.12) and an exchange of the summation order that

$$S(\mathcal{A}, \mathcal{P}) = \sum_{a \in \mathcal{A}} \sum_{d|(a,P)} \mu(d)$$

$$\le \sum_{a \in \mathcal{A}} \left(\sum_{\substack{d_1|P_1, d_1|a \\ \nu(d_1) \le k_1}} \mu(d_1) \right) \cdots \left(\sum_{\substack{d_r|P_r, d_r|a \\ \nu(d_r) \le k_r}} \mu(d_r) \right) \qquad (12.13)$$

$$= \sum_{\substack{d_1, \ldots, d_r \\ d_j|P_j, \nu(d_j) \le k_j}} \mu(d_1) \cdots \mu(d_r) \, \#\{a \in \mathcal{A} : d_1 \cdots d_r \mid a\}.$$

The possibility of choosing different even k_j's in this formulation leads to better results than those obtained by applying (12.10) with just one even number k.

On the other hand, Brun's estimate (12.10) gives a lower bound just by taking ℓ odd, while no such simple device works here. We outline a path to lower estimates, based on the following inequality.

Lemma 12.6 *Suppose that $0 \le x_j \le y_j$ for $1 \le j \le r$. Then*

$$x_1 \cdots x_r \ge y_1 \cdots y_r - \sum_{\ell=1}^{r} (y_\ell - x_\ell) \prod_{\substack{j=1 \\ j \ne \ell}}^{r} y_j.$$

Proof. The inequality holds (with equality) for $r = 1$. For $r > 1$, the result follows inductively from

$$y_1 \cdots y_r - x_1 \cdots x_r$$

$$= (y_1 \cdots y_{r-1} - x_1 \cdots x_{r-1})y_r + x_1 \cdots x_{r-1}(y_r - x_r)$$

$$\leq (y_1 \cdots y_{r-1} - x_1 \cdots x_{r-1})y_r + y_1 \cdots y_{r-1}(y_r - x_r). \qquad \square$$

One can establish a lower bound by applying the last inequality with

$$x_j = \sum_{\substack{d|(a,P_j)}} \mu(d), \quad y_j = \sum_{\substack{d|(a,P_j) \\ \nu(d) \leq k_j}} \mu(d), \quad 1 \leq j \leq r,$$

and taking all k_j even. By (12.10) and Corollary 12.5 with the condition $p(d) = p(n)$ dropped, we have

$$0 \leq y_\ell - x_\ell \leq \sum_{\substack{d|(a,P_\ell) \\ \nu(d) = k_\ell + 1}} 1.$$

It follows from Lemma 12.6 that

$$\sum_{d|(a,P)} \mu(d) \geq \prod_{j=1}^{r} \sum_{\substack{d|(a,P_j) \\ \nu(d) \leq k_j}} \mu(d) - \sum_{\ell=1}^{r} \sum_{\substack{d|(a,P_\ell) \\ \nu(d) = k_\ell + 1}} 1 \prod_{\substack{j=1 \\ j \neq \ell}}^{r} \sum_{\substack{d|(a,P_j) \\ \nu(d) \leq k_j}} \mu(d).$$

We thus obtain the Brun-Hooley lower bound

$$S(\mathcal{A}, \mathcal{P}) = \sum_{a \in \mathcal{A}} \sum_{d|(a,P)} \mu(d)$$

$$\geq \sum_{\substack{d_1,\ldots,d_r \\ d_j | P_j, \nu(d_j) \leq k_j}} \mu(d_1) \cdots \mu(d_r) \#\{a \in \mathcal{A} : d_1 \cdots d_r \mid a\}$$

$$- \sum_{\ell=1}^{r} {\sum}^{*} \mu\left(\frac{d_1 \cdots d_r}{d_\ell}\right) \#\{a \in \mathcal{A} : d_1 \cdots d_r \mid a\},$$

with $\sum^{*}$ extending over $d_1, \ldots, d_r$ satisfying

$$d_j \mid P_j \quad \text{and} \quad \nu(d_j) \leq k_j, \quad \text{for} \quad j \neq \ell,$$

$$d_\ell \mid P_\ell \quad \text{and} \quad \nu(d_\ell) = k_\ell + 1.$$

An explicit Brun-Hooley lower bound is stated in Problem 12.1 below.

At this point, we return to establishing an upper bound for $S(\mathcal{A}, \mathcal{P})$, starting from (12.13) and the hypothesized relation (12.7). We have

$$
S(\mathcal{A}, \mathcal{P}) \leq \sum_{\substack{d_1,\ldots,d_r \\ d_j | P_j,\, \nu(d_j) \leq k_j}} \frac{\mu(d_1)\omega(d_1)}{d_1} \cdots \frac{\mu(d_r)\omega(d_r)}{d_r} X
$$

$$
+ \sum_{\substack{d_1,\ldots,d_r \\ d_j | P_j,\, \nu(d_j) \leq k_j}} |r_{\mathcal{A}}(d_1 \cdots d_r)|
$$

$$
=: X\Sigma + R, \quad \text{say.}
$$

We estimate Σ with the aid of Corollary 12.4. For k_j even we have

$$
\sum_{\substack{d|P_j \\ \nu(d) \leq k_j}} \frac{\mu(d)\omega(d)}{d} = \sum_{d|P_j} \frac{\mu(d)\omega(d)}{d}
$$

$$
+ \sum_{\substack{d|P_j \\ \nu(d)=k_j+1}} \frac{\omega(d)}{d} \prod_{\substack{p|P_j \\ p<p(d)}} \left(1 - \frac{\omega(p)}{p}\right).
$$

To simplify calculations, we replace the last product by 1 and obtain

$$
\sum_{\substack{d|P_j \\ \nu(d) \leq k_j}} \frac{\mu(d)\omega(d)}{d} \leq \sum_{d|P_j} \frac{\mu(d)\omega(d)}{d} + \sum_{\substack{d|P_j \\ \nu(d)=k_j+1}} \frac{\omega(d)}{d}.
$$

Now

$$
\sum_{d|P_j} \frac{\mu(d)\omega(d)}{d} = \prod_{p|P_j} \left(1 - \frac{\omega(p)}{p}\right) =: V_j
$$

and, upon noting possible permutations among $k_j + 1$ distinct prime factors of P_j, we obtain

$$
\sum_{\substack{d|P_j \\ \nu(d)=k_j+1}} \frac{\omega(d)}{d} \leq \frac{1}{(k_j+1)!} \left\{ \sum_{p \in \mathcal{P}_j} \frac{\omega(p)}{p} \right\}^{k_j+1},
$$

provided that $\nu(P_j) \geq k_j + 1$ (and the left-hand sum is empty otherwise). Finally, by a familiar estimate,

$$\sum_{p \in \mathcal{P}_j} \frac{\omega(p)}{p} \leq \sum_{p \in \mathcal{P}_j} \log\left(1 - \frac{\omega(p)}{p}\right)^{-1} = \log V_j^{-1} =: L_j.$$

We find that

$$\sum_{\substack{d|P_j \\ \nu(d) \leq k_j}} \frac{\mu(d)\omega(d)}{d} \leq V_j + \frac{L_j^{k_j+1}}{(k_j+1)!} = V_j\left\{1 + \frac{e^{L_j} L_j^{k_j+1}}{(k_j+1)!}\right\}.$$

In some applications we take $k_r = \infty$ and replace the last inequality by

$$\sum_{d|P_r} \mu(d)\,\omega(d)/d =: V_r.$$

Thus we obtain

$$\Sigma \leq \prod_{j=1}^{r} V_j\left\{1 + \frac{e^{L_j} L_j^{k_j+1}}{(k_j+1)!}\right\} \leq \prod_{j=1}^{r}\left\{V_j \exp\left(\frac{e^{L_j} L_j^{k_j+1}}{(k_j+1)!}\right)\right\}.$$

We now have our main estimate:

Theorem 12.7 *Suppose that $\mathcal{P}$ is the set of all primes in some interval $[2, z)$. Let $r \geq 1$ and $\{z_j\}$ satisfy $2 = z_{r+1} < z_r < \cdots < z_1 = z$. Set*

$$\mathcal{P}_j := \mathcal{P} \cap [z_{j+1}, z_j) \quad \text{and} \quad P_j := \prod_{p \in \mathcal{P}_j} p, \quad j = 1, \ldots, r.$$

Suppose $\mathcal{A}$ is a sequence and ω a multiplicative function for which (12.7) holds. Let $k_1, \ldots, k_r$ be any sequence of even nonnegative integers (possibly with $k_r = \infty$). Define

$$L_j := \sum_{p \in \mathcal{P}_j} \log\left\{1 - \frac{\omega(p)}{p}\right\}^{-1}, \quad E := \sum_{j=1}^{r} \frac{L_j^{k_j+1}}{(k_j+1)!} e^{L_j};$$

in case $k_r = \infty$, define the rth term in the sum for E to be 0. Also, let

$$V(\mathcal{P}) := \prod_{p \in \mathcal{P}}\left(1 - \frac{\omega(p)}{p}\right), \quad R := \sum_{\substack{d_1, \ldots, d_r \\ d_j|P_j, \nu(d_j) \leq k_j}} |r_{\mathcal{A}}(d_1 \cdots d_r)|.$$

Then

$$S(\mathcal{A}, \mathcal{P}) \leq X V(\mathcal{P}) \exp E + R.$$

Often, we can easily estimate R using

Lemma 12.8 *Assume that the conditions of the previous theorem hold and that, in addition, $|r_{\mathcal{A}}(d)| \leq \omega(d)$ for all $d \mid \prod_{j=1}^{r} P_j$. Then*

$$R < \prod_{j=1}^{r} z_j^{k_j} / V(\mathcal{P}).$$

If we take $k_r = \infty$, then

$$R < \prod_{p \in \mathcal{P}_r} \{1 + \omega(p)\} \prod_{j=1}^{r-1} z_j^{k_j} / V(\mathcal{P}).$$

Proof. If $d \mid P_j$ and $\nu(d) \leq k_j$, then $d \leq z_j^{k_j}$ and we have

$$\sum_{\substack{d \mid P_j \\ \nu(d) \leq k_j}} \omega(d) \leq z_j^{k_j} \sum_{d \mid P_j} \frac{\omega(d)}{d}$$

$$= z^{k_j} \prod_{p \in \mathcal{P}_j} \left(1 + \frac{\omega(p)}{p}\right) \leq z^{k_j} \prod_{p \in \mathcal{P}_j} \left(1 - \frac{\omega(p)}{p}\right)^{-1}.$$

If $k_r = \infty$, we have the identity

$$\sum_{\substack{d \mid P_r \\ \nu(d) < \infty}} \omega(d) = \prod_{p \mid P_r} (1 + \omega(p)).$$

Now $|r_{\mathcal{A}}(d_1 \cdots d_r)| \leq \omega(d_1) \cdots \omega(d_r)$, and the result follows upon multiplying together the factors. Note that we may insert $\prod_{p \in \mathcal{P}_r} (1 - \omega(p)/p)^{-1}$ into the formula when $k_r = \infty$, since this factor is at least 1. $\qquad \square$

PROBLEM 12.1 With the notation of Theorem 12.7, set

$$R' := z\left(\prod_{j=1}^{r} z_j^{k_j}\right) E/V(\mathcal{P}).$$

Use Lemma 12.6 and an argument analogous to the proof of Theorem 12.7 to establish the lower bound estimate

$$S(\mathcal{A}, \mathcal{P}) \geq XV(\mathcal{P})(1 - \exp E) - R - R'.$$

12.5   The large sieve

In this and the following sections we establish an arithmetical form of the so-called large sieve. This term commonly designates results involving a mean square trigonometric estimate as well as the arithmetic formulation we present. The trigonometric inequality has significant applications for character sum and Dirichlet series estimates as well as for the arithmetic problems we shall study.

Theorem 12.9   (Montgomery). *For each prime p let $J(p)$ denote the union of $\omega(p)$ distinct residue classes modulo p. Let M be an integer and N a positive integer. Let $\mathcal{N}$ be a set of integers in the interval $[M, M + N]$ which lie in none of the residue classes $J(p)$. Then, for each $X \geq 1$, the inequality*

$$|\mathcal{N}| \leq (N + X^2)/Q(X)$$

holds, where

$$Q(X) := \sum_{q \leq X} \mu^2(q) \prod_{p|q} \frac{\omega(p)}{p - \omega(p)} . \tag{12.14}$$

Note that the symbol $Q(X)$, which is commonly used in this topic, is not the counting function of squarefree integers.

Theorem 12.9 produces rather good estimates regardless of whether the average value of $\omega(p)$ is large or small. In contrast, the sieve estimates of Brun, Selberg, or Hooley deteriorate when the average number of residue classes being sieved is large.

Suppose, as an extreme case, that $\omega(p_0) = p_0$ for some p_0, i.e. *all* residue classes modulo p_0 have been removed. In this case $|\mathcal{N}| = 0$, $Q(X) = \infty$ (provided that $X \geq p_0$), and the theorem holds trivially. *We shall henceforth assume that $\omega(p) < p$ for all primes p.*

The following result is apparently more general than Theorem 12.9; however, each is easily deducible from the other.

Theorem 12.10   *Let $J(p)$, $\omega(p)$, M, N, X, and $Q(X)$ be as in Theorem 12.9. For $M \leq n \leq M+N$ let a_n be real or complex numbers satisfying $a_n = 0$ if $n \in J(p)$ for some p. Then*

$$\left| \sum_{n=M}^{M+N} a_n \right|^2 \leq \frac{N + X^2}{Q(X)} \sum_{n=M}^{M+N} |a_n|^2.$$

PROBLEM 12.2   Show that Theorem 12.10 is equivalent to Theorem 12.9.

PROBLEM 12.3   Show that Theorem 12.9 holds if $N = 0$.

12.6   An extremal majorant

We shall establish the large sieve inequality by an elegant method of S. Graham and J. Vaaler which extends Selberg's original λ method to effectively handle cases of arbitrary $\omega(p) < p$. The argument uses a class of entire functions having the properties described in the following

Lemma 12.11   *Let $M \in \mathbb{R}$, $N \in \mathbb{Z}^+$, $0 < \delta < 1/2$, and $e(y) = e^{2\pi i y}$. There exists a continuous nonnegative function G of bounded total variation on $\mathbb{R}$ having the additional properties*

$$G(t) \geq 1, \quad M \leq t \leq M + N, \tag{12.15}$$

$$\int_{-\infty}^{\infty} G(t)\, dt = N + \delta^{-1}, \tag{12.16}$$

$$(\mathfrak{F}G)(x) := \int_{-\infty}^{\infty} G(t)e(-tx)\, dt = 0, \quad x \in \mathbb{R}, \ |x| \geq \delta. \tag{12.17}$$

Proof.   We normalize the problem by showing, for given $\ell > 0$, that there exists a continuous nonnegative function f of bounded total variation on $\mathbb{R}$ having the additional properties

$$f(u) \geq 1, \quad 0 \leq u \leq \ell, \tag{12.15'}$$

$$\int_{-\infty}^{\infty} f(u)\, du = \ell + 1, \tag{12.16'}$$

$$\mathfrak{F}f = 0 \text{ on } (-\infty, -1] \cup [1, \infty). \tag{12.17'}$$

The assertion of the lemma follows from the existence of such a function f. Indeed, given δ and N, we set $\ell = \delta N$ and $G(t + M) = f(\delta t)$. It is easy to verify that such a function G satisfies the conclusion of the lemma.

Remark 12.12   The relation (12.16') is extremely sharp; if f satisfies (12.15') and (12.17'), then it follows from the Poisson summation formula (cf. Appendix) that

$$\ell + 1 = \int_{-\infty}^{\infty} f = (\mathfrak{F}f)(0) = \sum_{n=-\infty}^{\infty} f(n) \geq \sum_{0 \leq n \leq \ell} f(n) \geq [\ell] + 1. \quad (12.18)$$

Returning to the proof of the lemma, we define $B : \mathbb{C} \to \mathbb{C}$ by

$$B(z) := \left(\frac{\sin \pi z}{\pi}\right)^2 \left\{ \frac{2}{z} + \sum_{k=0}^{\infty} \frac{1}{(k-z)^2} - \sum_{k=1}^{\infty} \frac{1}{(k+z)^2} \right\}. \quad (12.19)$$

We shall show that

$$f(u) := \frac{1}{2} B(u) + \frac{1}{2} B(\ell - u), \quad u \in \mathbb{R},$$

is continuous, nonnegative, of bounded total variation on $\mathbb{R}$, and satisfies (12.15'), (12.16'), and (12.17').

If M is a positive number and $|z| \leq M/2$, then

$$|(k \pm z)^{-2}| \leq (k - k/2)^{-2} = 4k^{-2}$$

for any $k \geq M$. By the Weierstrass M-test, both series in (12.19) converge uniformly on compact subsets of $\mathbb{C}$. The apparent singularities of B, at integral values of z, are removable, and thus B is an entire function.

The identity

$$\pi^2 \csc^2 \pi z = \sum_{k=-\infty}^{\infty} (k - z)^{-2}$$

and simple manipulations yield

$$B(z) = 1 + 2\left(\frac{\sin \pi z}{\pi}\right)^2 \left\{ \frac{1}{z} - \sum_{k=1}^{\infty} (k+z)^{-2} \right\}. \quad (12.20)$$

Now

$$\frac{1}{z} - \sum_{k=1}^{\infty} \frac{1}{(k+z)^2} = \sum_{k=1}^{\infty} \left\{ \frac{1}{z+k-1} - \frac{1}{z+k} - \frac{1}{(z+k)^2} \right\}$$

$$= \sum_{k=1}^{\infty} \frac{1}{(z+k-1)(z+k)^2},$$

and hence

$$B(z) = 1 + 2\left(\frac{\sin \pi z}{\pi}\right)^2 \sum_{k=1}^{\infty} (z + k - 1)^{-1}(z + k)^{-2}. \qquad (12.21)$$

We derive estimates for $B(u)$ for $u \geq 0$ from these representations. The last formula for B implies that $B(u) \geq 1$ for all $u \geq 0$ and that $B(u) = 1$ at all nonnegative integers. The representation (12.20) and the inequalities

$$0 < \frac{1}{u} - \sum_{k=1}^{\infty} \frac{1}{(k+u)^2} \leq \frac{1}{u} - \int_{u+1}^{\infty} t^{-2} \, dt = \frac{1}{u} - \frac{1}{u+1} < \frac{1}{u^2},$$

valid for $u > 0$, yield

$$1 \leq B(u) \leq 1 + 2\left(\frac{\sin \pi u}{\pi u}\right)^2, \quad u \geq 0.$$

Also, we see from the definition of B that

$$B(u) + B(-u) = 2\left(\frac{\sin \pi u}{\pi u}\right)^2. \qquad (12.22)$$

This identity and the preceding inequalities for $B(u)$ for $u > 0$ yield

$$-1 \leq B(u) \leq -1 + 2\left(\frac{\sin \pi u}{\pi u}\right)^2, \quad u < 0.$$

In particular, $B(u) = -1$ for any negative integer u.

Now we choose

$$f(u) = \frac{1}{2} B(u) + \frac{1}{2} B(\ell - u).$$

Since B is entire, f is certainly continuous. We deduce from the inequalities of the last two paragraphs that

$$1 \leq f(u) \leq 1 + \left(\frac{\sin \pi u}{\pi u}\right)^2 + \left(\frac{\sin \pi(\ell - u)}{\pi(\ell - u)}\right)^2, \quad \text{if } 0 \leq u \leq \ell, \qquad (12.23)$$

and

$$0 \leq f(u) \leq \left(\frac{\sin \pi u}{\pi u}\right)^2 + \left(\frac{\sin \pi(\ell - u)}{\pi(\ell - u)}\right)^2, \quad \text{if } u < 0 \text{ or } u > \ell. \qquad (12.24)$$

Thus f is nonnegative on $\mathbb{R}$ and satisfies (12.15').

In order to prove that f is of bounded variation, it suffices to show that

$$\int_{-\infty}^{\infty} |B'(t)| dt < \infty.$$

Since B is entire, B' is certainly bounded on the compact interval $[-1, 1]$. To estimate B' on $[1, \infty)$ we differentiate (12.20), obtaining

$$B'(u) = \frac{4}{\pi} \sin \pi u \cos \pi u \sum_{k=1}^{\infty} (u + k - 1)^{-1}(u + k)^{-2}$$

$$+ 2 \left(\frac{\sin \pi u}{\pi} \right)^2 \left\{ -\frac{1}{u^2} + 2 \sum_{k=1}^{\infty} (k + u)^{-3} \right\}.$$

Thus, for $1 \leq u < \infty$ we have

$$|B'(u)| \leq \frac{4}{\pi} \left\{ \frac{1}{u(u+1)^2} + \int_1^{\infty} \frac{dt}{(u+t-1)(u+t)^2} \right\}$$

$$+ \frac{2}{\pi^2} \left\{ \frac{1}{u^2} + 2 \int_0^{\infty} \frac{dt}{(t+u)^3} \right\} = O(u^{-2}).$$

For $u \leq -1$ we differentiate formula (12.22) to obtain

$$B'(u) = B'(-u) + \frac{4 \sin \pi u \cos \pi u}{\pi u^2} - \frac{4 \sin^2 \pi u}{\pi^2 u^3}$$

$$= O(u^{-2}), \quad -\infty < u \leq -1.$$

Thus $|B'|$ is integrable, and so f is of bounded total variation on $\mathbb{R}$.

The inequalities (12.23), (12.24) imply that $\int_{-\infty}^{\infty} f < \infty$; we determine the exact value of this integral. For $T > \ell$,

$$2 \int_{-T}^{T} f(t) \, dt = \int_{-T}^{T} B(t) \, dt + \int_{-T}^{T} B(\ell + t) \, dt$$

$$= 2 \int_{-T}^{T} B(t) \, dt + \int_{T}^{T+\ell} B(t) \, dt - \int_{-T}^{-T+\ell} B(t) \, dt$$

$$=: I + II - III, \quad \text{say}.$$

Now

$$I = 2 \int_0^{T} \{ B(t) + B(-t) \} \, dt = 4 \int_0^{T} \left\{ \frac{\sin \pi t}{\pi t} \right\}^2 dt \to 2$$

as $T \to \infty$ (cf. (11.11)). The inequalities leading up to (12.23) imply that $B(t) \to 1$ as $t \to \infty$ and $B(t) \to -1$ as $t \to -\infty$. Thus $II - III \to 2\ell$ as $T \to \infty$, and (12.16') holds.

It remains to show that $(\mathfrak{F}f)(x) = 0$ for $|x| \geq 1$. To do this we first estimate $B(z)$ for z complex and $|z|$ large. For $\Re z \geq 0$, we split the sum occurring in equation (12.21) into three parts, namely the term with $k = 1$, the terms with $2 \leq k \leq |z|$ and the terms with $k > |z|$.

For $2 \leq k \leq |z|$, we have $|z + k| > |z + k - 1| \geq |z|$ and hence

$$\left| \sum_{2 \leq k \leq |z|} (z + k - 1)^{-1}(z + k)^{-2} \right| < \frac{|z|}{|z|^3} = \frac{1}{|z|^2}.$$

For $k > |z|$, we have $|z + k| > |z + k - 1| \geq k - 1$ and hence

$$\left| \sum_{k > |z|} (z + k - 1)^{-1}(z + k)^{-2} \right| < \sum_{k > |z|} \frac{1}{(k - 1)^3} \ll \frac{1}{|z|^2}$$

for $|z| \geq 3$, say. Combining the estimates, we find that

$$B(z) - 1 \ll |\sin \pi z|^2 / |z|^2$$

for $|z| \geq 3$ and $\Re z \geq 0$. By (12.22), in the same range,

$$1 + B(-z) \ll |\sin \pi z|^2 / |z|^2.$$

These estimates imply that

$$B(z) \ll 1 + |\sin \pi z|^2 / |z|^2$$

for all $z \in \mathbb{C}$, and thus on this set

$$f(z) = \frac{1}{2}\{B(z) + B(\ell - z)\} \ll 1 + \left| \frac{\sin \pi z}{z} \right|^2 + \left| \frac{\sin \pi(\ell - z)}{\ell - z} \right|^2.$$

This bound can be written more simply by noting that

$$|\sin w| \leq \exp |\Im w|$$

and, for $|z| \geq 2\ell + 5$, that $|z - \ell| \geq |z| - \ell \geq |z|/2$. Thus

$$f(z) \ll \exp(2\pi |\Im z|) / |z|^2, \quad |z| \geq 2\ell + 5.$$

Now consider $(\mathfrak{F}f)(x)$ for $x \geq 1$. By Cauchy's theorem we have

$$\int_{-T}^{T} f(t)e(-tx)dt = \int_{\substack{|z|=T \\ \Im z \leq 0}} f(z)e(-zx)dz.$$

If $T \geq 2\ell + 5$ we get

$$\int_{-T}^{T} f(t)e(-tx)\,dt \ll \frac{\pi T}{T^2} \to 0$$

as $T \to \infty$, uniformly in $x \geq 1$. It follows that

$$(\mathfrak{F}f)(x) := \int_{-\infty}^{\infty} f(t)e(-tx)dt = 0, \quad x \geq 1.$$

Similarly, $(\mathfrak{F}f)(x) = 0$ for $x \leq -1$ by taking the semicircular contour in the upper half plane. This establishes (12.17′).   □

As an immediate application of the preceding lemma, we next establish the rather surprising result that if r is an integer that is not too large, then the sum of G taken over any residue class modulo r is the same.

Lemma 12.13   *Let G be as in Lemma 12.11. Let r be an integer satisfying $1 \leq r \leq 1/\delta$. Then for any integer a,*

$$\sum_{n \equiv a(r)} G(n) = \frac{N + \delta^{-1}}{r}.$$

Proof.   Let $F(x) = G(rx + a)$. Then, if ℓ and m run over all integers,

$$\sum_{n \equiv a(r)} G(n) = \sum_{m} G(rm + a) = \sum_{m} F(m) = \sum_{\ell} \widehat{F}(\ell)$$

by the Poisson summation formula. A change of variables shows that

$$\widehat{F}(t) = (1/r)e(at/r)(\mathfrak{F}G)(t/r),$$

so

$$\sum_{n \equiv a(r)} G(n) = \frac{1}{r} \sum_{\ell} e(a\ell/r)(\mathfrak{F}G)(\ell/r).$$

Since $r \leq 1/\delta$, all terms on the right hand side of the last equation vanish except the term with $\ell = 0$, which gives $(\mathfrak{F}G)(0)/r = (N + \delta^{-1})/r$.   □

12.7   Proof of Theorem 12.9

We begin by extending the definition of J given in Theorem 12.9 to square-free integers d by taking

$$J(d) := \bigcap_{p \mid d} J(p), \ d > 1, \text{ and } J(1) := \mathbb{Z}.$$

By the Chinese remainder theorem, there are $\prod_{p \mid d} \omega(p)$ distinct residue classes modulo d in $J(d)$ for d squarefree and $d > 1$. If we set $\omega(1) = 1$ and $\omega(p^\alpha) = \omega(p)$ for $\alpha > 1$ (we use only squarefree integers in what follows, but it is convenient to have $\omega(p^\alpha) = 0$ whenever $\omega(p) = 0$), then ω can be extended to $\mathbb{Z}^+$ as a multiplicative function.

Following the idea of Selberg's sieve method, we let $\{\lambda_d\}$, $d = 1, 2, \ldots$ be real numbers to be specified later, subject to the two conditions

$$\lambda_1 = 1, \tag{12.25}$$

$$\lambda_d = 0 \text{ if } d > X \text{ or if } \mu(d) = 0. \tag{12.26}$$

Then

$$|\mathcal{N}| \le \sum_{n=M}^{M+N} \left(\sum_{\substack{d \\ n \in J(d)}} \lambda_d \right)^2,$$

for if n lies in no residue class $J(p)$, we get a contribution only from $d = 1$.

Take G to be as in Lemma 12.11 with $\delta = 1/X^2 < 1/2$. (Note that the theorem is trivially true for $1 \le X < 2$.) Then

$$|\mathcal{N}| \le \sum_n G(n) \left(\sum_{\substack{d \\ n \in J(d)}} \lambda_d \right)^2 = \sum_{d,e} \lambda_d \lambda_e \sum_{n \in J(d) \cap J(e)} G(n).$$

Now $J(d) \cap J(e) = J([d,e])$, where $[d,e]$ denotes the least common multiple of d and e, and by applying Lemma 12.13 to each of the $\omega([d,e])$ residue classes, we obtain

$$|\mathcal{N}| \le (N + X^2) \sum_{d,e} \lambda_d \lambda_e \frac{\omega([d,e])}{[d,e]} .$$

If $\omega(p) = 0$ for some prime p dividing d or e, then clearly $\omega([d,e]) = 0$. Thus we may restrict the summation here to pairs d, e such that $\omega(p) > 0$

for all primes p dividing $[d, e]$. Let $\mathcal{M}$ denote the multiplicative semigroup generated by the primes p for which $\omega(p) > 0$. Here we use the notation $\sum^*$ to indicate a summation restricted to elements of $\mathcal{M}$. With this notation, we finish the proof by choosing $\{\lambda_d\}$ to minimize

$$S := \sum_{d,e}^* \lambda_d \lambda_e \frac{\omega([d,e])}{[d,e]} \tag{12.27}$$

subject to the conditions (12.25) and (12.26). We show that $S = 1/Q(X)$ holds for such a choice of $\{\lambda_d\}$.

For any multiplicative function f we have

$$f([a,b]) \, f((a,b)) = f(a)f(b)$$

for all $a, b \in \mathbb{Z}^+$ (cf. (2.12)) and thus

$$S = \sum_{d,e}^* \lambda_d \lambda_e \frac{\omega(d)\omega(e)(d,e)}{de \, \omega((d,e))} \, .$$

Next, we define an arithmetic function g by setting

$$(g*1)(r) = r/\omega(r), \quad r \in \mathcal{M},$$

and $g(r) = \infty$ otherwise; g is an (extended real valued) multiplicative function and satisfies

$$g(r) = \prod_{p|r} \frac{p - \omega(p)}{\omega(p)}$$

for r squarefree. With this definition

$$Q(X) = \sum_{r \leq X} \mu^2(r)/g(r).$$

Now

$$S = \sum_{d,e}^* \frac{\lambda_d \, \omega(d)}{d} \frac{\lambda_e \omega(e)}{e} \sum_{\substack{r|d \\ r|e}} g(r).$$

Exchanging the summation order and taking account of (12.26), we get

$$S = \sum_r^* \mu^2(r)g(r)y_r^2,$$

where

$$y_r = \sum_{d \equiv 0(r)} \frac{\lambda_d \omega(d)}{d} = \sum_t \lambda_{rt} \frac{\omega(rt)}{rt} .$$

Note that the preceding formula gives $y_r = 0$ if $r \notin \mathcal{M}$. Further, values of $\omega(rt)$ are irrelevant for rt not squarefree, since $\lambda_{rt} = 0$ in this case. Also $y_r = 0$ for $r > X$, since $\lambda_{rt} = 0$ when $rt > X$.

By Möbius inversion (with finite sums!)

$$\sum_r \mu(r) y_{rd} = \sum_r \mu(r) \sum_t \lambda_{rdt} \frac{\omega(rdt)}{rdt}$$

$$= \sum_s \lambda_{ds} \frac{\omega(ds)}{ds} \sum_{r|s} \mu(r) = \frac{\lambda_d \omega(d)}{d} .$$

In particular,

$$\lambda_1 = \sum_r \mu(r) y_r = 1. \tag{12.28}$$

We minimize the positive definite quadratic form S subject to the two conditions on $\{\lambda_d\}$. Rewriting (12.28) as

$$1 = \sum_r{}^* \mu(r) y_r = \sum_{r \leq X}{}^* \frac{|\mu(r)|}{\sqrt{g(r)}} \cdot \mu(r) \sqrt{g(r)} y_r,$$

we apply the Cauchy-Schwarz inequality to obtain

$$1 \leq \sum_{r \leq X}{}^* \frac{\mu^2(r)}{g(r)} \sum_{r \leq X}{}^* \mu^2(r) g(r) y_r^2, \tag{12.29}$$

i.e. $Q(X)S \geq 1$. Taking y_r so that

$$\mu(r) \sqrt{g(r)} y_r = C|\mu(r)|/\sqrt{g(r)}, \quad r \leq X, \ r \in \mathcal{M},$$

for a suitable constant C yields equality in (12.29). By (12.28) again,

$$1 = \sum_{r \leq X}{}^* \mu(r) y_r = C \sum_{r \leq X}{}^* |\mu(r)|/g(r) = CQ(X).$$

Thus the choice

$$y_r = \begin{cases} \mu(r)/\{g(r)Q(X)\}, & r \leq X, \\ 0, & r > X, \end{cases}$$

satisfies the conditions (12.25) and (12.26) on $\{\lambda_d\}$ and yields $S = 1/Q(X)$. Recalling that $|\mathcal{N}| \leq (N + X^2)S$, we obtain the claimed inequality of Theorem 12.9. $\square$

12.8   Notes

12.2 Some references for small sieves:

- H. Halberstam and H.-E. Richert, Sieve Methods, Academic Press, London, 1974
- A. Selberg, Lectures on Sieves, Collected Papers, vol. 2, Springer–Verlag, Berlin, 1991
- G. Greaves, Sieves in Number Theory, Erg. Math. u. ihrer Grenz. (3), vol. 43, Springer–Verlag, Berlin, 2001

12.4.   The material in this section comes from K. Ford and H. Halberstam, J. Number Theory, vol. 81 (2000), pp. 335–350, who coined the term "Brun-Hooley sieve." An upper bound of this type was discovered by C. Hooley, Acta Arith., vol. 66 (1994), pp. 359–368.

12.5. The mean square trigonometric inequality mentioned at the start of this section is discussed e.g., in H. L. Montgomery and R. C. Vaughan, Mathematika, vol. 20 (1973), pp. 119–134. Theorem 12.1 first appeared in Montgomery's paper, J. London Math. Soc., vol. 43 (1968), pp. 93–98.

12.3 was found (in a related form) independently by A. Beurling and A. Selberg. See the survey paper by Vaaler, Bull. A.M.S., New Series, vol. 12 (1985), pp. 183–216 and two papers by Graham and Vaaler in Trans. A.M.S., vol. 265 (1981), pp. 283–302 and in Topics in Classical Number Theory, Vol. I, (Proceedings of a conference held in Budapest, 1981), Colloq. Math. Soc. János Bolyai, 34, North-Holland, Amsterdam, 1984, pp. 599–615.

12.7.   The proof of the large sieve given here was suggested to us by S. W. Graham. It is based on his article with J. D. Vaaler in the János Bolyai volume cited just above. Selberg's sieve was introduced in Norske Vid. Selsk. Forh., vol. 19 (1947), pp. 64–67.

Chapter 13

Application of Sieves

13.1 A Brun-Hooley estimate of twin primes

In this chapter, we give some applications of the two sieve methods developed in the last chapter. One application, which we shall carry out using each method, shows that the twin primes constitute, in a certain sense, "zero percent" of all primes. In fact, we shall see that the proportion of twin primes is so small that the sum of their reciprocals converges, while, as we showed in (1.1), the sum of the reciprocals of all the primes diverges.

Theorem 13.1 *The counting function of the twin primes satisfies*

$$TP(x) \ll x/\log^2 x, \quad x \ge x_0. \tag{13.1}$$

Also,

$$\sum_{p,\, p+2 \text{ prime}} 1/p < \infty.$$

Proof. We take $\mathcal{A} = \{n(n+2) : 1 \le n \le N\}$, with $N = [x]$, which we suppose to be at least 2000. Also, take $\mathcal{P}$ to be all primes smaller than $z = N^{1/9}$ and $P = \prod_{p \in \mathcal{P}} p$. Now

$$TP(x) < S(\mathcal{A}, \mathcal{P}) + z,$$

and we estimate $S(\mathcal{A}, \mathcal{P})$ using Theorem 12.7 and Lemma 12.8. With the notation of the theorem and lemma, (12.7) holds with $X = N$, $\omega(2) = 1$, and $\omega(p) = 2$ for all $p \in [3, z)$. Also, $|r_{\mathcal{A}}(d)| \le \omega(d)$ holds for all $d \mid P$ by

313

(12.8). We have

$$S(\mathcal{A}, \mathcal{P}) \leq \frac{1}{2} N \prod_{3 \leq p < z} \left(1 - \frac{2}{p}\right) \cdot \exp E + R.$$

The product above has the order of magnitude

$$\prod_{p < z} \left(1 - \frac{1}{p}\right)^2 \sim \frac{e^{-2\gamma}}{\log^2 z} = \frac{c}{\log^2 N}. \tag{13.2}$$

Thus it suffices to show for a suitable partition $\{\mathcal{P}_j\}$ of $\mathcal{P}$ and choice of "Brun numbers" $\{k_j\}$ that

$$E \quad \text{is bounded} \tag{13.3}$$

and

$$R \ll N / \log^2 N. \tag{13.4}$$

With $z = N^{1/9}$, define a real sequence $\{z_j\}$ by

$$\log z_j = 2^{1-j} \log z, \quad 1 \leq j \leq r - 1,$$

$$z_r = \log \log N, \quad \text{and} \quad z_{r+1} = 2.$$

(Note that $\log \log 2000 > 2$.) Here r is taken so that

$$z^{2^{1-r}} \leq \log \log N < z^{2^{2-r}} = z_{r-1}.$$

For $1 \leq j \leq r - 1$, Lemma 4.10, Mertens' estimate for $\sum 1/p$, yields

$$L_j = - \sum_{z_{j+1} \leq p < z_j} \log\left(1 - \frac{2}{p}\right) \leq \sum_{z_{j+1} \leq p < z_j} \left(\frac{2}{p} + \frac{B}{p^2}\right)$$

$$\leq 2 \log \frac{\log z_j}{\log z_{j+1}} + \frac{B'}{\log z_r} \leq 2 \log 2 + \frac{B'}{\log \log \log 2000} =: L,$$

an absolute constant.

With the choice $k_j = 2j$ for $j = 1, \ldots, r - 1$, and $k_r = \infty$, we get

$$E < \sum_{j=1}^{\infty} \frac{L^{2j+1}}{(2j+1)!} e^L < e^{2L} \ll 1,$$

and thus (13.3) holds.

By Lemma 12.8 and the estimate (13.2) for $V(\mathcal{P})$,

$$R \ll 3^{\pi(z_r)} \prod_{j=1}^{r-1} z_j^{k_j} (\log N)^2.$$

Now

$$\sum_{j=1}^{r-1} k_j \log z_j = \sum_{j=1}^{r-1} 2j \cdot 2^{1-j} \log z$$

$$< 2 \log \left(N^{1/9} \right) \sum_{j=1}^{\infty} j(1/2)^{j-1} = (8/9) \log N,$$

and it follows that

$$R < 3^{\log \log N} \log^2 N \cdot N^{8/9} < (\log N)^4 N^{8/9} = o(N/\log^2 N);$$

thus (13.4) also holds. This establishes the estimate $TP(x) \ll x/\log^2 x$.

The second assertion of the theorem follows at once upon expressing the given sum in the Stieltjes form $\int_1^\infty u^{-1} \, dTP(u)$, integrating by parts, and using the preceding upper bound for TP. Alternatively, we can see that the given sum is finite by noting that its restriction to each interval $2^n < p < 2^{n+1}$ is less than $2^{-n} TP(2^{n+1}) = O(n^{-2})$.

13.2 The Brun-Titchmarsh inequality

In §9.6 we established the P.N.T. for arithmetic progressions in the form

$$\pi(x; q, a) = \{1 + o(1)\} \, (\operatorname{li} x)/\varphi(q)$$

for $(q, a) = 1$ and $x \to \infty$. One can obtain better error estimates by studying the zeros of the functions $L(\cdot, \chi)$ for the $\varphi(q)$ residue characters $\chi \pmod q$ and applying the Perron inversion formula. An unconditional estimate is given in the following

Theorem 13.2 (Siegel-Walfisz). *Let $(a, q) = 1$ and $q < (\log x)^N$, where N is any (fixed) positive number. There exists a positive (absolute) constant c such that*

$$\pi(x; q, a) = \frac{\operatorname{li} x}{\varphi(q)} + O_N\{x \exp(-c \log^{1/2} x)\}.$$

While this is a deep theorem, it has several limitations:

(1) It gives no information if q is large relative to x.
(2) It is not useful for estimating differences $\pi(x + y; q, a) - \pi(x; q, a)$ if $y/\{\varphi(q) \log x\}$ is small compared with $x \exp(-c \log^{1/2} x)$.
(3) The bound implied by the O_N cannot be computed at present, so the theorem is not useful for numerical calculation.

We shall not give a proof of this theorem here. Instead, we can obtain numerical upper bounds in a wide variety of cases by means of

Theorem 13.3 (Brun-Titchmarsh inequality). *Let q and a be positive integers with $(q, a) = 1$. Let $x \geq 0$ and $y \geq 4q$. Then*

$$\pi(x + y; q, a) - \pi(x; q, a) \leq \frac{2y}{\varphi(q) \log(y/q)} \left\{ 1 + O\left(\frac{\log \log(y/q)}{\log(y/q)}\right) \right\}.$$

Proof. Let us consider the set of integers of the form $nq + a$ satisfying $x < nq + a \leq x + y$ and delete those which are congruent to $0 \,(\text{mod} \, p)$ for some $p \leq x$. The number Z of survivors provides an upper estimate for $\pi(x + y; q, a) - \pi(x; q, a)$.

We recast this argument slightly to apply Theorem 12.9 conveniently. Let

$$M = \left[\frac{x - a}{q} + 1\right] \quad \text{and} \quad M + N = \left[\frac{x + y - a}{q}\right] \leq M + \left[\frac{y}{q}\right]$$

and let $\mathcal{N}$ denote the set of Z integers n in the interval $[M, M+N]$ satisfying $nq \not\equiv -a \,(\text{mod} \, p)$ for all $p \leq x$.

If $p \mid q$, then, since $(a, q) = 1$, the relation $nq \not\equiv -a \,(\text{mod} \, p)$ holds for all integers n, and we need sieve no residue classes for this prime p. Thus $\omega(p) = 0$ if $p \mid q$. Also, we take $\omega(p) = 0$ for all primes $p > x$, since we do not wish to sieve with respect to these primes. It remains to consider primes $p \leq x$, $p \nmid q$. Since $p \nmid q$, the congruence $nq \equiv -a \,(\text{mod} \, p)$ has exactly one solution $n \,(\text{mod} \, p)$. In this case we delete one residue class modulo p, i.e. take $\omega(p) = 1$ for $p \nmid q$, $p \leq x$.

It follows from Theorem 12.9 that for any $X \geq 1$, we have

$$Z \leq (y/q + X^2)/Q(X), \quad Q(X) = \sum_{\ell \leq X} \mu^2(\ell) \prod_{p \mid \ell} \frac{\omega(p)}{p - \omega(p)}.$$

It is convenient for the estimation of $Q(X)$ to assume that $X \leq x$. This ensures that $\omega(p) = 1$ if $p \nmid q$ and $p \leq X$. With this condition, we have

$$Q(X) = \sum_{\substack{\ell \leq X \\ (\ell,q)=1}} \mu^2(\ell) \prod_{p|\ell} \frac{1}{p-1}$$

$$= \sum_{\substack{\ell \leq X \\ (\ell,q)=1}} \mu^2(\ell) \prod_{p|\ell} \left(\frac{1}{p} + \frac{1}{p^2} + \cdots\right) = \sum_{\substack{(m,q)=1 \\ \mathrm{sqf}(m) \leq X}} \frac{1}{m},$$

where $\mathrm{sqf}(m)$ denotes the squarefree part of m. Thus

$$Q(X) \geq \sum_{\substack{m \leq X \\ (m,q)=1}} \frac{1}{m}$$

$$= \prod_{p|q} \left(1 - \frac{1}{p}\right) \prod_{p|q} \left(1 + \frac{1}{p} + \frac{1}{p^2} + \cdots\right) \sum_{\substack{m \leq X \\ (m,q)=1}} \frac{1}{m}$$

$$\geq \prod_{p|q} \left(1 - \frac{1}{p}\right) \sum_{j \leq X} \frac{1}{j} > \frac{\varphi(q)}{q} \log X.$$

Thus, for $1 \leq X \leq x$ we have

$$Z \leq \left(\frac{y}{q} + X^2\right) \Big/ \left(\frac{\varphi(q)}{q} \log X\right).$$

If $y \leq x^2$, then we choose

$$X^2 = y/(q \log(y/q)) < x^2$$

and get

$$\pi(x+y;q,a) - \pi(x;q,a) \leq \frac{2y}{\varphi(q)} \frac{(1 + 1/\log(y/q))}{(\log(y/q) - \log\log(y/q))}$$

$$\leq \frac{2y}{\varphi(q) \log(y/q)} \left\{1 + O\left(\frac{\log\log(y/q)}{\log(y/q)}\right)\right\}.$$

Thus the claimed inequality holds for $4q \leq y \leq x^2$. This is the most significant case of the theorem.

The rest of the argument, where $y > x^2$, still requires some care. We give two different arguments according to the size of q.

Case 1. If $4q > x + y - \sqrt{y}$, then we have

$$\frac{1}{2}y \leq y - \sqrt{y} < y - \sqrt{y} + x < 4q,$$

and thus $y < 8q$. We estimate $\pi(x + y; q, a) - \pi(x; q, a)$ trivially in this case by counting *all* integers $m \equiv a \pmod q$ in $(x, x + y]$:

$$\pi(x + y; q, a) - \pi(x; q, a) \leq [y/q + 1] \leq 8.$$

Recalling that $y \geq 4q$ by hypothesis, we can write

$$8 \leq \frac{2y}{\varphi(q) \log(y/q)} \left\{ 1 + O\left(\frac{\log \log(y/q)}{\log(y/q)} \right) \right\}$$

for a suitable O constant, since $1.38 < \log y/q < 2.08$.

Case 2. It remains to treat $y > x^2$ and $x + y - \sqrt{y} \geq 4q$. We write

$$\pi(x + y; q, a) - \pi(x; q, a) \leq \pi(x + y; q, a)$$

$$\leq \pi(x + y; q, a) - \pi(\sqrt{y}; q, a) + \sqrt{y}/q + 1. \qquad (13.5)$$

Let $\Delta = x + y - \sqrt{y}$, the difference in arguments in the last expression. We have $4q \leq \Delta \leq y$, and hence the main line of reasoning applies with Δ in place of y and $\sqrt{y}$ in place of x. We get

$$\pi(x + y; q, a) - \pi(\sqrt{y}; q, a) \leq \frac{\Delta/q}{\log(\Delta/q)} \frac{2q}{\varphi(q)} \left\{ 1 + O\left(\frac{\log \log(\Delta/q)}{\log(\Delta/q)} \right) \right\}.$$

We can replace Δ by y in the last expression, since $y/q \geq \Delta/q \geq 4$ and each of $t \mapsto t/\log t$ and $t \mapsto t \log \log t / \log^2 t$ is an increasing function for $t > e$. For the second claim, set $t = e^u$ and form

$$\{ e^u u^{-2} \log u \}' = u^{-3} e^u \{ 1 + (u - 2) \log u \}.$$

The last expression is clearly nonnegative for $u \geq 2$. For $1 < u < 2$ we have

$$1 + (u - 2) \log u > 1 - (2 - u)(u - 1) > 0.$$

Thus

$$\pi(x + y; q, a) - \pi(\sqrt{y}; q, a) \leq \frac{2y}{\varphi(q) \log(y/q)} \left\{ 1 + O\left(\frac{\log \log(y/q)}{\log(y/q)} \right) \right\},$$

and by (13.5), the theorem holds also for $y > x^2$. □

PROBLEM 13.1   Let $(k, \ell) = 1$ and $Z(x; k, \ell)$ denote the number of positive integers not exceeding x, *all* of whose prime factors are congruent to ℓ modulo k. Let $H(x; k, \ell)$ denote the sum of the reciprocals of the positive integers q whose squarefree part is at most x and *none* of whose prime factors is congruent to ℓ modulo k. Show that $Z(x)H(\sqrt{x}) \leq 2x$ for all $x \geq 1$ and that $Z(x) = O\{x \log^{-\delta} x\}$, where $\delta = 1 - \varphi(k)^{-1}$.

13.3   Primes represented by polynomials

Dirichlet's famous theorem on primes in arithmetic progressions (Theorem 9.1) can be rephrased as follows: If f is a first degree polynomial in one variable with relatively prime integer coefficients and positive leading coefficient, then $f(n)$ is a prime for infinitely many positive integers n. No result of this type has been established for polynomials of higher degree, although it is easy to make reasonable conjectures.

For example, we do not know whether $n^2 + 1$ is prime for infinitely many positive integers n. (Equivalently, we can ask: Are there infinitely many primes p such that $p - 1$ is a perfect square?) More generally, if f is a nonconstant polynomial over $\mathbb{Z}$ which has positive leading coefficient and is irreducible over $\mathbb{Q}$, are there infinitely many positive integers n such that $f(n)$ is prime? We can restate the question in a quantitative form: What is the approximate size of

$$F(N) := \#\{n \in [1, N] : f(n) \text{ is prime}\}?$$

Of course, if the coefficients of f have a common divisor exceeding 1 or, more generally, if $f(n)$ is divisible by some prime p_0 for all n, then $f(n)$ is composite for all but a finite number of values of n. For example, $f(n) = n^2 - 3n + 4$ is irreducible but prime only for $n = 1$ and $n = 2$; otherwise it is properly divisible by 2. Thus, we exclude from consideration polynomials f satisfying $f(n) \equiv 0 \pmod{p_0}$ for some prime p_0 and all $n \in \mathbb{Z}^+$.

As a first stab at making a heuristic formula for $F(N)$, we simply appeal to the P.N.T. For large values of n, among the integers around $f(n)$, primes occur with relative frequency $1/\log f(n)$. If f has leading coefficient $C \geq 1$ and degree h, then for large t,

$$f(t) = Ct^h\{1 + O(1/t)\}$$

and so

$$\log f(t) = h \log t + \log C + O(1/t).$$

Thus a first approximation to $F(n)$ might be

$$\int_a^N \frac{dt}{\log f(t)} = \int_2^N \frac{dt}{h \log t + \log C} + O(1) \sim \frac{1}{h} \operatorname{li}(N). \qquad (13.6)$$

Here a is taken large enough so that $\log f(t) \geq 1$ for $t \geq a$.

A moment's thought shows that this formula will not always approximate $F(N)$. For the polynomial $f(n) = n^2 - 3n + 4$ we have $F(N) = 2$ for all $N \geq 2$ while the formula is asymptotic to $N/(2 \log N)$. The formula fails to take account of the possibility that an integer of the form $f(n)$ may be divisible by some primes with a relative frequency different from that of random integers. Indeed, random integers are divisible by a prime p with a relative frequency $1/p$, while on the other hand, integers of the form $f(n)$ are divisible by p with a relative frequency $\omega_0(p)/p$, where $\omega_0(p)$ equals the number of solutions of the congruence $f(x) \equiv 0 \pmod{p}$. That is, $\omega_0(p)$ is the number of distinct elements x in a complete residue system modulo p for which $f(x) \equiv 0 \pmod{p}$. For example, $n^2 + 1$ is never divisible by a prime $p \equiv 3 \pmod 4$, but if $p \equiv 1 \pmod 4$, then $n^2 + 1$ is divisible by p with relative frequency $2/p$.

The respective relative frequencies for which nondivisibility by p occurs are $1 - 1/p$ for random integers and $1 - \omega_0(p)/p$ for integers of the form $f(n)$. Since the argument on which (13.6) was based ignored this distinction, we remedy this oversight by multiplying by $(1 - \omega_0(p)/p)(1 - 1/p)^{-1}$ for each prime p. Thus a more reasonable heuristic formula for $F(N)$ is

$$F(N) \approx c(f) \int_2^N \frac{dt}{h \log t + \log C}, \qquad (13.7)$$

where

$$c(f) := \prod_p \left\{ \left(1 - \frac{\omega_0(p)}{p} \right) \left(1 - \frac{1}{p} \right)^{-1} \right\}. \qquad (13.8)$$

The right hand side of (13.7) is the commonly accepted heuristic formula for $F(N)$; we shall show in Lemma 13.5 that the product in (13.8) converges. Formula (13.7) has provided a remarkably good approximation where it has been tested numerically. In the special case $f(n) = n^2 + 1$, the conjectural formula (13.7) becomes

$$F(N) \approx \frac{1}{2} c(f) \int_2^N \frac{dt}{\log t}, \qquad (13.9)$$

where

$$c(f) = \prod_{p>2} \left\{ 1 - \frac{(-1)^{(p-1)/2}}{p-1} \right\} \doteq 1.372813. \qquad (13.10)$$

While it has not been proved that an asymptotic relation actually holds in (13.7) or even that F is an unbounded function when $\omega_0(p) < p$ for all primes p, sieve methods do enable us to show that $F(N)$ cannot be of a larger order of magnitude than the right side of (13.7). Specifically, we use Theorem 12.9 to prove the following.

Theorem 13.4 *Suppose f is a nonconstant polynomial in one variable with integer coefficients and positive leading coefficient and that f is irreducible over the field of rationals. Let $\omega_0(p)$ denote the number of solutions of the congruence $f(x) \equiv 0 \pmod{p}$, and suppose that $\omega_0(p) < p$ for all primes p. Let*

$$F(N) = \#\{n \in [1, N] : f(n) \text{ is prime}\}.$$

Then, for N large, we have

$$F(N) \le 2c(f)N/\log N + O(N \log \log N/\log^2 N),$$

where $c(f)$ is defined by (13.8).

Lemma 13.5 *Under the hypotheses of the preceding theorem, the product defining $c(f)$ converges and is non-zero.*

Proof. We have the identity

$$\left(1 - \frac{1}{p}\right)^{-1} \left(1 - \frac{\omega_0(p)}{p}\right) = 1 - \frac{\omega_0(p) - 1}{p} - \frac{\omega_0(p) - 1}{p(p-1)}.$$

It is known that, for all but finitely many primes p, $\omega_0(p)$ is the number of prime ideals of first degree dividing p in the ring of integers of the algebraic number field generated by a zero of f. (Cf. Theorem 8.1 in Henry B. Mann, Introduction to Algebraic Number Theory, Ohio State University Press, Columbus, 1955.) Accordingly, $\sum_p (\omega_0(p) - 1)/p$ converges by Formula 120 of E. Landau, J. reine angew. Math., vol. 125 (1902), pp. 64–188; also in [LanC] vol. 1, pp. 201–325. Now $\omega_0(p) \le \deg f$, so that

$$\sum_p (\omega_0(p) - 1)p^{-1}(p-1)^{-1}$$

converges absolutely. By the above identity, the product defining $c(f)$ converges. Since $\omega_0(p) < p$ for all primes p, $c(f) \ne 0$.   □

Lemma 13.6   *Under the hypotheses of Theorem 13.4, put*

$$a_n := \mu^2(n) \prod_{p|n} \omega_0(p)\{1 - \omega_0(p)/p\}^{-1}.$$

Then for large X we have

$$\sum_{n \le X} a_n n^{-1} = c(f)^{-1} \log X + O(1).$$

Remark 13.7   If $f(x) = ax + b$, where $a > 0$ and $(a,b) = 1$, we have $\omega_0(p) = 0$ if $p \mid a$ and $\omega_0(p) = 1$ if $p \nmid a$. In this case the lemma asserts that

$$\sum_{\substack{n \le X \\ (n,a)=1}} \frac{\mu^2(n)}{\varphi(n)} = \frac{\varphi(a)}{a} \log X + O(1),$$

which of course can be obtained directly (cf. Problem 3.25).

Proof.   For $\Re s > 1$ we have

$$\sum_{n=1}^{\infty} \frac{a_n}{n^s} = \prod_p \left\{1 + \frac{\omega_0(p)}{p^s}\left(1 - \frac{\omega_0(p)}{p}\right)^{-1}\right\} = \sum_{m=1}^{\infty} \frac{\alpha_m}{m^s} \cdot \prod_p \left(1 - \frac{\omega_0(p)}{p^s}\right)^{-1},$$

where

$$\sum_{m=1}^{\infty} \frac{\alpha_m}{m^s} = \prod_p \left\{1 + \frac{\omega_0(p)}{p^s}\left(1 - \frac{\omega_0(p)}{p}\right)^{-1}\right\}\left\{1 - \frac{\omega_0(p)}{p^s}\right\}$$

$$= \prod_p \left\{1 + \frac{\omega_0(p)}{p^s}\frac{\omega_0(p)}{p - \omega_0(p)} - \frac{\omega_0(p)^2}{p^{2s}}\frac{p}{p - \omega_0(p)}\right\}$$

and $\sum \alpha_m m^{-s}$ converges absolutely for $\Re s > 1/2$. Similarly

$$\prod_p \left(1 - \frac{\omega_0(p)}{p^s}\right)^{-1} = \sum_{m=1}^{\infty} \frac{\beta_m}{m^s} \cdot \prod_p \left(1 - \frac{1}{p^s}\right)^{-\omega_0(p)},$$

where

$$\sum_{m=1}^{\infty} \frac{\beta_m}{m^s} = \prod_p \left\{ \left(1 - \frac{1}{p^s}\right)^{\omega_0(p)} \left(1 - \frac{\omega_0(p)}{p^s}\right)^{-1} \right\}$$

and $\sum \beta_m m^{-s}$ converges absolutely for $\Re s > 1/2$.

Further

$$\prod_p \left(1 - \frac{1}{p^s}\right)^{-\omega_0(p)} = \sum_{m=1}^{\infty} \frac{\delta_m}{m^s} \cdot \zeta_f(s),$$

where $\zeta_f(s)$ is the Dedekind zeta function of the algebraic number field generated by a zero of f and $\sum \delta_m m^{-s}$ converges for $\Re s > 1/2$. Putting these results together we find that

$$\sum_{n=1}^{\infty} \frac{a_n}{n^s} = \sum_{j=1}^{\infty} \frac{\alpha_j}{j^s} \cdot \sum_{k=1}^{\infty} \frac{\beta_k}{k^s} \cdot \sum_{\ell=1}^{\infty} \frac{\delta_\ell}{\ell^s} \cdot \zeta_f(s) = \sum_{m=1}^{\infty} \frac{\epsilon_m}{m^s} \cdot \zeta_f(s), \qquad (13.11)$$

where $\sum \epsilon_m m^{-s}$ converges absolutely for $\Re s > 1/2$. If $\zeta_f(s) = \sum b_n n^{-s}$, then by Weber's theorem

$$\sum_{n \le x} b_n = Bx + O(x^\theta), \qquad (13.12)$$

where B is a positive constant depending on f and θ is some constant in $(0, 1)$ depending on the degree of f. (Cf. Satz 210 of E. Landau, Einführung in die elementare und analytische Theorie der algebraischen Zahlen und der Ideale, 2nd ed., Leipzig, Teubner, 1927.)

Now by (13.11), (13.12), and the stability theorem (Th. 3.29) we have

$$\sum_{n \le x} a_n = Ax + O(x^{\theta'}),$$

where A is a positive constant and $\theta' = \max(\theta, 1/2 + \epsilon)$. It follows from the (proof of the) Dirichlet-Dedekind theorem (Th. 1.8) that

$$A = \lim_{s \to 1+} (s-1) \sum_{n=1}^{\infty} a_n n^{-s} = \lim_{s \to 1+} \zeta(s)^{-1} \sum_{n=1}^{\infty} a_n n^{-s}.$$

But

$$\zeta(s)^{-1} \sum_{n=1}^{\infty} a_n n^{-s}$$

$$= \prod_p \left\{ \left(1 - \frac{1}{p^s}\right)\left(1 + \frac{\omega_0(p)(1 - \omega_0(p)/p)^{-1}}{p^s}\right) \right\}$$

$$= \exp \sum_p \sum_{m=1}^{\infty} \left\{ -\frac{1}{mp^{ms}} + (-1)^{m+1}\frac{\omega_0(p)^m(1 - \omega_0(p)/p)^{-m}}{mp^{ms}} \right\}$$

$$= \exp \left\{ \sum_p \frac{\omega_0(p) - 1}{p^s} + \sum_p \sum_{m=1}^{\infty} \frac{\eta(p, m)}{p^{ms}} \right\},$$

where $\sum_p \sum_{m=1}^{\infty} \eta(p, m)p^{-ms}$ converges absolutely for $\Re s > 1/2$.

By the continuity theorem for Dirichlet series (Cor. 6.16)

$$\lim_{s \to 1+} \sum_p \frac{\omega_0(p) - 1}{p^s} = \sum_p \frac{\omega_0(p) - 1}{p}$$

and

$$\lim_{s \to 1+} \sum_p \sum_{m=1}^{\infty} \frac{\eta(p, m)}{p^{ms}} = \sum_p \sum_{m=1}^{\infty} \frac{\eta(p, m)}{p^m}.$$

Then, by the continuity of the exponential function, we have

$$A = \exp \left\{ \sum_p \frac{\omega_0(p) - 1}{p} + \sum_p \sum_{m=1}^{\infty} \frac{\eta(p, m)}{p^m} \right\}$$

$$= \prod_p \left\{ \left(1 - \frac{1}{p}\right)\left(1 + \frac{\omega_0(p)(1 - \omega_0(p)/p)^{-1}}{p}\right) \right\}$$

$$= \prod_p \left\{ \left(1 - \frac{1}{p}\right)\left(1 - \frac{\omega_0(p)}{p}\right)^{-1} \right\} = 1/c(f).$$

The result of the lemma now follows from the partial summation formula

$$\sum_{n \le X} a_n n^{-1} = X^{-1} \sum_{n \le X} a_n + \int_1^x u^{-2} \left(\sum_{n \le u} a_n \right) du. \qquad \square$$

Proof of Theorem 13.4. Taking $X = (N/\log N)^{1/2}$ and $\omega(p) = \omega_0(p)$, we apply Theorem 12.9. We find

$$N + X^2 = N(1 + 1/\log N)$$

and

$$Q(X) = \sum_{n \leq X} a_n n^{-1} = c(f)^{-1} \log X + O(1)$$

$$= \frac{1}{2} c(f)^{-1} (\log N - \log \log N) + O(1),$$

so that

$$\frac{N + X^2}{Q(X)} = 2c(f)N/\log N + O(N \log \log N / \log^2 N).$$

Thus the theorem is proved. $\qquad\qquad\square$

13.4   A uniform two residue sieve estimate

Here we shall give an upper estimate for the number of primes $p \leq N$ for which $|ap + k|$ is also prime. This formulation includes two famous unsolved problems. If we take $a = 1$ and $k = 2$, we obtain an upper estimate for the number of pairs of twin primes p, $p + 2$ with $p \leq N$. The bound we obtain also enables us to take a small step towards solving the Goldbach conjecture, which asserts that every even integer $n > 2$ is representable as the sum of two primes. We shall consider these problems further in the next section.

Theorem 13.8   *Let $N \geq 3$ and let a and k be arbitrary nonzero integers. Then*

$$\sum_{\substack{p \leq N \\ |ap + k| \ prime}} 1 \leq \left\{ 16 + O\left(\frac{\log \log N}{\log N} \right) \right\} D(ak) N / \log^2 N.$$

Here

$$D(m) = \prod_{p > 2} \{1 - (p-1)^{-2}\} \cdot \prod_{\substack{p \mid m \\ p > 2}} \frac{p-1}{p-2}.$$

The constant implied by the O is absolute, and the estimate is uniform in a, k, and N.

Remarks 13.9 1. The constant 16 can be replaced by $8 + \epsilon$, for any $\epsilon > 0$, (cf. [HlRi], Th. 3.12) by using the Selberg sieve and the Bombieri-Vinogradov theorem. Heuristic arguments suggest that the result is valid with 16 replaced by 2 (cf. §13.6).

2. If $(a, k) > 1$, then $|ap + k|$ can be prime for at most one prime p. Also, if a and k are of the same parity, then $2 \mid (ap + k)$ for all odd primes p, and $|ap + k|$ can be prime for at most one odd prime p. We shall henceforth assume that a and k are relatively prime and of opposite parity; in particular that ak is even.

Let us now set up our estimation problem in a sieve framework. Let Y be a parameter, $1 < Y < N$. Let

$$\mathcal{N} = \{n \in [1, N]: \text{all prime divisors of } n(an + k) \text{ exceed } Y\}.$$

Then

$$\sum_{\substack{p \leq N \\ |ap + k| \text{ prime}}} 1 \leq \pi(Y) + |\mathcal{N}|, \tag{13.13}$$

and we estimate $|\mathcal{N}|$ by Theorem 12.9. For each $p \leq Y$ we take

$$J(p) = \{n : n(an + k) \equiv 0 \,(\mathrm{mod}\, p)\}.$$

The congruence $n(an + k) \equiv 0 (\mathrm{mod}\, p)$ has two solutions modulo p if $p \nmid ak$, namely $n \equiv 0 \,(\mathrm{mod}\, p)$ and n satisfying $an \equiv -k \,(\mathrm{mod}\, p)$, and has only one solution if $p \mid ak$, namely $n \equiv 0 \,(\mathrm{mod}\, p)$. Thus we take

$$\omega(p) = \begin{cases} 2, & \text{if } p \nmid ak \text{ and } p \leq Y, \\ 1, & \text{if } p \mid ak \text{ and } p \leq Y, \\ 0, & \text{if } p > Y. \end{cases}$$

We call this a two residue sieve because two residue classes modulo p are being deleted for most primes p not exceeding the parameter Y.

For any $X > 1$ we have

$$|\mathcal{N}| \leq (N + X^2)/Q(X), \tag{13.14}$$

where

$$Q(X) := \sum_{m \leq X} \mu^2(m) \prod_{p|m} \frac{\omega(p)}{p - \omega(p)} .$$

It is convenient to take $Y = X$ (which in turn is still to be specified in terms of N). We must obtain a lower bound for $Q(X)$ which is uniform in a and k. This we do in the following two lemmas.

Lemma 13.10 *Let $X \geq 1$ and ak be even. Let $\prod^*_{p|Z}$ denote a product taken over odd primes $p \mid Z$. Then*

$$Q(X) \prod^*_{p|ak} \frac{p-1}{p-2} \geq \sum_{n \leq X} \mu^2(n) \prod^*_{p|n} \frac{2}{p-2} .$$

Proof. The left hand side equals

$$\prod^*_{p|ak} \frac{p-1}{p-2} \sum_{m \leq X} f(m),$$

where $f(m) = \mu^2(m) \prod_{p|m} \omega(p)/\{p - \omega(p)\}$. The function f has the following three properties: it is multiplicative, $f(2) = 1$, and f vanishes at any multiple of 4. Thus we can write

$$\sum_{m \leq X} f(m) = \sum_{\substack{m \leq X \\ m \text{ odd}}} f(m) + \sum_{\substack{m \leq X \\ 2\|m}} f(m) = \sum_{\substack{m \leq X \\ m \text{ odd}}} f(m) + \sum_{\substack{m \leq X/2 \\ m \text{ odd}}} f(m).$$

A similar partition applies to the right hand side of the stated inequality, so it suffices to prove that

$$\prod^*_{p|ak} \frac{p-1}{p-2} \sum_{\substack{m \leq X \\ m \text{ odd}}} \mu^2(m) \prod_{p|m} \frac{\omega(p)}{p - \omega(p)} \geq \sum_{\substack{n \leq X \\ n \text{ odd}}} \mu^2(n) \prod^*_{p|n} \frac{2}{p-2} . \qquad (13.15)$$

Each odd positive squarefree integer m can be expressed uniquely in the form $m = m_1 m_2$, where $m_1 \mid ak$, $(m_2, ak) = 1$, and m_1 and m_2 are both odd and positive. We use m_1 and m_2 in this way throughout the argument.

Thus we write the left hand side of (13.15) as

$$\prod_{p|ak}^{*} \frac{p-1}{p-2} \sum_{\substack{m \leq X \\ m = \bar{m}_1 m_2}} \mu^2(m) \prod_{p|m_1} \frac{1}{p-1} \prod_{p|m_2} \frac{2}{p-2}$$

$$= \sum_{m_1 m_2 \leq X} \mu^2(m_1 m_2) \prod_{p|m_1} \frac{p-1}{p-2} \prod_{\substack{p|ak \\ p \nmid m_1}}^{*} \frac{p-1}{p-2} \prod_{p|m_1} \frac{1}{p-1} \prod_{p|m_2} \frac{2}{p-2}$$

$$= \sum_{m_1 m_2 \leq X} \mu^2(m_1 m_2) \prod_{p|m_2} \frac{2}{p-2} \prod_{p|m_1} \frac{1}{p-2} \prod_{\substack{p|ak \\ p \nmid m_1}}^{*} \left(1 + \frac{1}{p-2}\right).$$

We expand the last product and multiply each of the resulting terms by $\prod_{p|m_1} (p-2)^{-1}$. Then the last two products together become

$$\sum \mu^2(d) \prod_{p|d} \frac{1}{p-2},$$

where this sum extends over odd multiples d of m_1 which divide ak.

If we make the above substitution and rearrange terms we find that the left side of (13.15) equals

$$\sum_{\substack{d|ak \\ d \text{ odd}}} \mu^2(d) \prod_{p|d} \frac{1}{p-2} \sum_{m_2 \leq X} \mu^2(m_2) \prod_{p|m_2} \frac{2}{p-2} \sum_{\substack{m_1|d \\ m_1 \leq X/m_2}} 1.$$

The innermost sum equals $2^{\omega(d)}$ provided that $d \leq X/m_2$. If we restrict the sum over m_2 to satisfy $m_2 \leq X/d$, then the left hand side of (13.15) is at least as large as

$$\sum_{\substack{d|ak \\ d \text{ odd}}} \sum_{m_2 \leq X/d} \mu^2(dm_2) \prod_{p|m_2} \frac{2}{p-2} \prod_{p|d} \frac{2}{p-2}.$$

The last expression equals the right hand side of (13.15), because dm_2 ranges over all odd integers not exceeding X. $\square$

Lemma 13.11   *Let $X \geq 2$. Then*

$$\sum_{n \leq X} \mu^2(n) \prod_{\substack{p|n \\ p>2}} \frac{2}{p-2} = \prod_{p>2} \left(1 + \frac{1}{p(p-2)}\right) \left\{(1/4) \log^2 X + O(\log X)\right\}.$$

Proof. Let $f(n)$ denote the summand on the left side of the above equation. The function f is multiplicative and we can write

$$\sum_{n\leq X} f(n) = \sum_{\substack{n\leq X \\ n \text{ odd}}} f(n) + \sum_{\substack{n\leq X \\ 2\|n}} f(n)$$

$$= \sum_{\substack{n\leq X \\ n \text{ odd}}} f(n) + \sum_{\substack{n\leq X/2 \\ n \text{ odd}}} f(n). \qquad (13.16)$$

It suffices to estimate $\sum f(n)$ for $n \leq X$ and n odd.

Let $f_1(n) = nf(n)$ if n is odd and 0 if n is even. We shall estimate $\sum_{n\leq X} f_1(n)$ and apply partial summation. We estimate the last sum by representing f_1 as $1 * 1 * g$, for some arithmetic function g, and applying (the method of) Theorem 3.29, the stability theorem.

For $\Re s > 1$ introduce the D.s.

$$\sum_{n=1}^{\infty} f_1(n)n^{-s} = \prod_{p>2} \left\{ 1 + \frac{2p^{1-s}}{p-2} \right\}$$

$$= \zeta^2(s)(1-2^{-s})^2 \prod_{p>2}(1-p^{-s})^2 \left(1 + \frac{2p^{1-s}}{p-2} \right)$$

$$= \zeta^2(s)(1-2\cdot 2^{-s}+2^{-2s})\times$$

$$\prod_{p>2} \left\{ 1 + \frac{4}{p-2}p^{-s} + \frac{(-2-3p)}{p-2}p^{-2s} + \frac{2p}{p-2}p^{-3s} \right\}.$$

From this equation we deduce that $f_1 = 1*1*g$, where g is multiplicative, $g(2) = -2$; $g(4) = 1$; $g(2^\alpha) = 0$, $\alpha \geq 3$; and for all primes $p \geq 3$,

$$g(p) = \frac{4}{p-2}; \quad g(p^2) = \frac{-2-3p}{p-2}; \quad g(p^3) = \frac{2p}{p-2};$$

and $g(p^\alpha) = 0$ for all $\alpha \geq 4$. The D.s. for g converges absolutely for any abscissa greater than $1/2$.

We obtain by Lemma 3.1 and Corollary 3.32

$$F_1(X) := \sum_{n \le X} f_1(n) = \sum_{n \le X} 1 * 1 * g(n)$$

$$= \sum_{n \le X} N_2(X/n)g(n)$$

$$= \sum_{n \le X} \left\{ \frac{X}{n} \log \frac{X}{n} + O\left(\frac{X}{n}\right) \right\} g(n)$$

$$= X \log X \sum_{n \le X} \frac{g(n)}{n} + O\left\{ X \sum_{n \le X} |g(n)| \frac{(1 + \log n)}{n} \right\}.$$

Now we have

$$\sum_{n > X} \left| \frac{g(n)}{n} \right| \le X^{-1/3} \sum_{n > X} |g(n)| n^{-2/3} = O(X^{-1/3})$$

and

$$\sum_{n=1}^{\infty} |g(n)| \frac{(1 + \log n)}{n} = \sum_{n=1}^{\infty} \frac{|g(n)|}{n^{2/3}} \frac{(1 + \log n)}{n^{1/3}} < \infty.$$

Thus

$$F_1(X) = X \log X \cdot \sum_{n=1}^{\infty} g(n)/n - X \log X \cdot \sum_{n > X} \frac{g(n)}{n} + O(X)$$

$$= \widehat{G}(1) X \log X + O(X),$$

where $\widehat{G}$ is the D.s. of g.

Partial summation yields

$$\sum_{\substack{n \le X \\ n \text{ odd}}} f(n) = \int_{1/2}^{X} u^{-1} dF_1(u) = \frac{F_1(X)}{X} + \int_{1/2}^{X} \frac{F_1(u)}{u^2} \cdot du$$

$$= \frac{1}{2} \widehat{G}(1) \log^2 X + O(\log X).$$

By the product formula for $\widehat{G}$ we get

$$\widehat{G}(1) = \frac{1}{4} \prod_{p > 2} (1 - p^{-1})^2 \left(1 + \frac{2}{p-2}\right) = \frac{1}{4} \prod_{p > 2} \left\{1 + \frac{1}{p(p-2)}\right\}.$$

Thus by the above and (13.16) we have

$$\sum_{n \le X} f(n) = \frac{1}{8} \prod_{p>2} \left\{ 1 + \frac{1}{p(p-2)} \right\} \left\{ \log^2 X + \log^2 \frac{X}{2} \right\} + O(\log X),$$

which yields the desired estimate. □

Proof of Theorem 13.8. The second factor in the definition of $D(m)$ is always at least 1. Thus, for all $m \ge 1$ we have

$$D(m) \ge \prod_{p>2} \left\{ 1 + \frac{1}{p(p-2)} \right\}^{-1} > \exp\left\{ -\sum_{p>2} \frac{1}{p(p-2)} \right\}$$

$$> \exp\left\{ -\sum_{n \ge 1} \frac{1}{(2n-1)(2n+1)} \right\} = \exp\left(-\frac{1}{2} \right) > \frac{1}{2}.$$

If we set $X = (N/\log N)^{1/2}$ and apply (13.13), (13.14), and Lemmas 13.10 and 13.11, we have

$$\sum_{\substack{p \le N \\ |ap+k| \text{ prime}}} 1 \le \frac{N+X^2}{Q(X)} + X \le \frac{(N+X^2)D(ak)}{(1/4)\log^2 X + O(\log X)} + X$$

$$\le \frac{N(1+1/\log N)D(ak)}{(1/16)(\log N - \log\log N)^2 + O(\log N)} + 2D(ak)\left(\frac{N}{\log N} \right)^{\frac{1}{2}}$$

$$\le \left\{ 16 + O\left(\frac{\log\log N}{\log N} \right) \right\} \frac{D(ak)N}{\log^2 N}. \qquad \square$$

13.5   Twin primes and Goldbach's problem

Theorem 13.8 gives an immediate upper bound for $TP(x)$, the number of pairs of twin primes p, $p+2$ satisfying $p \le x$. We have

Theorem 13.12   (Brun-Selberg). *For $x \to \infty$,*

$$TP(x) \le \left\{ 16 \prod_{p>2} \left(1 - (p-1)^{-2} \right) + o(1) \right\} x/\log^2 x.$$

Proof.   Take $a = 1$ and $k = 2$ in Theorem 13.8 to get the claimed bound on TP. □

PROBLEM 13.2   Let $U(t)$ denote the upper density of the set of integers n for which $p_{n+1} - p_n < t \log n$. Using Theorem 13.8 (or otherwise) show that there exists a constant K such that $U(t) \le Kt$ for all $t > 0$.

We say that an even integer has the *Goldbach property* (briefly: is Goldbach) if it is representable in at least one way as a sum of two primes. The Goldbach conjecture is that every even number exceeding 2 has this property. We shall show that a positive proportion of the even integers have the Goldbach property. This result was first established by Schnirelmann, and it was the starting point of his proof that every integer exceeding one is representable as a sum of at most A primes, where A is some absolute constant. Let $G(x)$ denote the number of even integers not exceeding x which have the Goldbach property.

Theorem 13.13   *There exists a positive constant α such that for all sufficiently large x, $G(x) > \alpha x$.*

Proof.   For n even and $n > 4$ let

$$P(n) = \#\{p \le n : n - p \text{ is prime}\},$$

and $P(n) = 0$ otherwise. By the Cauchy-Schwarz inequality we have

$$\left\{\sum_{n \le x} P(n)\right\}^2 \le G(x)\left\{\sum_{n \le x} P(n)^2\right\}.$$

On the one hand

$$\sum_{n \le x} P(n) = \sum_{n \le x} \sum_{p+p'=n}^* 1 = \sum_{p+p' \le x}^* 1$$

$$\ge \left\{\sum_{p \le x/2}^* 1\right\}^2 = \frac{(1+o(1))x^2}{4 \log^2 x},$$

where $*$ denotes omission of the prime $p = 2$.

On the other hand we have by Theorem 13.8, with $k = n$ (even) and $a = -1$ that

$$P(n) \le (16 + o(1)) \prod_p^* \{1 - (p-1)^{-2}\} \prod_{p|n}^* \frac{p-1}{p-2} \frac{n}{\log^2 n}.$$

Since $n/\log^2 n$ is increasing for $n \geq 8$, we have

$$\sum_{n \leq x} P(n)^2 \leq (256 + o(1)) \prod_p{}^* \{1 - (p-1)^{-2}\}^2 \frac{x^2}{\log^4 x} \sum_{n \leq x} \prod_{p|n}{}^* \left(\frac{p-1}{p-2}\right)^2.$$

Let f be the multiplicative function defined for $n \geq 2$ by

$$f(n) = \prod_{p|n}{}^* \left(\frac{p-1}{p-2}\right)^2.$$

If we set $g = f * \mu$, then g is multiplicative and for each prime p we have $g(p) = f(p) - 1$. We have $g(2) = 0$, and if $p > 2$, then $g(p) = (2p - 3)(p - 2)^{-2}$. Also, $g(p^\alpha) = 0$ for all primes p and all $\alpha \geq 2$. Thus

$$\sum_{n \leq x} f(n) = \sum_{n \leq x} (1 * g)(n) = \sum_{n \leq x} \left[\frac{x}{n}\right] g(n) \leq x \sum_{n=1}^{\infty} \frac{g(n)}{n}$$

$$= x \prod_{p>2} \left\{1 + \frac{g(p)}{p}\right\} = x \prod_{p>2} \left\{1 + \frac{2p-3}{p(p-2)^2}\right\}.$$

It follows that

$$\sum_{n \leq x} P(n)^2 \leq (256 + o(1)) \prod_{p>2} \{1 - (p-1)^{-2}\}^2 \left\{1 + \frac{2p-3}{p(p-2)^2}\right\} \frac{x^3}{\log^4 x}$$

$$= (256 + o(1)) \prod_{p>2} \left\{\frac{p^3 - 3p^2 + 3p}{p^3 - 3p^2 + 3p - 1}\right\} \frac{x^3}{\log^4 x},$$

and so we have finally

$$G(x) \geq \left(\frac{1}{4096} + o(1)\right) \prod_{p>2} \left\{1 - \frac{1}{p(p^2 - 3p + 3)}\right\} x. \qquad \square$$

PROBLEM 13.3 Improve the constant achieved in Theorem 13.13 by the following devices. (1) Apply Hölder's inequality in the "balanced" form

$$\sum_{n \leq x} P(n) \leq \left\{\sum_{n' \leq x} \left(\frac{\sqrt{n'}}{\log n'}\right)^\alpha\right\}^{1/\alpha} \left\{\sum_{n \leq x} \left(\frac{P(n) \log n}{\sqrt{n}}\right)^\beta\right\}^{1/\beta}$$

with $\alpha > 1$, $\beta > 1$, $\frac{1}{\alpha} + \frac{1}{\beta} = 1$, and n' running through the Goldbach numbers. (2) Estimate the last factor by obtaining bounds for

$$\sum_{\substack{n \leq x \\ n \text{ even}}} \prod_{\substack{p \mid n \\ p > 2}} \left(\frac{p-1}{p-2}\right)^{\beta}$$

and applying partial summation. (3) Show that

$$\sum_{n \leq x} P(n) \sim \frac{x^2}{2 \log^2 x}.$$

(4) Obtain $G(x) := \sum_{n' \leq x} 1$ from $\sum (n')^{\alpha/2} (\log n')^{-\alpha}$ by partial summation. (Using $\alpha = 5/4$ and $\beta = 5$ we found this way that $G(x) > x/40$ for all sufficiently large x.)

13.6   A heuristic formula for twin primes

In §12.1 we mentioned the conjectural formula

$$TP(x) \sim Kx/\log^2 x$$

for a suitable constant K. Here we sketch a heuristic argument for this formula with an explicit expression for K and give some numerical data supporting it.

It is convenient to consider a somewhat more general problem: estimate the number $F_k(N)$ of integers $n \in [1, N]$ for which $f_1(n)$, $f_2(n)$, ..., $f_k(n)$ are simultaneously prime, where f_1, f_2, ..., f_k are distinct irreducible polynomials over $\mathbb{Z}$, each with a positive leading coefficient. The argument yielding the heuristic formula (13.7) can be extended to give a heuristic formula for $F_k(n)$. As before, we may assume, for $1 \leq i \leq k$ and all primes p, that $\omega_i(p) < p$, where $\omega_i(p)$ is the number of solutions of the congruence $f_i(x) \equiv 0 \pmod{p}$. Since distinct irreducible polynomials are relatively prime, the number of solutions $\omega(p)$ of the congruence

$$f_1(x) f_2(x) \cdots f_k(x) \equiv 0 \pmod{p}$$

satisfies

$$\omega(p) = \omega_1(p) + \omega_2(p) + \cdots + \omega_k(p)$$

for all but finitely many primes p. If f_i has degree h_i and leading coefficient $C_i =: \exp c_i$, an argument similar to that given in §13.3 yields the heuristic formula

$$F_k(N) \approx c(f_1, f_2, \ldots, f_k) \int_a^N \frac{dt}{\log f_1(t) \cdots \log f_k(t)}$$

$$= c(f_1, \ldots, f_k) \int_a^N \frac{dt}{(h_1 \log t + c_1) \cdots (h_k \log t + c_k)} + O(1)$$

$$\sim \frac{c(f_1, \ldots, f_k)}{h_1 \cdots h_k} \frac{N}{(\log N)^k},$$

where

$$c(f_1, \ldots, f_k) := \prod_p \left\{ (1 - 1/p)^{-k} (1 - \omega(p)/p) \right\}.$$

Here a is a positive number greater than the largest real zero of any of the polynomials $f_1, \ldots, f_k$. The convergence of the product defining $c(f_1, \ldots, f_k)$ follows from the fact that $\sum_p (\omega_i(p) - 1)/p$ converges for each i and hence that $\sum_p (\omega(p) - k)/p$ converges. The upper bound inequality analogous to that given in Theorem 13.4 is

$$F_k(N) \le 2^k k! \, c(f_1, \ldots, f_k) \left\{ 1 + o(1) \right\} N/(\log N)^k.$$

This estimate exceeds the heuristic formula by the factor $2^k k! \, h_1 h_2 \cdots h_k$.

The conjectured counting function of twin primes corresponds to the case $k = 2$, $f_1(x) = x$, and $f_2(x) = x + 2$. We find here that $\omega(2) = 1$ and $\omega(p) = 2$ for $p \ge 3$, so that

$$c(f_1, f_2) = 2 \prod_{p \ge 3} \left(1 - \frac{2}{p}\right)\left(1 - \frac{1}{p}\right)^{-2} = 2 \prod_{p \ge 3} \{1 - (p-1)^{-2}\} \doteq 1.32032,$$

and we have the conjectured formula of §12.1:

$$TP(x) \approx 2 \prod_{p \ge 3} \{1 - (p-1)^{-2}\} x/\log^2 x.$$

Thus the above heuristic formula for $F_k(N)$ reduces to the conjectural formula for the counting function of twin primes.

Comparing this formula with the upper estimate of Theorem 13.12 or the general upper estimate for $F_k(N)$ mentioned above, we see that the estimate is 8 times as large as the conjectured result.

We conclude this chapter with some numerical data of R. Brent on the distribution of twin primes [Math. Comp. 29 (1975), p. 51; ibid., 30 (1976), p. 379], which support the heuristic formula for $TP(x)$. Let $r(x)$ denote the integer nearest to

$$TP(x) - 2 \prod_{p \geq 3} \{1 - (p-1)^{-2}\} \int_2^x \log^{-2} t \, dt.$$

x	$TP(x)$	$r(x)$	$|r(x)|/TP(x)$
10^3	35	-11	$3.1 \cdot 10^{-1}$
10^4	204	-9	$4.4 \cdot 10^{-2}$
10^5	1224	-25	$2.0 \cdot 10^{-2}$
10^6	8169	-79	$9.6 \cdot 10^{-3}$
10^7	58980	226	$3.8 \cdot 10^{-3}$
10^8	440312	-56	$1.2 \cdot 10^{-4}$
10^9	3424506	-802	$2.3 \cdot 10^{-4}$
10^{10}	27412679	1262	$4.6 \cdot 10^{-5}$
10^{11}	224376048	7183	$3.2 \cdot 10^{-5}$

Table 13.1   TWIN PRIME DATA

Using explicit data up to 10^{11} and the conjectured asymptotic formula beyond that point, Brent has given a "conjectural evaluation" of the so-called Brun constant (cf. Theorem 13.1)

$$\sum_{p,\, p+2 \text{ prime}} \left(\frac{1}{p} + \frac{1}{p+2} \right). \tag{13.17}$$

He found that this number probably is close to 1.90216. The sum (13.17) taken over primes $p < 10^{11}$ is equal to $1.797904\ldots$.

PROBLEM 13.4   Let $B(x)$ denote the sum of (13.17) extended over twin primes $p, p+2$ with $p \leq x$. Calculate $B(100)$. Using the conjectural formula to estimate $B(x) - B(100)$, find an x which gives a value of $B(x)$ close to 1.6. (Explicit calculation yields $B(5741) \doteq 1.59993$ and $B(5849) \doteq 1.60027$.)

13.7   Notes

For more information on the material of this section and further applications of sieve methods, see [HlRi].

13.2. Proofs of Theorem 13.2 can be found in [Dav, §22] and [KarN, Ch. IX, §3]. Theorem 13.3 has been proved by H. L. Montgomery and R. C. Vaughan, Mathematika, vol 20 (1973), pp. 119–134, with no error term and subject only to the condition $y > q$. Their argument makes use of a weighted form of the large sieve and some numerical computations.

13.3. The heuristic formula (13.7) dates back at least to Hardy and Littlewood, Acta Math., vol. 44 (1922), pp. 1–70, particularly pp. 46–49; also in Collected Papers of G. H. Hardy, vol. 1, Oxford, 1966, pp. 561–630, particularly pp. 606–609. For general irreducible polynomials, such formulas were studied explicitly by P. T. Bateman and R. A. Horn, Math. Comp., vol. 16 (1962), pp. 363–367, although the idea was known earlier.

The conjecture that $n^2 + 1$ is prime for infinitely many integers n is sometimes called Landau's conjecture, since he discussed it in his address to the 1912 International Congress as one of the particularly challenging unsolved problems about prime numbers (the other three being the Goldbach conjecture, the infinitude of twin primes, and the existence of a prime between any two consecutive squares), cf. Proc., Fifth Int. Cong. of Math., vol. 1, Cambridge, 1913, pp. 93–108, particularly p. 106; also in [LanC], vol. 5, pp. 240–255, particularly p. 253, and in Jahresber. Deutsch. Math. Verein., vol. 21 (1912), pp. 208–228, particularly p. 224. While these four problems remain unsolved as stated, interesting partial results have been obtained for each of them. The nearest approach to Landau's conjecture to date is H. Iwaniec's result (cited below) that there are infinitely many integers n such that $n^2 + 1$ is the product of at most two primes.

The heuristic formula (13.9) for the counting function of integers for which $n^2 + 1$ is prime was tested by A. E. Western, Proc. Cambridge Philos. Soc., vol. 21 (1922), pp. 108–109. After the introduction of electronic computation, further such tests were made by D. Shanks, Math. Comp., vol. 13 (1959), pp. 78–86; vol. 14 (1960), pp. 320–332; and vol. 17 (1963), pp. 188–193. Similar data exist for other irreducible polynomials.

For $f(n) = n^2 + 1$, we know that

$$F(N) \le \{2 + o(1)\}c(f) \int_2^N \frac{dt}{\log t},$$

where $c(f)$ is given by (13.10), but no nontrivial lower estimate is known for $F(N)$. On the other hand, for $F^*(N)$ the number of integers n in $[1, N]$ for which $n^2 + 1$ has at most two prime factors, H. Iwaniec, Invent. Math., vol. 47 (1978), pp. 171–188, proved that

$$F^*(N) > \frac{c(f)}{155} \int_2^N \frac{dt}{\log t}$$

for all sufficiently large values of N.

13.4. Theorem 13.13 was obtained by L. Schnirelmann in 1930 using Brun's sieve method, which provides an upper bound for $P(n)$ but with a much larger constant than the number 16 occurring in Theorem 13.8. Schnirelmann's achievement was made widely known to western mathematicians through Landau's exposition in Nachr. Ges. Wiss. Göttingen 1930, pp. 255–276 (reprinted in [LanC], vol. 9, pp. 167–188); this work also is contained in Landau's book, Über einige neuere Fortschritte der additiven Zahlentheorie, Cambridge Tracts in Math. and Math. Physics, No. 35, Cambridge U. Press, 1937 (reprinted by Stechert–Hafner, New York, 1964). The idea of combining an upper bound for $P(n)$ with the Cauchy-Schwarz inequality may seem an obvious step, but it had eluded earlier sieve researchers.

13.5. It has been shown that almost all even numbers have the Goldbach property. Montgomery and Vaughan established in Acta Arith., vol. 27 (1975), pp. 353-370, that

$$\#\{n \le x : n \text{ even}, n \text{ not Goldbach}\} < x^{1-\delta}$$

for some effectively computable positive constant δ and all sufficiently large x. By arguments more elaborate than those we have used in proving Theorem 13.13, Vaughan has shown in Bull. London Math. Soc., vol. 8 (1976), pp. 245–250 and in J. reine angew. Math., vol. 290 (1977), pp. 93–108, that

$$\liminf_{x \to \infty} G(x)/x \ge 1/4 \text{ and } G(x) > x/26 \text{ for all } x \ge 4.$$

13.6. There is a nice discussion of the heuristic formula for the number of integers n in the interval $[2, N]$ for which both n and $n + d$ are primes in G. Pólya, Amer. Math. Monthly, vol. 66 (1959), pp. 375–384.

Appendix A

Results from Analysis and Algebra

A.1 Properties of real functions

The material in this section is intended for readers who are acquainted with the basic facts about Riemann-Stieltjes (R.S.) integrals. Those familiar with R.S. (or Lebesgue-Stieltjes) theory, product measures, and convolutions can omit this material. As noted in §3.3, we use the Pollard R.S. integral, whose definition uses refinements of partitions. Suppose that f is a left continuous function and g is right continuous on $[a, b]$ and that both functions have a discontinuity at a point $p \in (a, b]$. This common discontinuity presents no problem for the Pollard integral; the contribution to $\int_a^b f\, dg$ at p is $f(p-)$ times the jump of g at p.

Propositions from the text whose proofs are given in the Appendix have been restated and numbered as they first appeared.

A.1.1 Decomposition

Given a complex valued function f defined on a domain D, we can represent f in the form $\Re f + i \Im f$, with $\Re f$ and $\Im f$ each a real valued function on D. For any linear process, e.g. integration, we can study the two parts separately if it is convenient. Analytic properties, such as continuity, bounded variation, or harmonicity that f may enjoy are inherited by the real functions. (Of course, holomorphy is lost in the passage to real functions.)

Also, it is a familiar fact ([Apos], Th. 6.13) that a real function of bounded variation on an interval can be expressed as a difference of two nondecreasing functions.

Lemma A.1 *Let f be a real right continuous monotone increasing function on an interval $[a, b]$. Then f has at most a denumerable collection of discontinuities $\{x_i\}$ on $[a, b]$. Let $f_i := f(x_i) - f(x_i-)$, the "jump" of f at x_i, and let*

$$g(x) = \sum_{x_i \leq x} f_i.$$

Then $f - g$ is a continuous increasing function on $[a, b]$.

Proof. By [Apos, Th. 6.2], the collection of discontinuities of f on $[a, b]$ is denumerable. The function g is well defined, for by [Apos, Th. 6.1] the series $\sum f_i$ is (absolutely) summable. Also g is clearly increasing.

The functions f and g are each continuous from the right, and hence so too is their difference. As monotone functions, f and g each have limits from the left as well [Apos, Th. 4.51], and thus their difference also has such a limit. For $\xi \in (a, b]$ we must show that

$$\lim_{x \to \xi-} \{f(x) - g(x)\} = f(\xi) - g(\xi). \tag{A.1}$$

If ξ is a point of continuity of f, then it is a point of continuity of g as well and (A.1) holds. If $\xi = x_I$, a point of discontinuity of f, then

$$\lim_{x \to \xi-} f(x) = f(\xi) - f_I, \quad \lim_{x \to \xi-} g(x) = g(\xi) - f_I,$$

and again (A.1) holds.

Finally, we show that $f - g$ is monotone. Let $a \leq \xi < \eta \leq b$. By Theorem 6.1 of [Apos] again, $g(\eta-) - g(\xi) \leq f(\eta-) - f(\xi)$ or

$$f(\xi) - g(\xi) \leq f(\eta-) - g(\eta-) = f(\eta) - g(\eta). \qquad \square$$

A.1.2   Riemann-Stieltjes integrals

Theorem 3.7 *Let F and g be functions of bounded variation on $[a, b]$ and assume that F is continuous from the right at each point of $[a, b)$ and g is continuous from the left at each point of $(a, b]$. Then $\int_a^b g\,dF$ exists.*

Proof. Using the remark following the statement of Theorem 7.27 in [Apos], for g of bounded variation and F continuous on $[a, b]$, the integral $\int_a^b g\,dF$ exists.

Given F of bounded variation on $[a, b]$, by linearity we may suppose the function real and we can express it as $F_c + F^+ - F^-$, where F_c is continuous and $F^{\pm}$ are nondecreasing pure jump functions on $[a, b]$. Note that a pure jump function with only a finite number of discontinuities is called a step function. By linearity, it suffices to show that $\int_a^b g dF^+$ exists. To simplify notation we assume that $F^+ = F$.

In the special case

$$F(x) = \begin{cases} f_1, & a \le x < c, \\ f_2, & c \le x \le b, \end{cases}$$

for f_1, f_2 constants and g a left continuous function at c, $\int_a^b g \, dF$ exists by Theorem 7.9 of [Apos]. By linearity, this result extends to right continuous step functions F on $[a, b]$.

Given F a pure jump function having an infinite number of jumps on $[a, b]$ and a number $\epsilon > 0$, by the total variation condition there exists a finite collection of points $\{x_i\}$ on which all but at most ϵ of the variation of F occurs. If we subtract from F the step function with suitable jumps at the points $\{x_i\}$, the resulting function F_R has total variation less than ϵ on $[a, b]$. Then, for any partition $\{y_i\}$ of $[a, b]$ we have

$$\left| \sum_i g(\xi_i)\{F_R(y_{i+1}) - F_R(y_i)\} \right| < \epsilon \sup_{a \le x \le b} |g(x)|,$$

which can be made arbitrarily small. Thus $\int g dF$ exists. $\qquad \square$

PROBLEM A.1   Calculate $\int_0^1 g \, dF$ from the definition:
   a. $F(t) = t^2$, $g(t) = t$. Hint. Show that one can use partitions of the form $\{2^{-n} i\}$, $i = 0, 1, \ldots, 2^n$.
   b. $F(t) = [t + 1/2]$, $g(t) = -[-t + 1/2]$.

PROBLEM A.2   Let $F(t) = g(t) = [t + 1/2]$. Show that $\int_0^1 g \, dF$ does not exist.

Lemma 3.8   *Let F and g satisfy the hypotheses of the preceding theorem. For $a < x < b$ let $\varphi(x) = \int_a^x g dF$. Then φ is continuous from the right and $\varphi(x) - \varphi(x-) = g(x)\{F(x) - F(x-)\}$.*

Proof.   Since F is of bounded variation on $[a, b]$, again supposing F to be real, we express it as $F^+ - F^-$, the difference of two nondecreasing functions. By linearity, we can suppose that F itself is nondecreasing.

To show φ continuous from the right, take $a \leq x < y < b$ and write

$$|\varphi(y) - \varphi(x)| \leq M\{F(y) - F(x)\},$$

where $M = \sup_{a \leq t \leq b} |g(t)|$, a finite number. By the continuity hypothesis,

$$\lim_{y \to x+} \{F(y) - F(x)\} = 0,$$

and thus $\varphi(y) \to \varphi(x)$.

To calculate $\varphi(x) - \varphi(x-)$, let $a \leq y < x < b$ and form

$$\varphi(x) - \varphi(y) - g(x)(F(x) - F(y)) = \int_y^x (g - g(x)) \, dF. \tag{A.2}$$

By the continuity hypothesis, $g(t) - g(x) \to 0$ for $y < t < x$ as $y \to x-$. Taking limits as $y \to x-$, the right hand side of (A.2) goes to zero and we get the claimed formula. $\qquad\qquad\qquad\qquad\qquad\qquad\qquad\qquad\qquad\qquad\qquad\square$

A.1.3   Integrators

For convenience, we restate here the function H defined by equation (3.10): For $x < 1$, take $H(x) := 0$, and for $x \geq 1$, set

$$H(x) := \int_{1-}^x G(x/t) dF(t). \tag{A.3}$$

Lemma 3.16   *Let $F, G \in \mathcal{V}$ and let H be defined by (A.3). Then*

$$H_v(x) \leq \int_{1-}^x G_v(x/t) dF_v(t) \leq G_v(x) F_v(x),$$

and hence H is loc. B.V.

Proof.   Recall that, for any $t > s \geq 1$, we have

$$|G(t) - G(s)| \leq G_v(t) - G_v(s).$$

Let $1 \leq x < y$ and form

$$H(y) - H(x) = \int_{1-}^y \{G(y/t) - G(x/t)\} \, dF(t);$$

note that $G(x/t) = 0$ for $t > x$. Then

$$|H(y) - H(x)| \le \int_{1-}^{y} \{G_v(y/t) - G_v(x/t)\} \, dF_v(t)$$

$$= \int_{1-}^{y} G_v(y/t) dF_v(t) - \int_{1-}^{x} G_v(x/t) \, dF_v(t).$$

For any $x \ge 1$ and any partition of $[1 - \epsilon, x]$ we obtain

$$\sum_i |H(t_{i+1}) - H(t_i)| \le \int_{1-}^{x} G_v(x/t) \, dF_v(t),$$

since the sum of estimates for $|H(t_{i+1}) - H(t_i)|$ is telescoping. Taking the supremum of $\sum_i |H(t_{i+1}) - H(t_i)|$ over all partitions, we obtain the first inequality of the lemma. The second inequality follows immediately from the monotonicity of F_v and G_v, whence H is loc. B.V. $\qquad\square$

Recall that in §3.4.1 we defined S_x to be the empty set for $x < 1$ and

$$S_x := \{(u, v) \in (\tfrac{1}{2}, \infty) \times (\tfrac{1}{2}, \infty) : uv \le x\}, \qquad x \ge 1.$$

Lemma 3.18 *Each set S_x has content. Let F and $G \in V$ and let H be defined on $[1, \infty)$ by equation (A.3) and $H(x) = 0$ for $x < 1$. Then*

$$(dF \times dG)(S_x) = H(x).$$

Proof. If $x < 1$, then $S_x = \phi$, which has content, and $(dF \times dG)(S_x) = 0 = H(x)$. Henceforth, we suppose that $x \ge 1$.

We may assume that each of F, G is nondecreasing. Since the integral defining $H(x)$ exists, for given $\epsilon > 0$ we can find a partition P of $[1/2, x]$ with the following property: if $P' = \{t_i\}_{i=0}^{N}$ is any refinement of P and $\{t_i^*\}$ is any set of evaluation points with $t_i \le t_i^* \le t_{i+1}$, then

$$\left| \sum_{i=0}^{N-1} G\left(\frac{x}{t_i^*}\right) dF((t_i, t_{i+1})) - \int_{1-}^{x} G\left(\frac{x}{t}\right) dF(t) \right| < \epsilon.$$

If we take $t_i^* = t_i$, then the sum is $(dF \times dG)(U)$, where U is a finite union of rectangles covering S_x. If $t_i^* = t_{i+1}$, then the sum is $(dF \times dG)(T)$, where T is a finite union of rectangles contained in S_x. Since the inner and outer product integrators can be made arbitrarily close, S_x has content and, further, $(dF \times dG)(S_x) = H(x)$. $\qquad\square$

Lemma 3.20   *Let F, $G \in \mathcal{V}$, let $H(x) = 0$ for $x < 1$, and let $H(x)$ be defined by equation (A.3) for $x \geq 1$. Then H is continuous from the right and $H \in \mathcal{V}$.*

Proof.   Again, we may assume that F, $G \uparrow$. Let $1 \leq x < y \leq 2x$. Given $\epsilon > 0$, we approximate G on $[1/2, 2x]$ by a right continuous step function G_S with steps of size g_i at points

$$1 = t_1 < t_2 < \cdots < t_n = x < \cdots < t_m = 2x$$

such that $|G(t) - G_S(t)| < \epsilon$ on $[1/2, 2x]$. (For $1/2 \leq t < 1$, we take $G_S(t) := 0 = G(t)$.) Now

$$H(y) - H(x) = \int_{1-}^{y} G_S(y/t)\, dF(t) - \int_{1-}^{x} G_S(x/t)\, dF(t)$$

$$+ \int_{1-}^{y} \{G(\tfrac{y}{t}) - G_S(\tfrac{y}{t})\}\, dF(t) - \int_{1-}^{x} \{G(\tfrac{x}{t}) - G_S(\tfrac{x}{t})\}\, dF(t).$$

By the uniform approximation of G_S to G, the last two integrals are each at most $\epsilon F(y)$; integration by parts of the first two integrals yields

$$\int_{1-}^{y} F(y/t)\, dG_S(t) - \int_{1-}^{x} F(x/t)\, dG_S(t)$$

$$= \int_{1-}^{x} \{F(y/t) - F(x/t)\}\, dG_S(t) + \int_{x}^{y} F(y/t)\, dG_S(t)$$

$$= I + J, \quad \text{say.}$$

For y close to $x+$, we have $J = 0$, for then G_S is constant on $[x, y]$. Also

$$I = \sum_{i=1}^{n} \{F(y/t_i) - F(x/t_i)\} g_i,$$

and each term of this sum goes to zero as $y \to x+$. Thus $H(y) - H(x) \to 0$ as $y \to x+$.

Finally, $H \in \mathcal{V}$, since (i) the support of H is contained in $[1, \infty)$, (ii) H is continuous from the right at each point, and (iii) H is locally of bounded variation, by Lemma 3.15. $\square$

Let φ be a loc. B.V. function which is continuous from the left on $(0, \infty)$, let F and $G \in \mathcal{V}$ and let $0 < a < b < \infty$. Define

$$\int_a^b \varphi\,(dF * dG) := \int_a^b \varphi\,dH,$$

where $dH = dF * dG$ is an integrator associated with $H \in \mathcal{V}$.

Lemma 3.23 *Let F, G, φ, a, and b be as above. Then*

$$\int_a^b \varphi(dF * dG) = \int_{s=1-}^b \left\{ \int_{t=a/s}^{b/s} \varphi(st)dG(t) \right\}dF(s). \tag{A.4}$$

Proof. Suppose first that $\varphi = \chi_{(c,d]}$, the indicator function of an interval $(c, d]$. If $(a, b] \cap (c, d] = \phi$, then each side of (A.4) is 0 and there is nothing more to show. Otherwise, let $(a, b] \cap (c, d] = (e, f]$. Then

$$\int_a^b \varphi(dF * dG) = \int \chi_{(a,b]}\chi_{(c,d]}(dF * dG) = \int \chi_{(e,f]}dH = H(f) - H(e)$$

$$= \iint_{e<st\leq f} dF(s)dG(t) = \int_s \int_t \chi_{(e,f]}(st)dG(t)dF(s)$$

$$= \int_s \int_t \chi_{(a,b]}(st)\chi_{(c,d]}(st)dG(t)dF(s)$$

$$= \int_{s=1-}^b \int_{t=a/s}^{b/s} \chi_{(c,d]}(st)dG(t)dF(s).$$

By linearity, the formula holds for any left continuous step function φ, and then by uniform approximation, it extends to any left continuous function φ that is locally B.V. $\qquad\square$

Lemma 3.24 *Convolution of integrators is associative.*

Proof. It suffices to show for all $x \geq 1$ that

$$\int_{1-}^x dF * (dG * dH) = \int_{1-}^x (dF * dG) * dH.$$

By (3.13), commutativity of convolution, and Lemma 3.23, we obtain

$$\int_{1-}^{x} dF * (dG * dH) = \int_{1-}^{x} F\left(\frac{x}{u}\right)(dH * dG)(u)$$

$$= \int_{t=1-}^{x} \int_{s=1-}^{x/t} F\left(\frac{x}{st}\right) dH(s) dG(t).$$

Integrating the inner integral by parts (as in passing from (3.10) to (3.11)) we find that

$$\int_{s=1-}^{x/t} F\left(\frac{x/t}{s}\right) dH(s) = \int_{s=1-}^{x/t} H\left(\frac{x/t}{s}\right) dF(s),$$

and thus

$$\int_{t=1-}^{x} \int_{s=1-}^{x/t} F\left(\frac{x}{st}\right) dH(s) dG(t) = \int_{t=1-}^{x} \int_{s=1-}^{x/t} H\left(\frac{x}{st}\right) dF(s) dG(t).$$

The roles of F and H have now been exchanged, and so

$$\int_{1-}^{x} dF * (dG * dH) = \int_{1-}^{x} dH * (dG * dF),$$

which is equivalent to what we want to show. $\square$

A.2 The Euler gamma function

Here are gamma relations we use. Proofs of these formulas are given e.g. in [Car], [Con], [TiTF], and [Olv].

- Euler integral:

$$\Gamma(z) = \int_{0}^{\infty} e^{-t} t^{z-1} dt \quad (\Re z > 0)$$

- Product formula:

$$\Gamma(z) = e^{-\gamma z} z^{-1} \prod_{n=1}^{\infty} \left(1 + \frac{z}{n}\right)^{-1} e^{z/n} \quad (z \in \mathbb{C} \setminus \{0, -1, -2, \dots\})$$

- Recurrence formula: $\quad \Gamma(z+1) = z\,\Gamma(z) \quad (z \in \mathbb{C} \setminus \{0, -1, -2, \dots\})$

- Reflection formula:

$$\Gamma(z)\Gamma(1-z) = \frac{\pi}{\sin \pi z} \quad (z \in \mathbb{C} \smallsetminus \mathbb{Z})$$

- Multiplication formula:

$$\Gamma\left(\frac{z}{2}\right)\Gamma\left(\frac{z+1}{2}\right) = \sqrt{\pi}\, 2^{1-z}\Gamma(z) \quad (z \in \mathbb{C} \smallsetminus \{0,-1,-2,\dots\})$$

- Stirling's formula: Let $\delta \in (0, \pi/2)$. Then

$$\text{Log } \Gamma(z) = \log\sqrt{2\pi} + (z-1/2)\,\text{Log } z - z + O(1/|z|)$$

holds uniformly in the region

$$\{z \in \mathbb{C} : |\arg z| \le \pi - \delta, |z| \ge 1\}$$

Here Log is the branch of the logarithm that is positive on the positive real axis.

A.3 Poisson summation formula

Theorem A.2 *Suppose that f is a continuous and piecewise differentiable function on $\mathbb{R}$ and that $\int_{-\infty}^{\infty}(|f| + |f'|) < \infty$. Then $\sum_{-\infty}^{\infty} f(n)$ exists and equals*

$$\lim_{N\to\infty} \sum_{n=-N}^{N} (\mathfrak{F}f)(n),$$

where

$$(\mathfrak{F}f)(x) := \int_{-\infty}^{\infty} f(t)e^{-2\pi i x t}\, dt.$$

Proof. First we show that f vanishes at $\pm\infty$. We have

$$f(y) - f(x) = \int_{x}^{y} f'(t)\, dt \to 0$$

as $x,\, y \to +\infty$ or $x,\, y \to -\infty$. By the Cauchy convergence criterion, $\lim_{x\to\infty} f(x)$ and $\lim_{x\to-\infty} f(x)$ exist. Further, these limits are zero, since $\int_{-\infty}^{\infty} |f|$ exists.

Next,

$$\lim_{X,Y\to\infty} \int_{-X}^{Y} f(t) \, d(t - [t]) =$$

$$\lim_{X,Y\to\infty} \left\{ (t - [t] - \frac{1}{2}) f(t) \Big|_{-X}^{Y} - \int_{-X}^{Y} (t - [t] - \frac{1}{2}) f'(t) \, dt \right\}.$$

Since $\int |f|$ and $\int |f'|$ exist and f vanishes at $\pm\infty$, the last formula implies that $\sum_{-\infty}^{\infty} f(n)$ exists and satisfies

$$\sum_{-\infty}^{\infty} f(n) = \int_{-\infty}^{\infty} f(t) \, dt + \int_{-\infty}^{\infty} (t - [t] - \frac{1}{2}) f'(t) \, dt. \tag{A.5}$$

We study the last integral. By Lemma 7.15, $t - [t] - \frac{1}{2}$, the negative of the sawtooth function (except on integers), is represented by its Fourier series

$$-\sum_{n=1}^{\infty} \frac{1}{\pi n} \sin 2\pi n t$$

except at integer values of t, where the series assumes the value 0 while the original function assumes its right hand limit $-1/2$.

Thus

$$\int_{-\infty}^{\infty} f'(t)(t - [t] - \frac{1}{2}) \, dt = -\int_{-\infty}^{\infty} f'(t) \lim_{N\to\infty} \sum_{0<|n|\le N} \frac{e^{2\pi i n t}}{2\pi i n} \, dt$$

$$= \lim_{N\to\infty} \sum_{0<|n|\le N} -\frac{1}{2\pi i n} \int_{-\infty}^{\infty} f'(t) e^{2\pi i n t} \, dt. \tag{A.6}$$

The exchange of the sum and integral is valid, since

$$\int_{|t|>T} O(1) |f'(t)| \, dt \to 0 \text{ as } T \to \infty$$

and the symmetric partial sums of the Fourier series converge to the sawtooth function boundedly on $\mathbb{R} \setminus \mathbb{Z}$ and uniformly on closed intervals that do not contain integers.

Finally, we integrate by parts the integrals on the right side of (A.6) and obtain

$$\int_{-\infty}^{\infty} f'(t)e^{2\pi int}dt = -2\pi in \int_{-\infty}^{\infty} f(t)e^{2\pi int}dt.$$

It follows from (A.5) that

$$\sum_{-\infty}^{\infty} f(n) = \int_{-\infty}^{\infty} f(t)dt + \lim_{N\to\infty} \sum_{0<|n|\le N} \int_{-\infty}^{\infty} f(t)e^{2\pi int}dt$$

$$= \lim_{N\to\infty} \sum_{|n|\le N} (\mathfrak{F}f)(-n) = \lim_{N\to\infty} \sum_{|n|\le N} (\mathfrak{F}f)(n). \qquad \square$$

Theta functions occur in diverse areas of mathematics, e.g. the theory of elliptic modular functions, a proof of the functional equation of $\zeta(s)$, and a solution of the heat equation. Here we introduce one of these functions, which we shall call *the* theta function. It is defined for $y > 0$ by

$$\vartheta(y) := \sum_{n=-\infty}^{\infty} e^{-\pi n^2 y}.$$

In the following problem we present the theta functional equation.

PROBLEM A.3 Prove that $\vartheta(y) = (1/\sqrt{y})\,\vartheta(1/y)$ for $y > 0$. Hint. Apply Theorem A.2 with $f(t) = \exp(-\pi t^2 y)$ and observe that

$$\int_{-\infty}^{\infty} e^{-\pi(u+iv)^2} du$$

is a positive number that is independent of v.

A.4 Basis theorem for finite abelian groups

An abelian group is a group H whose composition operation satisfies $xy = yx$ for any two elements x, y of H (here the composition is indicated by juxtaposition). If $y_1, y_2, \ldots, y_n$ are elements of an abelian group H, let us denote by $\langle y_1, y_2, \ldots, y_n \rangle$ the set of elements of the form $y_1^{k_1} y_2^{k_2} \cdots y_n^{k_n}$, where $k_1, k_2, \ldots, k_n$ range over the integers. If $H = \langle y_1, y_2, \ldots, y_n \rangle$, we say that $y_1, y_2, \ldots, y_n$ *generate* H or that $y_1, y_2, \ldots, y_n$ *form an n-tuple of generators of* H. Let e denote the identity element of H. If $y_1, y_2, \ldots, y_n$

is an n-tuple of generators of H, then e, y_1, $y_2, \ldots, y_n$ is an $(n+1)$-tuple of generators of H. Conversely, if y_1, $y_2, \ldots, y_n$ is an n-tuple of generators of H and if $y_1 = e$, then $y_2, \ldots, y_n$ is an $(n-1)$-tuple of generators of H.

If an n-tuple of generators y_1, $y_2, \ldots, y_n$ for the abelian group H has the additional property that the assumption $y_1^{a_1} y_2^{a_2} \cdots y_n^{a_n} = e$ implies $y_1^{a_1} = y_2^{a_2} = \cdots = y_n^{a_n} = e$, then we say that y_1, $y_2, \ldots, y_n$ form a *basis* of H. The object of this section is to prove that every abelian group with a finite number of elements has a basis. It is immediate that if y_1, $y_2, \ldots, y_n$ form a basis for H and if $y_1 = y_2 = \cdots = y_r = e$ for some $r \in [1, n)$, then $y_{r+1}, \ldots, y_n$ form a basis for H. We begin with a preliminary assertion.

Theorem A.3   *If H is a finite abelian group with $H = \langle y_1, y_2, \ldots, y_n \rangle$ and if c_1, $c_2, \ldots, c_n$ are integers with $\gcd(c_1, c_2, \ldots, c_n) = 1$, then there exist elements y_1', $y_2', \ldots, y_n'$ of H such that $H = \langle y_1', y_2', \ldots, y_n' \rangle$ with $y_1' = y_1^{c_1} y_2^{c_2} \cdots y_n^{c_n}$.*

Proof.   We proceed by induction on the sum $|c_1| + |c_2| + \cdots + |c_n| = s$. The assertion is immediate if $s = 1$, for then one of c_1, $c_2, \ldots, c_n$ is ± 1 and all the others are zero. So suppose $s > 1$. Then at least two of the numbers c_1, $c_2, \ldots, c_n$ are nonzero, say c_1 and c_2. We may suppose that $0 < |c_2| \le |c_1|$. Let $\delta := \operatorname{sgn}(c_1/c_2)$. Now $y_2 = (y_2 y_1^{\delta}) y_1^{-\delta}$ and so

$$H = \langle y_1, y_2, \ldots, y_n \rangle = \langle y_1, y_2 y_1^{\delta}, y_3, \ldots, y_n \rangle.$$

Now $\gcd(c_1 - \delta c_2, c_2, \ldots, c_n) = 1$ and

$$|c_1 - \delta c_2| + |c_2| + \cdots + |c_n| < s.$$

By induction, there exist elements y_1', $y_2', \ldots, y_n'$ of H such that $H = \langle y_1', y_2', \ldots, y_n' \rangle$ and

$$y_1' = y_1^{c_1 - \delta c_2} (y_2 y_1^{\delta})^{c_2} y_3^{c_3} \cdots y_n^{c_n} = y_1^{c_1} y_2^{c_2} y_3^{c_3} \cdots y_n^{c_n}. \qquad \square$$

Theorem A.4   *Every finite abelian group H has a basis.*

Proof.   We must show that, for some positive integer n, there exist elements y_1, $y_2, \ldots, y_n$ of H such that

  (1) $H = \langle y_1, y_2, \ldots, y_n \rangle$
  (2) $y_1^{a_1} y_2^{a_2} \cdots y_n^{a_n} = e$ for integers a_1, $a_2, \ldots, a_n$ implies that $y_1^{a_1} = y_2^{a_2} = \cdots = y_n^{a_n} = e$.

Since H is finite, for some n there is an n-tuple $x_1, x_2, \ldots, x_n$ of elements of H which generates H. For $x \in H$ let $f(x)$ be the order of x, i.e., the smallest positive integer g such that $x^g = e$. Now in the set of all n-tuples of elements of H which generate H, we say that

$$(y_1, y_2, \ldots, y_n) < (z_1, z_2, \ldots, z_n)$$

if this relation holds lexicographically with respect to order, i.e. if $f(y_1) < f(z_1)$ or if $f(y_1) = f(z_1)$ and $f(y_2) < f(z_2)$ or if $f(y_1) = f(z_1)$, $f(y_2) = f(z_2)$, and $f(y_3) < f(z_3)$, etc. This relation is transitive, and descending chains are finite. Thus there is an n-tuple $(y_1, y_2, \ldots, y_n)$ that generates H and is minimal in the sense that there is no other n-tuple $(z_1, z_2, \ldots, z_n)$ with $H = \langle z_1, z_2, \ldots, z_n \rangle$ and

$$(z_1, z_2, \ldots, z_n) < (y_1, y_2, \ldots, y_n).$$

We claim then that (2) also holds. Suppose to the contrary that (2) fails, say $y_1^{a_1} y_2^{a_2} \cdots y_n^{a_n} = e$ but not all of the elements $y_i^{a_i}$ are equal to e. We may suppose that $0 \le a_i < f(y_i)$, so that $y_i^{a_i} = e$ only when $a_i = 0$. Thus we are assuming that not all the a_i are zero. Suppose $a_1 = a_2 = \cdots = a_{\ell-1} = 0$ but $a_\ell > 0$. Let $d = \gcd(a_\ell, a_{\ell+1}, \ldots, a_n)$, so that

$$a_\ell = db_\ell, \ a_{\ell+1} = db_{\ell+1}, \ \ldots, \ a_n = db_n,$$

where $\gcd(b_\ell, b_{\ell+1}, \ldots, b_n) = 1$. By Theorem A.3 there exist elements $y_\ell', y_{\ell+1}', \ldots, y_n'$ of H such that

$$y_\ell' = y_\ell^{b_\ell} y_{\ell+1}^{b_{\ell+1}} \cdots y_n^{b_n} \quad \text{and} \quad \langle y_\ell, y_{\ell+1}, \ldots, y_n \rangle = \langle y_\ell', y_{\ell+1}', \ldots, y_n' \rangle.$$

Now

$$y_\ell'^d = y_\ell^{db_\ell} y_{\ell+1}^{db_{\ell+1}} \cdots y_n^{db_n} = y_\ell^{a_\ell} y_{\ell+1}^{a_{\ell+1}} \cdots y_n^{a_n} = y_1^{a_1} y_2^{a_2} \cdots y_n^{a_n} = e.$$

Therefore $f(y_\ell') \le d \le a_\ell < f(y_\ell)$. But

$$H = \langle y_1, \ldots, y_{\ell-1}, y_\ell', y_{\ell+1}', \ldots, y_n' \rangle$$

and

$$(y_1, \ldots, y_{\ell-1}, y_\ell', \ldots, y_n') < (y_1, \ldots, y_{\ell-1}, y_\ell, \ldots, y_n),$$

contradicting the minimality of $(y_1, \ldots, y_n)$. Thus (2) must hold. $\qquad \Box$

Corollary A.5   *If y_1, $y_2, \ldots, y_n$ form a basis for the finite abelian group H, then every element of H is expressible uniquely in the form $y_1^{j_1} y_2^{j_2} \cdots y_n^{j_n}$, where $0 \le j_i < f_i$, with f_i the order of y_i.*

Our proof of Theorem A.4 may produce a basis which begins with several occurrences of the identity; these can of course be deleted.

Bibliography

[Apos]   T. M. Apostol, *Mathematical analysis, 2nd ed.*, Addison–Wesley, Reading, Mass.-London-Don Mills, Ont., 1974.

[BaD]   P. T. Bateman and H. G. Diamond, "Asymptotic distribution of Beurling's generalized prime numbers," in *Studies in number theory*, W. J. LeVeque, ed., Math. Assoc. Amer., 1969, 152–210.

[Car]   C. Carathéodory, *Theory of functions of a complex variable*, transl. from German by F. Steinhart, vol. 1, Chelsea Pub. Co., New York, 1954.

[ChanI]   K. Chandrasekharan, *Introduction to analytic number theory*, Springer–Verlag, Berlin, 1968.

[ChanA]   K. Chandrasekharan, *Arithmetical functions*, Springer–Verlag, Berlin, 1970.

[Con]   J. B. Conway, *Functions of one complex variable, 2nd ed.*, Springer–Verlag, New York, 1978.

[CrPo]   R. Crandall and C. Pomerance, *Prime numbers, a computational perspective*, Springer–Verlag, New York, 2001.

[Dav]   H. Davenport, *Multiplicative number theory, 3rd ed.*, revised and with a preface by H. L. Montgomery, Graduate Texts in Mathematics, 74, Springer-Verlag, New York, 2000.

[Edw]   H. M. Edwards, *Riemann's zeta function*, Academic Press, New York, 1974.

[HlRi]   H. Halberstam and H.-E. Richert, *Sieve methods*, Academic Press, London, 1974.

[HarD]   G. H. Hardy, *Divergent series*, Oxford University Press, Oxford, 1949.

[HarR]   G. H. Hardy, *Ramanujan, Twelve lectures on subjects suggested by his life and work*, Cambridge University Press, Cambridge, 1940. Reprinted by Chelsea Pub. Co., New York, 1959.

[HaWr] G. H. Hardy and E. M. Wright, *An introduction to the theory of numbers, 5th ed.*, Oxford University Press, Oxford, 1979.

[Ing]   A. E. Ingham, *The distribution of prime numbers*, Cambridge Tracts in Math. and Math. Physics, 30, Cambridge Univ. Press, Cambridge, 1932.

[Ivic] A. Ivić, *The Riemann zeta–function*, Wiley–Interscience, New York, 1985.

[KarN] A. A. Karatsuba, *Basic analytic number theory*, transl. from the second Russian ed. and with a preface by M. B. Nathanson, Springer–Verlag, Berlin, 1993.

[KarVo] A. A. Karatsuba and S. M. Voronin, *The Riemann zeta–function*, transl. from Russian by N. Koblitz, Walter de Gruyter & Co, Berlin, 1992.

[LanH] E. Landau, *Handbuch der Lehre von der Verteilung der Primzahlen, 2nd ed.*, with an appendix by P. T. Bateman, Chelsea Pub. Co., New York, 1953.

[LanC] E. Landau, *Collected works*, Thales Verlag, Essen, 1987.

[LanV] E. Landau, *Vorlesungen über Zahlentheorie*, Hirzel, Leipzig, 1927. Reprinted by Chelsea Pub. Co., New York, 1947.

[Mat]   G. B. Mathews, *Theory of numbers, part 1*, Deighton Bell, Cambridge, 1892. Reprinted by Chelsea Pub. Co, New York, 1961.

[Mur]   M. R. Murty, *Problems in analytic number theory*, Graduate Texts in Mathematics, 206, Springer–Verlag, New York, 2001.

[Nark] W. Narkiewicz, *The development of prime number theory*, Springer Monographs in Math., Springer–Verlag, Berlin, 2000.

[NZM]   I. Niven, H. S. Zuckerman, and H. L. Montgomery, *An introduction to the theory of numbers, 5th ed.*, Wiley, New York, 1991.

[Olv]   F. W. J. Olver, *Asymptotics and special functions*, Academic Press, New York, 1974.

[Ten]   G. Tenenbaum, *Introduction to analytic and probabilistic number theory*, Cambridge Studies in Adv. Math., 46, Cambridge Univ. Press, Cambridge, 1995.

[TiTF] E. C. Titchmarsh, *Theory of functions, 2nd ed.*, Oxford Univ. Press, Oxford, 1939.

[TiHB] E. C. Titchmarsh, *The theory of the Riemann zeta–function, 2nd ed.*, edited and with a preface by D. R. Heath-Brown, Oxford Univ. Press, Oxford, 1986.

[WW]   E. T. Whittaker and G. N. Watson, *A course of modern analysis, 4th ed.*, Cambridge Univ. Press, Cambridge, 1927 (reprinted 1962).

Index of Names and Topics

Index of Symbols

Printed in the United States
By Bookmasters